International and Cultural Psychology

More information about this series at http://www.springer.com/series/6089

Jason E. Warnick • Dan Landis
Editors

Neuroscience in Intercultural Contexts

Springer

Editors
Jason E. Warnick
Department of Behavioral Sciences
Arkansas Tech University
Russellville, AR, USA

Dan Landis
University of Hawaii
Hilo, HI, USA

ISSN 1574-0455 ISSN 2197-7984 (electronic)
International and Cultural Psychology
ISBN 978-1-4939-2259-8 ISBN 978-1-4939-2260-4 (eBook)
DOI 10.1007/978-1-4939-2260-4

Library of Congress Control Number: 2015936329

Springer New York Heidelberg Dordrecht London

Printed on acid-free paper

Springer Science+Business Media LLC New York is part of Springer Science+Business Media (www.springer.com)

We dedicate this book to the memory of the late Nalini Ambady, a scholar whose significant research was the inspiration for this book. We lost her far too soon; she had many wonderful contributions to make to the field. May others continue in her footsteps.

Jason Warnick would like to dedicate this work to his wife Kyla and their two sons, Ian and Eli.

Dan Landis dedicates, always to Rae, who over 54 years has never lost faith that she made the right decision in the Spring of 1960.

Foreword: A Dedication to Nalini Ambady

The editors of this book have generously decided to dedicate it to the memory of Nalini Ambady who died from acute myelogenous leukemia on October 28, 2013. One of the pioneers of the field that has come to be known as cultural neuroscience, Nalini was originally commissioned to script the foreword for this edited volume. She was a prolific researcher whose work spanned cultural psychology, nonverbal behavior, stereotyping and discrimination, and cognitive neuroscience. She was renowned for her creativity as an experimentalist, for her habit of publishing groundbreaking work, and for her warmth, kindness, and grace as an individual. Yet her path was circuitous, unconventional, and characterized by the same intuitive reasoning that her research program would eventually describe as one of the best methods by which people gain insight about the world.

Nalini was born on March 20, 1959, in Kerala, India. She spent much of her early life moving around India as her family followed her father's various assignments and postings as a member of India's military forces. After attending the Lawrence School, Lovedale, as a teenager, where she served as Head Girl, she was a student at Lady Shri Ram College for Women, part of the University of Delhi, where she received a Bachelor's degree. A twist of fate then brought her to psychology. Having completed her undergraduate education, Nalini's parents were eager to arrange a marriage for her. Feeling the impending pressure that an arrangement would be reached, Nalini decided that her best strategy to delay would be to continue her studies. She therefore decided to apply to available graduate programs in psychology, a topic that had always interested her, and thought it best to go abroad to spurn her parents' efforts at her marriage. Well past the usual North American deadlines for graduate programs, Nalini found one appealing program whose application deadline had not passed: the terminal Master's program at the College of William and Mary in Williamsburg, Virginia. Nalini's hand-written application was successful and her maneuver to delay her arranged marriage would come to change her life in ways she did not expect.

Although she received generous aid and scholarships from William and Mary to attend the graduate program, the trip from India was well beyond her family's means. Her father therefore sold his Vespa motorcycle to cover the cost of her plane fare to Virginia. Nalini reported not the smoothest transition to life in the West. Although she rarely wore a sari, she happened to choose one when dressing the day she left India. She was greeted by a high-ranking administrator from the college at the airport when she arrived and later reflected on how exotic she must have seemed to him in the context of quaint Williamsburg, Virginia. Later, at a dinner of the students and faculty, Nalini was shocked to find that the party's guests had neglected the head of the whole-cooked fish presented on the buffet. A delicacy in her experience, she quickly snatched it up. Returning to the party, she was met with confusion and some horror from her North American hosts, for whom the fish's head is most typically discarded. Nalini's greatest challenge in her early years as a graduate student, however, was her unfamiliarity with the emerging technology of computers. Nalini had never learned to type and managed to survive for the first few years of graduate school drafting all of her papers and manuscripts in what she recounted as a beautiful and painfully trained penmanship. Eventually, her advisors made it clear to her that she would need to learn typing, as she would not be able to continue submitting hand-written manuscripts to academic journals. Even until her death, though, typing was not a skill to which Nalini took well. This led her to greatly prefer phone conversations over email exchanges, a curiosity in her behavior that most of her students and colleagues noticed but for which most never knew the cause. Her modal reply to an email of any length consisted of two words: "call me."

Studying psychology at William and Mary sparked an interest in Nalini that had previously not been kindled. She therefore decided to continue her education in psychology by pursuing a Ph.D. By this time, her parents were not as dogged in arranging a marriage for her back in India, but it did not hurt to protect her cause by staying in North America. She, hence, ventured north to Harvard University. Her graduate days at Harvard were challenging as she met with some bad luck in the lab and had difficulties with her initial supervisor. Nalini eventually was taken under the wing of Bob Rosenthal—famous for his work on self-fulfilling prophecies, statistical methodology, and the study of accuracy from nonverbal cues. One of Bob's best-known studies was on what became dubbed the Pygmalion Effect. In short summary, he found that teachers led to believe that some of their students would "bloom" over the course of the school year actually achieved greater success, presumably because the teachers began to behave differently toward those students in a way that encouraged their development and success. Nalini followed on this theme of Bob's earlier work on teacher–student interactions in her dissertation work. Nalini went to classrooms video-recording instructors while they were teaching. Her goal was to obtain clips of the teachers that she could then use to code aspects of their nonverbal behavior to see whether their success (measured in the form of the students' evaluations) could be gleaned from their nonverbal expressions.

For her study, Nalini needed segments of video in which the instructors were alone. That is, they could not be obstructed by interactions with students. Nalini was crestfallen to discover that in the hours of videotape, none of the clips of teachers alone exceeded more than 2 min. Such clips were far too short to code the nonverbal cues that she intended. Although she saw her dissertation as another failed attempt, Bob encouraged her to press on with the shorter clips. This single serendipitous event may have changed Nalini's life more than any other. Forced to use the briefer clips, Nalini stumbled upon the phenomenon that would become the signature of her career: the concept of thin slices of behavior. Extracting 10 s from the first, middle, and last 10 min of each teacher's instruction, Nalini was able to find that judgments of the teachers from these extremely brief, disjointed segments allowed for accurate prediction of their students' evaluations. A second study showed that these judgments predicted principal's evaluations of teachers' effectiveness as well. Nalini had happened upon an effect of social perception that others and she herself would have previously considered unthinkable. She pushed this even further by reducing the clips to as brief as 2-s segments cropped from the original 10-s clips. There, she found that even just those 2 s of viewing time allowed for judgments of the teachers that significantly corresponded with measures of their success (Ambady & Rosenthal, 1993).

Nalini had struck scientific gold. Her discouragement about research throughout graduate school finally gave way and was now made enthusiastic by the taste of some success. Not only had she uncovered the interesting phenomenon that teachers' success can be judged from such minimal information, she had developed a new methodology for research in person perception that would transform the field even more than it had her personal outlook. This work earned her a dissertation award from Division 5 of the American Psychological Association (Evaluation, Measurement, and Statistics) as well as the Behavioral Science Research Prize, shared with Bob, from the American Association for the Advancement of Science.

Her spirits buoyed, Nalini stayed on at Harvard after receiving her Ph.D. in 1991 to finish up her work on thin slices as a postdoctoral fellow with Bob until 1993. She then took her first faculty position at the College of the Holy Cross not far away in Worcester, Massachusetts. She was only at Holy Cross for a year before returning to Harvard as an assistant professor in 1994, where she spent roughly the next decade of her career. Although graduate school had presented challenges for Nalini, life as a junior faculty at Harvard lived up to its reputation as arduous. Much to her parents' delight, Nalini had met and married a talented Indian law student while a graduate student. She and her husband soon had two daughters who, as infants, were in regular attendance at Nalini's lab meetings and were often found sleeping in her office while she worked.

Despite the pressures that came with being junior faculty and a new mother, Nalini often reflected on those early years as some of the very best of her life. In particular, she found the camaraderie and support of her colleagues to be a rich soil in which grew close lifelong friendships. Not only did Nalini find a niche in her personal life during those first few years on the faculty, but professionally she was blossoming as well. In 1998, she received an Early Career Development (CAREER) award from the National Science Foundation and was the first psychologist to receive the Presidential Early Career Award for Scientists and Engineers (PECASE), awarded by U.S. President Bill Clinton. Following in the steps of her mentor, Nalini taught graduate statistics to students who today constitute some of the best and brightest scholars in the field. Nalini was an excellent teacher and received a teaching award from Harvard for her skill in the classroom. In terms of research, Nalini continued her work investigating thin slices but broadened her scope into new areas as well.

Nalini published her first paper on cross-cultural differences in 1996 (Ambady, Koo, Lee, & Rosenthal, 1996). This began a theme of research into cross-cultural differences in social cognition and social behavior that would become a hallmark of Nalini's career, with additional papers in this area due to be published even now after her death. It was this line of work that metamorphosed into her work on cultural neuroscience in later years. In the early time of her research career, however, Nalini also laid the ground for a wide array of effects that would come to distinguish her research in terms of creativity, impact, and breadth.

Some of Nalini's most notable work was on the effects of stereotypes on individuals' academic performance. Following the stereotype threat literature pioneered by Claude Steele and Joshua Aronson (Steele & Aronson, 1995), Nalini and her students showed that activating stereotypes about one's ethnic group or sex could boost performance as well as hinder it (Shih, Ambady, Richeson, Ambady, Fujita, & Gray, 2002; Shih, Pittinsky, & Ambady, 1999). This was a transformative finding that shifted the way researchers thought about the effects of stereotyping on performance and changed the subsequent research in the field on this topic.

Another incredibly influential vein in Nalini's research program was that on emotion recognition, particularly with regard to cross-cultural variation. Early work had shown that emotions were expressed and perceived relatively universally across cultures (Ekman & Friesen, 1971). In a landmark meta-analysis of almost 100 studies, Nalini found that there were, in fact, cultural differences in the magnitude of emotion recognition for in-group and out-group members, even if the overall outcome tended to be one that was accurate irrespective of who was expressing or perceiving (Elfenbein & Ambady, 2002a, 2002b). She went on to find that cultural familiarity seemed to be the mechanism driving this (Elfenbein & Ambady, 2003a, 2003b), and this line of work paved the way to another on what Nalini and her students referred to as "nonverbal accents" in the expression and judgment of emotional expressions (Marsh, Elfenbein, & Ambady, 2003) and other nonverbal behaviors (Marsh, Elfenbein, & Ambady, 2007). This work showed that very slight differences in the way that people express themselves give signs of their cultural background; for instance, how expressions of a broadly understood and universal emotion such as anger distinguish Japanese nationals from Japanese-Americans, or how differences in the way that Australians and Americans wave "hello" give away their nationality.

One of Nalini's best-known research areas was the study of how subtle nonverbal cues can allow for accurate judgments of sexual orientation. Beginning with a paper published in 1999, Nalini and her colleagues showed that thin slice clips as brief as 1 s in length (as well as still images taken from these clips) allowed perceivers to accurately judge men's and women's self-reported sexual orientation (Ambady, Hallahan, & Conner, 1999). This work continued later with a series of papers in which Nalini and her students explored the details and nuances of these effects (e.g., Freeman, Johnson, Ambady, & Rule, 2010; Rule, Ambady, Adams, & Macrae, 2008; Rule, Rosen, Slepian, & Ambady, 2011). One of these studies specifically examined the influence that targets' and perceivers' culture exerts on the accurate judgment of sexual orientation, finding that men's sexual orientation could be judged accurately from their faces largely independent of the culture of the person judging or being judged (Rule, Ishii, Ambady, Rosen, & Hallett, 2011).

Nalini began her foray into cognitive neuroscience research in 2003 as a collaborator on a project with her then post-doc, Reg Adams—one of the chapter authors in this book, on differences in amygdala response among individuals perceiving either direct or averted eye gaze in photos of others (Adams, Gordon, Baird, Ambady, & Kleck, 2003). From there, Nalini began applying cognitive neuroscience tools to her work on emotion recognition (Elfenbein, Mandal, Ambady, Harizuka, & Kumar, 2004; Mandal & Ambady, 2004) and stereotyping (Chiu, Ambady, & Deldin, 2004), which paved the way for a large research program aimed at understanding intergroup and cultural differences in the neural correlates of social perception.

It was around this time that Nalini left Harvard to continue her career at nearby Tufts University, just a few miles down the street. In 1999, Nalini had been promoted to associate professor at Harvard with an endowed chair named for John and Ruth Hazel. According to reports in Harvard's student newspaper, *The Harvard Crimson*, both the psychology department and dean of the faculty approved Nalini's tenured promotion to full professor in 2002 but was ultimately denied in a decision made by the then president Larry Summers (Vascellaro, 2002, 2003). Despite indications that Summers's decision might have been informed by a small minority of faculty members acting independently of the department to influence Summers, Nalini remained positive toward her colleagues and the support that she received from the majority of them before and after her departure from Harvard.

Disappointing as Summers's decision might have been, Nalini's departure from Harvard was bittersweet, as her move to Tufts would mark the beginning of a renaissance in her research program. Now a full professor with an endowed chair (Neubauer Faculty Fellow) and flush with grant support, Nalini's lab grew large—at one point numbering 4 postdoctoral fellows, 6 Ph.D. students, and over 30 undergraduate research assistants. The influx of new trainees added new directions to her research, inspired by her students' individual interests. She began a profusion of work on multiculturalism, specifically focused on implications for race, ethnicity, and the experience of people with multiracial backgrounds (e.g., Chiao, Heck, Nakayama, & Ambady, 2006; Pauker et al., 2009; Rattan & Ambady, 2013). In a related area of research, Nalini and her students began investigating the way that people's nonverbal responses to others shaped the impressions of third-party observers in a way that builds and maintains cultures of prejudice and discrimination (Weisbuch & Ambady, 2009; Weisbuch, Pauker, & Ambady, 2009). Meanwhile, three entirely new research tracks opened up in Nalini's lab: one on the effects of physical embodiment of psychological concepts on social perception (e.g., Slepian, Weisbuch, Rule, & Ambady, 2011), another on the dynamic and interactive nature of perception based on fluid and continuous theoretical models of cognition (see Freeman & Ambady, 2011 for review), and one on cues to personality and behavior present in social media (e.g., Ivcevic & Ambady, 2013).

Despite the growth into these new areas, however, Nalini still maintained active lines of inquiry on her long-standing topics of interest. She continued to publish novel and exciting work on emotion recognition (e.g., Weisbuch & Ambady, 2008) and extended this into new areas, such as considerations of how aging affects judgments of emotion (Krendl & Ambady, 2010). Of course, she continued to publish high-profile work on the accuracy of judgments based on thin slices of behavior, continuing within the theme of predicting success that she had begun with her initial dissertation work (e.g., Ambady, Krabbenhoft, & Hogan, 2006; Rule & Ambady, 2008), which included a book on the general topic entitled *First Impressions* (Ambady & Skowronski, 2008). She also continued to examine cultural differences in thought and behavior as they intersected with these new domains of research (e.g., Freeman, Ma, Han, & Ambady, 2013; Rule, Ambady, et al., 2010). Where her work really accelerated, however, was in social-cognitive neuroscience—particularly the application of cognitive neuroscience methods to answering questions about cultural differences in brain function and behavior.

Nalini's social neuroscience work followed several of the lines already laid out by her previous behavioral work. She published a flurry of studies examining the brain's role in thin slice judgments (e.g., Cloutier, Ambady, Meagher, & Gabrieli, 2012; Freeman, Schiller, Rule, & Ambady, 2010; Rule et al., 2011), the neural correlates of prejudice and stigma (e.g., Krendl, Kensinger, & Ambady, 2012; Krendl, Moran, & Ambady, 2013), and a host of studies following up on her initial work on the role of the amygdala and other subcortical structures in processing social cues from eye gaze (Adams et al., 2011, 2012).

The last of these served as a bridge into her work in cultural neuroscience. Her first cross-cultural neuroimaging study examined amygdala responses during perceptions of fear in the faces of cultural in-group and out-group members (Chiao, Iidaka, et al., 2008). She then extended this work to specifically considering the role that eye gaze plays in attenuating these cultural differences in amygdala response (Adams, Franklin, et al., 2010). Related to this, she and her students conducted an innovative and relatively groundbreaking study examining cross-cultural differences in superior temporal sulcus activity during mental state inferences from the eyes of cultural in-group and out-group members (Adams, Rule, et al., 2010). Not the least of this effort was the requirement to develop a version of the Reading the Mind in the Eyes Test (Baron-Cohen, Wheelwright, Hill, Raste, & Plumb, 2001) using East Asian stimuli that has since become widely used.

True to her start once more, Nalini and her lab also investigated cross-cultural differences in the neural substrates of thin slice judgments. In one study, she and her students explored cultural differences in amygdala responses to judgments predicting the electoral success of Japanese and American legislative political candidates (Rule, Freeman, et al., 2010). Another line of research exploring the brain's basis of inferences of judgments of dominance and submission (see also Chiao, Adams, Tse, Richeson, & Ambady, 2008) found that American and Japanese university students showed distinct responses when perceiving nonverbal displays of dominance and submission in reward-related areas of the brain (e.g., the caudate nucleus). This difference in response to dominant versus submissive bodily postures correlated with differences in cultural values between the USA and Japan in terms of adherence to individualist (dominant) and collectivist (submissive) behavior, as well as with individual differences in the endorsement of dominant versus submissive behavior and values (Freeman, Rule, Adams, & Ambady, 2009).

Perhaps even more influential than her original empirical work in the area of cultural neuroscience, Nalini also published a series of important review articles that tied together the findings of the multitude of researchers working in this emerging field. These chapters and review articles helped to introduce cultural neuroscience to researchers in both of its parent disciplines of cultural psychology (e.g., Ambady, 2011; Chiao & Ambady, 2007) and neuroscience (e.g., Freeman, Rule, & Ambady, 2009; Rule, Freeman, & Ambady, 2013), as well as for general audiences in psychology (Ambady & Bharucha, 2009).

After spending a year's sabbatical as a fellow at the Center for Advanced Study in the Behavioral Sciences at Stanford University from 2009 to 2010, Nalini permanently relocated from Tufts to the psychology department at Stanford University in the summer of 2011. Unfortunately, her time at Stanford would be short-lived. In November 2012, Nalini received the news that the leukemia she had initially and miraculously survived in 2004 (diagnosed just as she was arriving at Tufts) had returned. Faced with the need for a bone marrow transplant, a tide of outpouring from Nalini's former students, colleagues, and strangers in the field who were merely admirers of her work pooled together to launch an international campaign to find a donor and raise awareness of the need for South Asians in the international bone marrow registries. Despite their valiant and tireless efforts, a suitable donor was never found. Throughout rounds of debilitating chemotherapy and numerous physical setbacks, Nalini was working on her research to the very end. Some of her last days of consciousness were spent meeting with students and colleagues in her hospital room contemplating data, discussing changes in the field, and planning new studies with as much enthusiasm as she had at the peak of her health. Nalini loved her work and it was her passion for understanding human thought and behavior that allowed her to endure through so many trying times: from her early challenges in graduate school through her brutal fight with cancer.

Among all of her accomplishments, though, one of the things that mattered most to Nalini was her students. Nalini received five separate mentorship awards during her career. Much like her own mentor, Bob Rosenthal, Nalini thought of her students not just as apprentices or trainees but as an extension of her family. Her warmth and sincere interest in her students' lives, professional and personal, engendered a strong feeling of reciprocation among most of her students. At Tufts, her graduate students nicknamed her "Momma Ambady" and had shirts made that brandished a "Team Ambady" logo across the front. She showed support for her students' development as individuals and continued to maintain this support even in cases where her students decided that academia, teaching, or research was not for them. She believed it was important that every person follow his or her own personal passion and was happy to help with that however she could, regardless of where that path ultimately ended.

Although Nalini passed away at a time when cultural neuroscience was still developing as a discipline, her early contributions to the field helped substantially with those initial stages of growth. She therefore leaves behind a legacy in social psychology, cultural psychology, and the social and cultural neurosciences to which we are all beneficiaries. Her contributions to the field in terms of the findings that she generated, the training that she invested in some of the field's best researchers, and through her warmth and leadership are long lasting. We are left only to imagine how much more she would have given to the field and what unexpected new heights she might have reached had her prolific career not been curtailed so early. Yet she will live on not only in the memories of those who knew her personally but as each of us within the field stands upon the shoulders of her foundational work to help establish cultural neuroscience. Her life is an inspiring example of excellence in science and of how perseverance, passion, and unexpected luck can transform one's life and make an impact beyond what one might ever expect one could.

University of Toronto, Toronto, ON, Canada Nicholas O. Rule

References

Adams, R. B., Jr., Franklin, R. G., Jr., Kveraga, K., Ambady, N., Kleck, R. E., Whalen, P. J., et al. (2012). Amygdala responses to averted versus direct gaze fear vary as a function of presentation speed. *Social Cognitive and Affective Neuroscience, 7*, 568–577.

Adams, R. B., Jr., Franklin, R. G., Jr., Nelson, A. J., Gordon, H. L., Kleck, R. E., Whalen, P. J., et al. (2011). Differentially tuned responses to severely restricted versus prolonged awareness of threat: A preliminary fMRI investigation. *Brain and Cognition, 77*, 113–119.

Adams, R. B., Jr., Franklin, R. G., Jr., Rule, N. O., Freeman, J. B., Kveraga, K., Hadjikhani, N., et al. (2010). Culture, gaze, and the neural processing of fear expressions. *Social Cognitive and Affective Neuroscience, 5*, 340–348.

Adams, R. B., Jr., Gordon, H. L., Baird, A. A., Ambady, N., & Kleck, R. E. (2003). Effects of gaze on amygdala sensitivity to anger and fear faces. *Science, 300*, 1536.

Adams, R. B., Jr., Rule, N. O., Franklin, R. G., Jr., Wang, E. J., Stevenson, M. T., Yoshikawa, S., et al. (2010). Cross-cultural reading the mind in the eyes: An fMRI investigation. *Journal of Cognitive Neuroscience, 22*, 97–108.

Ambady, N. (2011). Culture, brain, and behavior. In M. Gelfand, C. Chiu, & Y. Hong (Eds.), *Advances in cultural psychology* (Vol. 2, pp. 53–90). Oxford: Oxford University Press.

Ambady, N., & Bharucha, J. (2009). Culture and the brain. *Current Directions in Psychological Science, 18*, 342–345.

Ambady, N., Hallahan, M., & Conner, B. (1999). Accuracy of judgments of sexual orientation from thin slices of behavior. *Journal of Personality and Social Psychology, 77*, 538–547.

Ambady, N., Koo, J., Lee, F., & Rosenthal, R. (1996). More than words: Linguistic and nonlinguistic politeness in two cultures. *Journal of Personality and Social Psychology, 5*, 996–1011.

Ambady, N., Krabbenhoft, M. A., & Hogan, D. (2006). The 30-sec sale: Using thin slice judgments to evaluate sales effectiveness. *Journal of Consumer Psychology, 16*, 4–13.

Ambady, N., & Rosenthal, R. (1993). Half a minute: Predicting teacher evaluations from thin slices of nonverbal behavior and physical attractiveness. *Journal of Personality and Social Psychology, 64*, 431–441.

Ambady, N., & Skowronski, J. (2008). *First impressions*. New York: Guilford.

Baron-Cohen, S., Wheelwright, S., Hill, J., Raste, Y., & Plumb, I. (2001). The "reading the mind in the eyes" test revised version: A study with normal adults, and adults with Asperger syndrome or high-functioning autism. *Journal of Child Psychology and Psychiatry, 42*, 241–251.

Chiao, J. Y., Adams, R. B., Tse, P., Richeson, J. A., & Ambady, N. (2008). Neural substrates for recognizing dominance and submission. *Group Processes and Intergroup Relations, 11*, 201–214.

Chiao, J. Y., & Ambady, N. (2007). Cultural neuroscience: Parsing universality and diversity across levels of analysis. In S. Kitayama & D. Cohen (Eds.), *The handbook of cultural psychology* (pp. 237–254). New York: Guilford Press.
Chiao, J. Y., Heck, H. E., Nakayama, K., & Ambady, N. (2006). Priming race in biracial observers affects visual search for Black and White faces. *Psychological Science, 17*, 387–392.
Chiao, J. Y., Iidaka, T., Gordon, H. L., Nogawa, J., Bar, M., Aminoff, E., et al. (2008). Cultural specificity in amygdala response to fear faces. *Journal of Cognitive Neuroscience, 20*, 2167–2174.
Chiu, P., Ambady, N., & Deldin, P. (2004). CNV in response to emotional in- and out-group stimuli differentiates high- and low-prejudiced individuals. *Journal of Cognitive Neuroscience, 16*, 1830–1839.
Cloutier, J., Ambady, N., Meagher, T., & Gabrieli, J. D. E. (2012). The neural substrates of person perception: Spontaneous use of financial and moral status knowledge. *Neuropsychologia, 50*, 2371–2376.
Ekman, P., & Friesen, W. V. (1971). Constants across cultures in the face and emotion. *Journal of Personality and Social Psychology, 17*, 124–129.
Elfenbein, H. A., & Ambady, N. (2002a). On the universality and cultural specificity of emotion recognition: A meta-analysis. *Psychological Bulletin, 128*, 203–235.
Elfenbein, H. A., & Ambady, N. (2002b). Is there an in-group advantage in emotion recognition? *Psychological Bulletin, 128*, 243–249.
Elfenbein, H. A., & Ambady, N. (2003a). When familiarity breeds accuracy: Cultural exposure and facial emotion recognition. *Journal of Personality and Social Psychology, 85*, 276–290.
Elfenbein, H. A., & Ambady, N. (2003b). Cultural similarity's consequences: A distance perspective on cross-cultural differences in emotion recognition. *Journal of Cross-Cultural Psychology, 34*, 92–109.
Elfenbein, H. A., Mandal, M. K., Ambady, N., Harizuka, S., & Kumar, S. (2004). Hemifacial differences in the in-group advantage in emotion recognition. *Cognition and Emotion, 18*, 613–629.
Freeman, J. B., & Ambady, N. (2011). A dynamic interactive theory of person construal. *Psychological Review, 118*, 1–33.
Freeman, J. B., Johnson, K. L., Ambady, N., & Rule, N. O. (2010). Sexual orientation perception involves gendered facial cues. *Personality and Social Psychology Bulletin, 36*, 1318–1331.
Freeman, J. B., Ma, Y., Han, S., & Ambady, N. (2013). Influences of culture and visual context on real-time social categorization. *Journal of Experimental Social Psychology, 49*, 206–210.
Freeman, J. B., Rule, N. O., Adams, R. B., Jr., & Ambady, N. (2009). Culture shapes a mesolimbic response to signals of dominance and subordination that associates with behavior. *NeuroImage, 47*, 353–359.
Freeman, J. B., Rule, N. O., & Ambady, N. (2009). The cultural neuroscience of social perception. *Progress in Brain Research, 178*, 191–201.
Freeman, J. B., Schiller, D., Rule, N. O., & Ambady, N. (2010). The neural origins of superficial and individuated judgments about ingroup and outgroup members. *Human Brain Mapping, 31*, 150–159.
Ivcevic, Z., & Ambady, N. (2013). Face to (Face)Book: The two faces of social behavior? *Journal of Personality, 81*, 290–301.
Krendl, A., & Ambady, N. (2010). Older adults' decoding of emotions: Role of dynamic versus static cues and age-related cognitive decline. *Psychology and Aging, 25*, 788–793.
Krendl, A. C., Kensinger, E. A., & Ambady, N. (2012). How does the brain regulate negative bias to stigma? *Social Cognitive and Affective Neuroscience, 7*, 715–726.
Krendl, A. C., Moran, J. M., & Ambady, N. (2013). Does context matter in evaluations of stigmatized individuals? An fMRI study. *Social Cognitive and Affective Neuroscience, 8*, 602–608.
Mandal, M. K., & Ambady, N. (2004). Laterality of facial expressions of emotion: Universal and culture-specific influences. *Behavioural Neurology, 15*, 23–34.
Marsh, A. A., Elfenbein, H. A., & Ambady, N. (2003). Nonverbal "accents": Cultural differences in facial expressions of emotion. *Psychological Science, 14*, 373–376.

Marsh, A. A., Elfenbein, H. A., & Ambady, N. (2007). Separated by a common language: Nonverbal accents and cultural stereotypes about Americans and Australians. *Journal of Cross-Cultural Psychology, 38*, 284–301.

Pauker, K., Weisbuch, M., Ambady, N., Sommers, S. R., Adams, R. B., Jr., & Ivcevic, Z. (2009). Not so Black and White: Memory for ambiguous group members. *Journal of Personality and Social Psychology, 96*, 795–810.

Rattan, A., & Ambady, N. (2013). Diversity ideologies and intergroup relations: An examination of colorblindness and multiculturalism. *European Journal of Social Psychology, 43*, 12–21.

Rule, N. O., & Ambady, N. (2008). The face of success: Inferences from Chief Executive Officers' appearance predict company profits. *Psychological Science, 19*, 109–111.

Rule, N. O., Ambady, N., Adams, R. B., Jr., & Macrae, C. N. (2008). Accuracy and awareness in the perception and categorization of male sexual orientation. *Journal of Personality and Social Psychology, 95*, 1019–1028.

Rule, N. O., Ambady, N., Adams, R. B., Jr., Ozono, H., Nakashima, S., Yoshikawa, S., et al. (2010). Polling the face: Prediction and consensus across cultures. *Journal of Personality and Social Psychology, 98*, 1–15.

Rule, N. O., Freeman, J. B., & Ambady, N. (2013). Culture in social neuroscience: A review. *Social Neuroscience, 8*, 3–10.

Rule, N. O., Freeman, J. B., Moran, J. M., Gabrieli, J. D. E., Adams, R. B., Jr., & Ambady, N. (2010). Voting behavior is reflected in amygdala response across cultures. *Social Cognitive and Affective Neuroscience, 5*, 349–355.

Rule, N. O., Ishii, K., Ambady, N., Rosen, K. S., & Hallett, K. C. (2011). Found in translation: Cross-cultural consensus in the accurate categorization of male sexual orientation. *Personality and Social Psychology Bulletin, 37*, 1449–1507.

Rule, N. O., Moran, J. M., Freeman, J. B., Whitfield-Gabrieli, S., Gabrieli, J. D. E., & Ambady, N. (2011). Face value: Amygdala response reflects the validity of first impressions. *NeuroImage, 54*, 734–741.

Rule, N. O., Rosen, K. S., Slepian, M. L., & Ambady, N. (2011). Mating interest improves women's accuracy in judging male sexual orientation. *Psychological Science, 22*, 881–886.

Shih, M., Ambady, N., Richeson, J. A., Fujita, K., & Gray, H. M. (2002). Stereotype performance boosts: The impact of self-relevance and the manner of stereotype activation. *Journal of Personality and Social Psychology, 83*, 638–647.

Shih, M., Pittinsky, T. L., & Ambady, N. (1999). Stereotype susceptibility: Identity salience and shifts in quantitative performance. *Psychological Science, 10*, 80–83.

Slepian, M. L., Weisbuch, M., Rule, N. O., & Ambady, N. (2011). Tough and tender: Embodied categorization of gender. *Psychological Science, 22*, 2–28.

Steele, C. M., & Aronson, J. (1995). Stereotype threat and the intellectual performance of African Americans. *Journal of Personality and Social Psychology, 69*, 797–811.

Vascellaro, J. E. (2002). Psychology professor denied tenure. *The Harvard Crimson*. Retrieved December 30, 2013, from http://www.thecrimson.com/article/2002/12/11/psychology-professor-denied-tenure-in-a/

Vascellaro, J. E. (2003). Innovative psychologist to leave for Tufts. *The Harvard Crimson*. Retrieved December 30, 2013, from http://www.thecrimson.com/article/2003/9/17/innovative-psychologist-to-leave-for-tufts/

Weisbuch, M., & Ambady, N. (2008). Affective divergence: Automatic responses to others' emotions depend on group membership. *Journal of Personality and Social Psychology, 95*, 1063–1079.

Weisbuch, M., & Ambady, N. (2009). Unspoken cultural influence: Exposure to and influence of nonverbal bias. *Journal of Personality and Social Psychology, 6*, 1104–1119.

Weisbuch, M., Pauker, K., & Ambady, N. (2009). The subtle transmission of race bias via televised nonverbal behavior. *Science, 326*, 1711–1714.

Preface

This book had two origination points. One point derived from the second editor (Landis) becoming convinced after 40 years in the field of intercultural relations research (as well as 35 years editing the major journal in the field) that a wall had been reached. It seemed that very little new was being produced in the research community. To be sure, populations changed (e.g., from Japanese or Israelis to South Africans, etc.) and there was some progress on the methodology (e.g., an emphasis on multilevel analysis), but still basically the same studies were being done over and over using the same measures and having the same theoretical basis, when there was one. The results, while often significant, had low variance accounts. So, perhaps really new methodologies as well as theories could help the field out of the doldrums.

The second origination point occurred when we ran across a brief article by Nalini Ambady describing the provocative results of neuroscientific studies of culture. Most of the studies we discovered were narrowly focused on finding cultural differences in brain function. That set us wondering how those differences might enable or degrade individuals' ability to function in different social and cultural settings. So, we set about contacting the leading workers in the field asking for contributions and further asking that they give specific consideration to the implications of their work for intercultural relations research. Most were enthusiastic about the idea, seeing it as a way of opening a new research avenue. We also encouraged that the researchers in brain science and intercultural research jointly produce the chapters of the book. That was true in some cases but not others. Hopefully, the future will see more joint efforts. We offer the joint editorship of this book as a model: a neuroscientist (JW) and an intercultural researcher (DL) working together to produce something that neither could have done alone.

We took the idea of the book to Tony Marsella, the editor of the International and Cultural Psychology series at Springer SBM, and he also was enthusiastic about the project. And, for this, we offer our gratitude and a heartfelt *Mahalo Nui Loa*.

There are others whose help and support made this book possible. In Russellville, Kyla Warnick was always able to breathe excitement into this project by asking wonderfully insightful questions and offering great discussions on the chapters. I can't thank her enough for the support she has provided throughout this project.

In Hilo, Rae M. Landis, who for 55 years has been a terrific supporter and prodder to compensate for her husband's inveterate tendency to procrastinate. She, above all, deserves credit for seeing this book finally completed.

We also acknowledge the important and significant contributions of Sharon Panulla, Execute Editor for Psychology at Springer SBM Publications. Her support and encouragement made this volume possible.

Russellville, AR, USA Jason E. Warnick
Hilo, HI, USA Dan Landis

Contents

Contributors

Anna Abraham Department of Psychology, School of Social, Psychological and Communication Sciences, Leeds Beckett University, Leeds, UK

Reginald B. Adams Jr. Department of Psychology, The Pennsylvania State University, University Park, PA, USA

Nalini Ambady (deceased)

Sergey Blok University of Maryland Center for Advanced Study of Language, College Park, MD, USA

Booz Allen Hamilton Inc., Rockville, MD, USA

Micaela Y. Chan Center for Vital Longevity, School of Behavioral and Brain Sciences, The University of Texas at Dallas, Dallas, TX, USA

Pin-Hao A. Chen Department of Psychological and Brain Sciences, Dartmouth College, Hanover, NH, USA

Emily L. Coderre Cognitive Neurology/Neuropsychology, Department of Neurology, John Hopkins University School of Medicine, Baltimore, MD, USA

Robert Doole Department of Biological Science and Technology, National Chiao Tung University, Hsinchu, Taiwan

Robert G. Franklin Jr. Department of Behavioral Sciences, Anderson University, Anderson, SC, USA

Jonathan B. Freeman Department of Psychology, New York University, New York, NY, USA

Sharon Glazer University of Maryland Center for Advanced Study of Language, College Park, MD, USA

Division of Applied Behavioral Sciences, University of Baltimore, Baltimore, MD, USA

Angela Gutchess Department of Psychology, Brandeis University, Waltham, MA, USA

Shihui Han Department of Psychology, PKU-IDG/McGovern Institute for Brain Research, Peking University, Beijing, People's Republic of China

Todd F. Heatherton Department of Psychological and Brain Sciences, Dartmouth College, Hanover, NH, USA

Chih-Mao Huang Department of Biological Science and Technology, National Chiao Tung University, Hsinchu, Taiwan

Laboratory for Cognitive Neuroscience, Institute of Linguistics, Academia Sinica, Taipei, Taiwan

Sarah Huff Department of Psychology, Brandeis University, Waltham, MA, USA

Dan Landis University of Hawaii, Keaau, HI, USA

Laura Ligouri Department of Psychology, Brandeis University, Waltham, MA, USA

Andrew M. Mathis University of Maryland Center for Advanced Study of Language, College Park, MD, USA

Booz Allen Hamilton Inc., Washington, DC, USA

Alissa J. Mrazek Department of Psychology, Northwestern University, Evanston, IL, USA

Jinkyung Na School of Behavioral and Brain Sciences, University of Texas at Dallas, Richardson, TX, USA

Nicholas O. Rule Department of Psychology, University of Toronto, Toronto, ON, Canada

Michael T. Stevenson Department of Occupational Therapy, Tufts University, Medford, MA, USA

Jason E. Warnick Department of Behavioral Sciences, Arkansas Tech University, Russellville, AR, USA

About the Editors

Jason E. Warnick, Ph.D. is an Associate Professor of Psychology in the Department of Behavioral Sciences at Arkansas Tech University and holds a doctorate in Experimental Psychology from the University of Mississippi. He is a fellow of the International Stress and Behavior Society and on the editorial board of seven peer-reviewed journals. In 2013, he was voted by his university's student body to receive the Professor of the Year Award, and in 2014, he received the Arkansas Tech University Faculty Excellence Award for Service. Additionally, he has received awards for academic advising from the National Academic Advising Association and for online course design from the Blackboard Catalyst Award Program. His research in the fields of neuroscience and sport psychology has been published in a variety of journals and edited books and has been featured in numerous media outlets.

Dan Landis (Ph.D, General-theoretical Psychology, 1963, Wayne State University) has held several academic and consulting positions, including Affiliate Professor of Psychology at the University of Hawaii (both Manoa and Hilo campuses), Professor Emeritus of Psychology, and Dean Emeritus, University of Mississippi. He is a past Chair of the Department of Psychology at Indiana-Purdue University, Indianapolis, Research Psychologist at Educational Testing Service, Senior Research Psychologist at the Franklin Institute Research laboratories, and Visiting Research Professors at the East-west Center and the Defense Equal Opportunity Management Institute. Dr. Landis is the author/coauthor of over 100 books, articles, technical reports, and presentations in areas such as the measurement of equal opportunity climate in military and civilian organizations, racial and gender discrimination, perception, statistics, sexual behavior and attitudes, and cross-cultural psychology and training. He is the founding Editor-in-Chief of the *International Journal of Intercultural Relations (1977–2011)*, the founding President of the International Academy for Intercultural Research (1997–2005), the coeditor/author of *Ethnic Conflict* (Sage, 1985), the coeditor/author of the three editions of the *Handbook of Intercultural Training* (1986, 1996, and 2004), and coeditor/author of *Handbook of Ethnic Conflict* (Springer, 2012). A Mandarin translation of the 2004 Handbook was published by Peking University Press in 2010. Dr. Landis is a Fellow of the American Psychological Association, a Fellow of the Society for the Psychological Study of Social Issues, and a Fellow of the Association for Psychological Science. He is listed in Who's Who in America and other biographical compendiums. In 1987, Dr. Landis, together with Dr. Mickey Dansby and Dr. Gloria Fisher, developed the Military Equal Opportunity Climate Survey (MEOCS) which, in various versions, is in wide use throughout the U.S. Department of Defense. During 1994–1996, Dr. Landis was appointed the first Shirley J. Bach Visiting Professor at the Defense Equal Opportunity Management Institute (DEOMI). During his tenure at DEOMI, he conducted further studies of equal opportunity climate in the U.S. Armed Forces as well as investigated racial disparities in the military justice system. In 2007, he was given a Lifetime Achievement Award by the International Academy for Intercultural Research, and in 2012, he was honored by the American Psychological Association with its award for Distinguished Contribution to the Internationalization of Psychology. Since 2005, he has been the Executive Director and Treasurer of the International Academy for Intercultural Research. In 2013, he was appointed the Editor for Behavioral Sciences of SpringerPlus, an open access, highly peer-reviewed, and online journal published by SpringerOpen. A fourth edition of the *Handbook of Intercultural Training* is now in progress with a publication date of 2015.

About the Authors

Anna Abraham is a Reader at the Leeds Beckett University in the UK and holds a doctorate in neuroscience from the Ruhr University Bochum in Germany. She is a psychologist and a cognitive neuroscientist whose research centers on the neurocognitive study of imagination, encompassing domains such as creativity, mental time travel, mental state reasoning, and the reality-fiction distinction.

Reginald B. Adams, Jr. Reg's work currently focuses on how we extract social and emotional meaning from nonverbal cues, particularly via the face. He seeks to better understand how multiple social messages (e.g., emotion, gender, race, age, etc.) combine and interact to form unified representations that guide our impressions of and responses to others. Although his questions are social psychological in origin, they draw heavily upon visual cognition and affective neuroscience to address social perception at the functional, cross-cultural, and neuroanatomical levels. He coedited the book *The Science of Social Vision*, at Oxford University Press, and has authored a wide range of related empirical articles. He received a SAGE Young Scholars Award from The Foundation for Personality and Social Psychology (2010) to support furthering these efforts. Reg earned his Ph.D. at Dartmouth College and is currently Associate Professor of psychology at Penn State University. Before joining the Penn State faculty, he was awarded a National Research Service Award (NRSA) to train as a postdoctoral fellow at Harvard and Tufts Universities working with Dr. Nalini Ambady and since has been funded by the National Science Foundation and the National Institutes of Mental Health and of Aging (NIH) to support his ongoing work on the compound nature of social sensory perception.

Nalini Ambady received her Ph.D. in social psychology from Harvard University and taught at Holy Cross College, Harvard University, where she was the John and Ruth Hazel Associate Professor of the Social Science, and Tufts University, where she was the Neubauer Faculty Fellow and Professor, before moving to Stanford University in 2011. Her research interests included examining the accuracy of social, emotional, and perceptual judgments, how personal and social identities affect cognition and performance, dyadic interactions (especially those involving status differentiated dyads), and nonverbal communication. She became particularly interested in applying innovative and integrative methods to examine such phenomena from multiple perspectives ranging from the biological to the sociocultural and connecting these insights to real-world problems. As part of this effort, while at Stanford she founded the Center for Social Psychological Answers to Real-World Questions (SPARQ), which continues the efforts to share social psychological insights with the people who work directly to improve society.

Sergey Blok is a behavioral researcher who specializes in the application of cognitive science principles to problems in health, usability, and personnel readiness. He holds a doctorate degree in Cognitive Psychology from Northwestern University and is a Lead Scientist at Booz Allen Hamilton. In his academic work, Dr. Blok has published on topics that include human reasoning, decision-making, and culture.

Micaela Y. Chan is currently a PhD student in the Cognition and Neuroscience program at the University of Texas Dallas and received her M.S. in Applied Cognition and Neuroscience at the University of Texas at Dallas and a B.Sc. in Psychology from the University of Illinois at Urbana-Champaign. Her current research interest is how large-scale functional brain networks changes over the adult life span and how these changes influence cognitive performance. Additionally, she is interested in further exploring how psychosocial (e.g., cultural surroundings and well-being) and neural (e.g., brain networks) variables interact and influence human behaviors.

Pin-Hao A. Chen is currently a PhD candidate in Cognitive Neuroscience at Dartmouth College. Before moving to the United States, he was a clinical psychologist in Taiwan. In 2011, he was sponsored by Fulbright scholarship and joined Dr. Heatherton's lab at Dartmouth College. He also received student travel award from International Cultural Neuroscience Consortium in 2012. He is especially interested in how to use brain-imaging techniques to understand the acculturation processes within immigrants. He has published a journal article in the journal *Culture and Brain* and had several under-review articles that are relevant to the issue of acculturation.

Emily L. Coderre is a postdoctoral fellow in the Department of Neurology at the Johns Hopkins University in Baltimore, MD. She holds an MSc in Cognitive Neuroscience and Neuroimaging and a PhD in Cognitive Psychology from the University of Nottingham in the UK. She is a member of a number of professional societies including the Cognitive Neuroscience Society and the Society for the Neurobiology of Language. She has received a number of awards for research support, including a travel prize to support a 6-month position as a guest researcher at the National Institutes of Health (NIH) during her graduate studies. Her research in cognitive neuroscience focuses on language processing in the brain, particularly in exceptional cases such as bilingualism and autism. She utilizes many techniques to investigate these topics, including electroencephalography (EEG) and functional magnetic resonance imaging (fMRI).

Robert Doole is a master's student at National Chiao Tung University, in Taiwan. His B.Sc. from Trent University, Canada, is in anthropology and psychology. He takes a multidisciplinary approach to the study of brains, consciousness, and cognition. Currently, he is working on operationalizing phenomenological theories of the mind for neurocognitive imaging experiments, as well as learning to engineer electronic brain interfaces and the dynamics which govern complex neural networks.

Robert G. Franklin, Jr. is an Assistant Professor of Psychology at Anderson University in Anderson, SC. He holds a doctorate in social psychology from Penn State University where he worked with Dr. Reg Adams, training broadly in the domain of social vision with a special focus on the relationship between mental state attributions and face memory. This work led to a prestigious university-wide *Alumni Association Dissertation Award.* In a separate but related line of work, he also received the Outstanding Publication by a Graduate Student Award from the Department of Psychology. Before arriving at Anderson, he was a postdoctoral research fellow working with Dr. Leslie Zebrowitz at Brandeis University, where he investigated the influence of aging on trait perceptions from facial cues. Broadly, his work now involves examining how social and cultural influences affect the cognitive and neural mechanisms underlying face perception.

Jonathan B. Freeman is an Assistant Professor of Psychology at New York University. He received his Ph.D. from Tufts University and was on the faculty at Dartmouth College before coming to NYU in 2014. Broadly, his research focuses on the cognitive and neural basis of person perception. He studies the mechanisms underlying basic ways we see and understand other people, including social categories and group membership (e.g., gender, race), personality traits, and emotion. Specifically, his work examines how social dimensions are perceived via multiple facial, vocal, bodily, and contextual cues and how such lower-level perceptual processes interact with higher-order social cognition and prior social and cultural knowledge to shape construals. He is additionally interested in how cognitive and neural dynamics during the person perception process predict downstream social behavior. His studies use a wide range of methodologies, including functional neuroimaging, electrophysiology, real-time behavioral techniques (e.g., computer mouse-tracking), and computational modeling. He is also the developer of the data collection and analysis software, MouseTracker, and has received fellowships and awards from the National Science Foundation, the National Institutes of Health, and the American Psychological Association.

Sharon Glazer is Professor and Chair of the Division of Applied Behavioral Sciences at the University of Baltimore since November 2013. Prior to that, she was a Research Professor at the University of Maryland Center for Advanced Study of

Language and Professor at San Jose State University. Dr. Glazer is also the Editor for the International Journal of Stress Management, on the editorial board for the International Journal of Intercultural Relations, Treasurer for the International Association for Cross-Cultural Psychology, and a member of the Advisory Board for the Institute for Cross-Cultural Management at Florida Institute of Technology. With the auspices of an Erasmus Mundus 3rd country scholar award, Dr. Glazer was a visiting professor to the University of Bologna and Rene Descartes University in Paris. She also taught at the University of Valencia, University of Barcelona, and University of Victoria in Wellington and was a Fulbright scholar at the Technical University of Budapest. Dr. Glazer's research revolves primarily around cross-cultural issues in organizational behavior, specifically occupational and organizational stress, global virtual teams, meaningfulness in life, social support, organizational commitment, values, and temporal orientation. Currently, Dr. Glazer is leading a cross-cultural study of leadership and stress.

Angela Gutchess is an Associate Professor of Psychology at Brandeis University, and she also has appointments in Neuroscience and the Volen Center for Complex Systems. She received her Ph.D. in Psychology from the University of Michigan. Her research investigates the influence of age and culture on memory and social cognition, using both behavioural and neuroimaging (fMRI) methods. She is interested in compensatory strategies and neural reorganization, particularly as an effect of context. Her cross-cultural collaboration on memory is funded by the National Science Foundation.

Dr. Shihui Han is a professor at the Department of Psychology and a principal investigator at PKU-IDG/McGovern Institute for Brain Research, Peking University, China. Dr. Han received a PhD in neuroscience from the University of Science and Technology of China. He held an appointment as a full professor of psychology at Peking University in 2001. He is the founding chief editor of a new journal "Culture and Brain" published by Springer and an associate editor of "Social Cognitive and Affective Neuroscience." He is a Fellow of the Association for Psychological Science. He is the recipient of an Outstanding Young Researcher award from National Natural Science Foundation of China in 2003 and a China Youth Science and Technology Award in 2004. His research aims to clarify social and biological factors that modulate the neural substrates of human social cognition. He has published over 150 empirical research papers as well as review articles in high-profile neuroscience and psychology journals.

Todd F. Heatherton is the Lincoln Filene Professor in Human Relations in the Department of Psychological and Brain Sciences at Dartmouth College. His research takes a social brain sciences approach, which combines theories and methods of evolutionary psychology, social cognition, and cognitive neuroscience to examine the neural underpinnings of social behavior. He is an associate editor of the Journal of Cognitive Neuroscience and serves on many editorial boards and grant review panels. He was elected president of the Society of Personality and Social Psychology in 2011 and has served on the executive committees of the Association of Researchers in Personality and the International Society of Self and Identity. He received the Award for Distinguished Service on Behalf of Social-Personality Psychology in 2005, was named to Thompson Reuters' ISI HighlyCited for Social Sciences in 2010, and received the Carol and Ed Diener Award for Outstanding Mid-Career Contributions to Personality Psychology in 2011. He is a fellow of the American Association for the Advancement of Science, the American Psychological Association, the Association for Psychological Science, the Society of Experimental Social Psychology, and the Society for Personality and Social Psychology.

Chih-Mao Huang is an Assistant Professor of the Department of Biological Science and Technology at National Chiao Tung University, Taiwan. He received his Ph.D. in psychology from the University of Illinois at Urbana-Champaign, USA, a master's degree in cognitive neuroscience from National Yang-Ming University, Taiwan, and bachelor's degree in life science, majoring in molecular biology, from National Tsing-Hua University, Taiwan. His fields of professional interest are cognitive neuroscience, cultural neuroscience, and cognitive aging.

Sarah Huff is a PhD student in Psychology at the University of Michigan. Prior to arriving at Michigan, she received her Master's Degree from Brandeis University and Bachelor's Degree from The Colorado College, both in Psychology. Her research focuses on self, identity, and culture. She is currently working on projects using neuroimaging (fMRI) to investigate how culture and genetic variations influence thinking about the self, emotion regulation, and making choices for the self and others. Sarah is also interested in how individuals manage multiply identities, especially different cultural identities. She recently spent the summer in Singapore as a National Science Foundation East Asia and Pacific Summer Institutes fellow studying multicultural identity and cultural adaptation.

Laura Ligouri is the Project Coordinator at the Saxelab Social Cognitive Neuroscience at MIT and is Co-Director of the Neuroscience and Social Conflict Initiative, a joint venture between MIT and Beyond Conflict. Prior to joining MIT, Laura completed her Master's degrees in both Psychology and Anthropology at Brandeis University. Laura's research addresses the underlying psychological structure of collective victimization and how this impacts dimensions of intergroup bias and violent conflict. Her work seeks to understand the bidirectional, mutual constitution of culture and neurobiological processes that give rise to mental representations of collective victimization and the way in which this translates to behavior.

Andrew M. Mathis is a consultant at Booz Allen Hamilton, holds a Bachelor's degree in Anthropology from Brown University, and is pursuing a Master's degree in Measurement, Statistics, & Evaluation from the University of Maryland, College

Park. He has contributed to research projects in anthropology, addictions psychology, and cross-cultural psychology and currently provides data analysis and IT expertise as a consultant.

Alissa J. Mrazek is a fourth year Ph.D. candidate in the Department of Psychology at Northwestern University and graduated summa cum laude with a BA in psychology and Spanish literature from Cornell University. Her research primarily focuses on empathy, motivation, and decision-making from both a social neuroscience and cross-cultural perspective. In 2013, she was an NSF fellow in the East Asia and Pacific Summer Institute spearheading a transnational experiment on social perception and cooperation. For her research on intergroup prosociality and variation in the serotonin transporter gene, she has received additional grants from the Dispute Resolution Research Center (2013, 2014) and awards from the International Cultural Neuroscience Consortium (2012, 2013). Additionally, she was accepted to and completed the Searle Teaching Certificate Program and received the Sales Teaching Fellowship in 2012 and 2013.

Jinkyung Na is an assistant professor in Psychological Sciences at the University of Texas at Dallas. He earned his bachelor's degree and MA in psychology from Seoul National University and a PhD in Social Psychology from the University of Michigan. Also, he did his postdoctoral training in the Center for Vital Longevity. He is a former recipient of the Brickman Memorial Prize and the Samsung Scholarship. His research focuses on the dynamic interplay between sociocultural factors and psychological processes.

Nicholas O. Rule is an Assistant Professor of Psychology at the University of Toronto where he also holds the Canada Research Chair in Social Perception and Cognition. He received a Ph.D. from Tufts University in 2010 under the mentorship of Nalini Ambady. In 2012, he received an Early Researcher Award from the Ministry of Research and Innovation of Ontario and in 2013 was awarded the Early Career Award from the International Social Cognition Network. His dissertation was finalist for the Harry and Pola Triandis Award from the International Association of Cross-Cultural Psychology and his research focuses on the behavioral and neural correlates of person perception with special emphasis on cross-cultural differences.

Michael T. Stevenson received his M.S. in social psychology from Penn State working with Dr. Reg Adams studying the effects of social categorization and social inclusiveness on emotion recognition accuracy across racial and cultural groups. He also pursued applied interests at Penn State examining the role of social information processing in parenting risk for child maltreatment working with Dr. Sandra Azar. He is currently coordinating a longitudinal project led by Dr. Linda Tickle-Degnen at Tufts University examining health quality of life in Parkinson's disease.

Chapter 1
Introduction and Rationale for This Book

Jason E. Warnick and Dan Landis

Abstract Warnick and Landis provide an overview of the major models of intercultural relations to serve as a foundation for the subsequent chapters in this text. The reader will gain a greater understanding of how the emerging field of cultural neuroscience can be applied to intercultural relations. Additionally, this chapter offers a brief guide to the book.

Each year, millions of people travel into nations and cultures that bear significant differences from their home milieus. The United Nations (Nations, 2013) estimates that in 2013 there were upwards of 232 million people living outside of their home cultures. Not surprising, the developed countries of Europe and North America receive the majority of immigrants, while Africa, Asia, and South America send the majority north. The 232 (3.2 million were tertiary students studying abroad (OECD, 2010)) figure includes not only refugees and immigrants seeking economic or political safety but also people traveling for more less short-term stays (e.g., managers of enterprises). For all of them, the experience can be exciting or terrifying or merely slightly exciting as opposed to slightly bothersome. For some, the new country becomes their permanent home; for others it is but a temporary way station in their life's journeys. Some never adjust to the new environment, huddling with peers from the same home country and rarely venturing out into the uncomfortable world (Kidder, 1977).

The purpose of this book is to hopefully bring together two research domains, which have heretofore remained distinct: intercultural relations and brain science. Intercultural relations has a rich history stretching back over more than half a century to at least the action research studies of Kurt Lewin (Lewin, 1946). Many

J.E. Warnick (✉)
Department of Behavioral Sciences, Arkansas Tech University,
407 West Q Street, Russellville, AR 72801, USA
e-mail: jwarnick@atu.edu

D. Landis
University of Hawaii, 17-123 Palula Place, Keaau, HI 96749, USA
e-mail: danl@hawaii.edu

J.E. Warnick, D. Landis (eds.), *Neuroscience in Intercultural Contexts*,
International and Cultural Psychology, DOI 10.1007/978-1-4939-2260-4_1

intercultural research scholars refer to the anthropological work of the Kluckhohns (Benedict, 1934; Boas, 1911; Kluckholn & Strodtbeck, 1961).[1] And, we are sure that even the ancient Greeks had something to say about the effect of moving into a new culture.

The study of brain functioning has a rich history that can be traced back to ancient Egypt. Papyrus scrolls dating to approximately 3000–2500 BCE describe the earliest known attempt to understand how injury to specific brain regions produce certain behavioral profiles (Feldman & Goodrich, 1999). This case study approach to understanding human brain functioning was the primary research method until the 1930s when electrical stimulation of the brain in conscious humans was tested (Penfield, 1958). Similar methods, while rare, are still conducted to this day and even led to advancements in the treatment of neuropsychiatric disorders (e.g., depression; (Mayberg et al., 2005)).

The contemporary study of brain functioning began with the advent of the positron emission tomography (PET) and functional magnetic resonance imaging (fMRI). These technologies allowed for the ability to see the movement of glucose and oxygen, respectively, and thus watch the metabolic activity in real time while the subjects were completing various tasks. The development of these technologies was followed by an explosion of neuroscientific inquiry in both academia (Legrenzi & Umilta, 2011) and in the popular media (Carr, 2010; Racine, Bar-Ilan, & Illes, 2006). This popularization was one of the driving forces that led the US Congress to declare the 1990s as the decade of the brain and provide increased funding toward neuroscientific research. In the following years, the popularity of neuroimaging has remained strong, while the research methods and the interpretation of neuroimaging results has been debated and refined (Lieberman, Berkman, & Wager, 2009; Vul, Harris, Winkielman, & Pashler, 2009). Social and cultural neuroscience is, by contrast, a fairly recent area of interest. Two recent reviews (Ambady & Bharucha, 2009; Cikara & Bavel, 2014) provide excellent summaries of the history and current status of the field.

While there is an obvious justification for bringing these two research domains together, they, until this book, remain separate. Indeed, the two major intercultural research journals in the field, *International Journal of Intercultural Relations* and the *Journal of Cross-cultural Psychology*, have not published a single paper making such a synthesis. We hope that, with this book, intercultural relations and neuroscientists will be encouraged to collaborate and that both fields will profit by such an interaction.

In this chapter, we will first provide some "war stories" that illustrate some of the issues involving in transitioning to a new culture. A review of the major theories of intercultural relations follows. These theories are derived from social psychology and communication research. We make no claim that the models we examine represent

[1] A good summary of the history of cross-cultural psychology is provided by Otto Klineberg (Klineberg, 1980).

the whole of intercultural relations research. Rather, they may be the ones most able to form a nexus with relevant and recent neuroscience. Then using a fairly complete model of the process of intercultural behavior, we use the existing neuroscience literature to suggest brain regions of interest (ROIs) that may be implicated for each component of the model. Finally, the plan of the book is presented.

1.1 The Vignettes

"James was a well-trained engineer of a leading manufacturer of large container/trailers. After success at the corporation's manufacturing US-based plant, he was promoted to lead manager of the entire European (including the then Soviet Union) area and was to be based in London. He was married and had two preteen children. The company paid all moving expenses, and the family took off for a nice house in a suburb of London. Now, James was a person who was accustomed to having things done (both at work and at home) in certain ways (e.g., requiring that dinner be on the table at a certain time) and did not take kindly to deviations. He would leave a restaurant if the food did not arrive within a narrow time band and would even send food back if not prepared to his exquisitely detailed specifications. But, he was an excellent engineer who had an extensive knowledge of the business so the company felt comfortable in the appointment. As soon as the family deplaned at Heathrow after the flight from New York and cleared customs, they went outside to find a taxi to take them to their new home. Of course, there was a queue at the taxi stand with a fairly longline of people waiting more or less patiently for the cars. James became impatient and dragging his family along pushed his way to the head of the line and jumped into the first taxi in-line. Before he could give the driver his destination, he was informed that the taxi was taken by the first person in-line. He became quite enraged and argued with the driver, to no avail. Finally, he exited the taxi and moved to the back of the line accompanied by the stares of the other people in the queue. This attitude and behavior was emblematic of his tenure in Europe. He could never understand that in many countries, business was never conducted during the first meeting. Indeed, it might not be conducted (in the formal sense) for several meetings. In other countries, there would be very little, if any, small talk with the business part launched before his rear end hit the chair. Despite all this, he completed his 3-year appointment and returned with his family to the United States. While he was not a favorite of his opposite numbers in the various countries, his company was a major player in the industry and provided an important product, so courtesy and business norms dictated a forbearance. For years afterward whenever another employee would comment on how cultural differences played an important part in gaining contracts, he would slough it off as a trivial part of doing business—culture, for him, was simply not important. Needless to say, his wife eventually divorced him, and he wound up a very lonely person estranged from his children"

Ralph Adams was working at his US company's mine site in Peru. Adams' Peruvian counterpart, Ignacia Dominquez, had been in the United States before Adams was assigned to Peru. Adams had not taken much time to spend with Dominguez in either country because of what he perceived to be scheduled and task priorities. Dominguez had invited him to dinner, but Adams had just not been able to find the time. The project that Adams and Dominguez were working on involved selecting a Peruvian contractor for a critical project. Adams wanted to use the US company's procedure for competitive selection of contractors. Dominguez had a very different approach. He said, 'I indent to use a contractor that I have worked with for a long time. I have a good relationship with the head of the company, who is my cousin, and I know he will do a good job.' Adams explained how important it was to get the best contractor at the lowest price. 'The US company's approach will allow us to look at several suppliers and pick the one that best meets our needs,' he said. Dominguez said he already knew the other Peruvian suppliers and assured Adams that his cousin's company was the best. After several more meetings, it was clear that Dominguez was simply not going to use the competitive bid process. Adams felt very uneasy about Dominguez and about his cousin's company.[2]

Aaron Jaminson recently moved to Hawaii Island (otherwise known as the Big Island). He bought a property in the Puna District were all the lots were 1 acre in size. The lots were zoned agricultural/residential. Aaron had come from Idaho and he was a train buff. More than a mere buff, he lived and breathed the history and kinds of trains. He had an extensive model train layout and could regal you with the provenance of all of the cars and engines in the layout—what railroad they depicted, where the cars and the engines had been built (e.g., the Baldwin works in Philadelphia), how much horsepower they generated, and so on. He could also cite chapter and verse about the use of trains in film (e.g., the New York Central Twentieth Century Limited run in Hitchcock's North by Northwest) and so on. Aaron decided to build a guesthouse on his property, sometimes in Hawaii referred to as an Ohana house. In-line with his love for trains, he designed the house to look like a caboose. He sent away for the plans and with those in hand, went down to the country planning department for approval to build. At the planning department, a local man looked over the plans and informed Aaron that they would not approve the project because, "We have no trains in Hawaii. And the last trains on Hawaii Island were destroyed in the 1946 Tsunami." Aaron explained that it is not a real train but merely looks like one and is actually a guesthouse. The inspector was unmoved, saying that it would not meet the standards set in the code. Aaron got progressively more agitated and his voice increased in volume and finally he was virtually shouting. In frustration, he stomped out muttering about "these hidebound Asians…"

When he got home, he was talking to his neighbor Greg who had lived in Hawaii for over three decades and was very knowledgeable about the local culture. Greg listened patiently to Aaron's tale of woe and then offered to go down to the building department and see what the problem was. Greg discovered that the major problem

[2] Vignette provided by Michael Tucker of Tucker International

was that Aaron had placed what looked like train wheels under the house, rather than securing the footings into cement-filled holes in the lava. Thus, the house would be unstable. Aaron had never given the inspector time to explain the problem. Greg explained that Aaron was new to the island and did not understand how business was done. The inspector said, "let me think about it."

A few days later, Aaron got a phone call from the inspector who said, "I've been thinking about your plans, and while we can't call it a caboose on the plans, how about we call it 'an agricultural processing shed?'" Aaron was quite pleased and proceeded to build his caboose. However, Aaron was never happy in Hawaii (too much regulation, he said), and within 3 years he put his property up for sale and moved back to Idaho where he could watch the trains go by day after day.[3]

Finally, a story is recanted about a newly arrived Middle Eastern foreign student at a medium-sized Southern university. Abdul had been to London and was familiar with the kinds of "service" ads that were often posted in the telephone booths. So, when he went into a booth in the small town abutting the university, he saw a notice "for a good time, call (phone number)." So, he called. The woman that answered seemed puzzled by the call and then became disturbed by the persistence of Abdul asking about the price for the "good time." As it turned out, the woman was the wife of the town's mayor. She kept the caller on the line while she sent a servant around the corner to the police station (I said it was a small town, remember?). She asked Abdul where he was and he told her. The next thing that occurred was that he was arrested and charged with several offenses. The upshot was that the university (which did not really have a foreign student advisor) could not protect him and he was sent packing back to his home country, hopefully a little more aware of some cultural differences.

In the next section, we will briefly summarize the major models of intercultural relations that have been proposed over the past half century. Most of these models will be incorporated in a larger model that we will use as the basis for developing some hypothetical relationships between the critical variables in intercultural relations and specific areas of the brain. We emphasize that the suggested relationships are to be subjected to empirical testing using such techniques as fMRI.

1.2 Models of Intercultural Interactions

1.2.1 The Two Equations of Triandis

In 1977, Harry Triandis proposed that social behavior (of which intercultural relations is a subset) could be described by a set of regression equations (Landis, Triandis, & Adamopoulous, 1978; Triandis, 1977). The first equation specified the relations between behavior, behavioral intentions, and habit:

[3] Vignette provided by Greg Trifonovitch

$$\text{ProbBeh} = F \times \left(w_1 H + w_2 \text{BehInt}\right)$$

where:

ProbBeh = probability that the behavior will occur.
H = past occurrence of the behavior.
BehInt = the intention to do the behavior.
Ws are individual difference regression coefficients.
F = facilitating conditions.

The second equation deals with the predictors of behavioral intentions (Landis, McGrew, & Triandis, 1975):

$$\text{BehInt} = w_1 A + w_2 S + w_3 \text{PcVc}$$

where:

A = affective response to the individuals who are the targets of the behavior (usually measured by semantic differential scales (Osgood, May, & Miron, 1975; Osgood, Suci, & Tannenbaum, 1957)).
S = social appropriateness of the behavior (e.g., roles and norms (Landis et al., 1978)).
PcVc = perceived consequences of the behavior.
w_1–w_3 are individual difference regression weights.

Triandis then adapted attribution theory to suggest that the task of a sojourner to a new culture was to create a set of "isomorphic attributions." Isomorphic attributions are those of the other in the social situation. These attributions would consist of roles, norms, affects, etc. and would lead to the development of appropriate behavioral intentions. To teach people isomorphic attributions, he and his colleagues developed a cognitive training technique: the culture assimilator or sensitizer (Albert, 1983; Bhawuk & Triandis, 1996; Cushner & Landis, 1996; Weldon, Carston, Rissman, Slobodin, & Triandis, 1975). This technique has generally shown itself to be the most valid and useful of the many training approaches available to date (Landis, Brislin, & Hulgus, 1985; Landis, Day, McGrew, Miller, & Thomas, 1976; Randolf, Landis, & Tzeng, 1977).

1.2.2 Gudykunst's AUM Model

Gudykunst (1988) defined (on page 73) the model as "…based on the assumption that managing uncertainty and anxiety is necessary for effective intergroup adjustment" (Gudykunst, 1988, 1995; Gudykunst, Guzley, & Hammer, 1996). Gudykunst then went on to specify a list of over 50 hypotheses that could be drawn from this simple model, all of which had operational grounding so that the testing was straightforward (Gao & Gudykunst, 1990; Gudykunst & Sudweeks, 1992). The full model is shown in Fig. 1.1 (Gao & Gudykunst, 1990). Gao and Gudykunst call the top left box social contact, the next one down is perceived similarity, and the last is cultural knowledge.

The model led to a set of assumptions (e.g., "uncertainty and anxiety are independent dimensions of intercultural adjustment"), axioms (e.g., "an increase in strangers' knowledge of the host culture will produce an increase in the accuracy of strangers' predictions and explanations of the behavior of host nationals" in other words, "isomorphic attributions"), and theorems (e.g., "an increase in the permanency of the strangers' stay in the host culture will produce an increase in the anxiety they experience upon entering the host culture"). Upon testing of alternative models, all of the left hand factors had significant weights to both of the intervening variables (anxiety and uncertainty reduction), but the impact on adaptation was indirect. Anxiety and uncertainty reduction were correlated (Gao & Gudykunst, 1990).

The two main variables in AUM theory were drawn from the previous work of Hofstede who, on the basis of an international data gathering in a single organization, suggested that cultures can be described along five dimensions: individualism/collectivisms (U. Kim, Triandis, Kagitcibasi, Choi, & Yoon, 1994), uncertainty avoidance, power distance, masculinity versus femininity, and Confucian dynamism (Hofstede, 1999). The second variable, anxiety, was adapted from the extensive literature on stress and coping in unfamiliar settings (Lazarus & Folkman, 1984; Ward, 2004). It is notable that AUM theory does not deal with the three other Hofstede dimensions, and it is possible that in some settings those other dimensions might be highly implicated in intercultural adaptation.

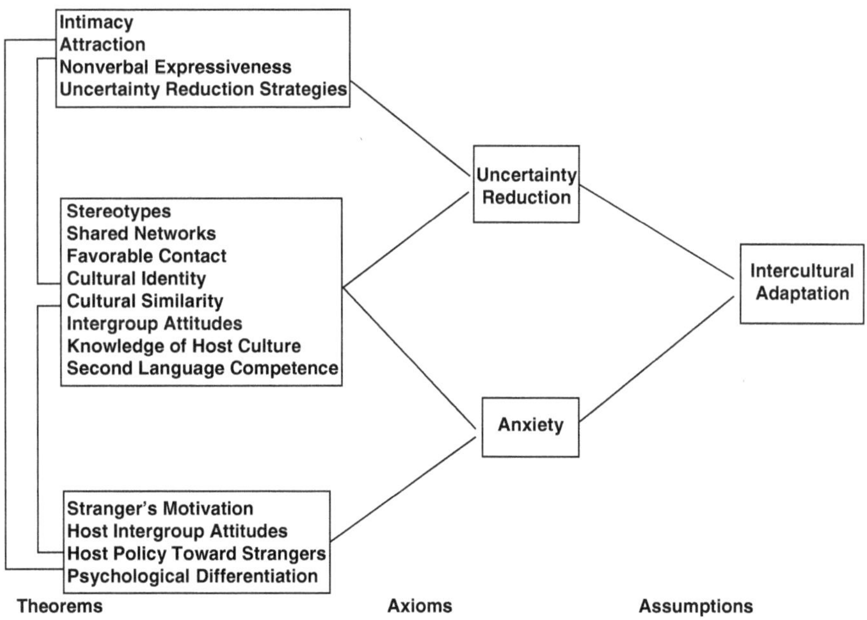

Fig. 1.1 Gudykunst's AUM model

1.2.3 Bennett's Ethnorelativism Stage Model (J. Bennett & Bennett, 2004; M. Bennett, 1993; Hammer, Bennett, & Wiseman, 2003)

Milton Bennett's Developmental Model of Intercultural Sensitivity (DMIS) (Fig. 1.2) outlines the stages that a person goes through from ethnocentrism to ethnorelativism. It is assumed that the sequence of stages is fairly fixed and is traversed in a given order. Hence, the model can be fairly termed a stage theory. That is, a given stage cannot be attained unless all previous stages have also been attained. As the figure below from Bennett and Bennett (2004) depicts, the most common starting point is denial in which the person holds on to the belief that any differences between cultural/racial groups are fictions (except when the "other" is seen as deficient in some way) perhaps promulgated by people with agendas (i.e., "some groups are naturally more able than others" or "**they** do well at sports but not at intellectually demanding professions," and "there are no difficulties between the groups" and, as a congressman asserted, "there is an urban culture that holds back minorities"). The next stage, defense, is one in which "…worldview is polarized into us—them distinctions, so the prevailing attitude is one of being under siege" (Bennett & Bennett, 2004, p. 154). An example is the belief, enunciated by at least one US congressman, that there is a "war against white people." And, "they are taking our jobs," a statement used to argue against allowing immigrants to enter the country.

Minimization, as a strategy for not recognizing differences is quite a bit more complex. For here, small differences are accepted, but the belief is that at some deep level those differences do not matter since all people are the same (i.e., "we are all made in God's image"). People at this stage are unlikely to resort to the extraordinary (e.g., violence) to maintain the status quo but may actually reach out to the "other" to bring them into some essential human group. However, the power of the dominant group will do little to change the fact of institutional privilege, since they are unlikely to recognize that such privilege even exists. Thus, people who rise to the top of organization have done so because they worked harder and deserved their positions, and, if the individuals of the nondominant group would do the same, they would also receive the same benefits.

Acceptance is, for Bennett, the first stage of the proceeding toward ethnorelativism. Here, the person accepts differences and also accords equality to the other. One is willing to accord other ethical systems the same status as one's own and therefore precludes making a choice among the positions. Within organizations, the value of diversity is acknowledged, and stringent efforts may be made to recruit the "others," but there may be little understanding of what to do once those people are in the company. So, as Bennett and Bennett (2004) note, "…they know how to 'talk the talk'…if not with much sophistication" (p. 156).

In adaption, there is a marked shifting of one's frame of reference. This is sometimes called "cultural empathy." This is similar, if not identical to Triandis' concept of "isomorphic attributions." That is, the person comes not only to appreciate cultural differences but also to actually see the world through the other's eyes. For the person, a critical issue will be the retention of a sense of self, while at the same time

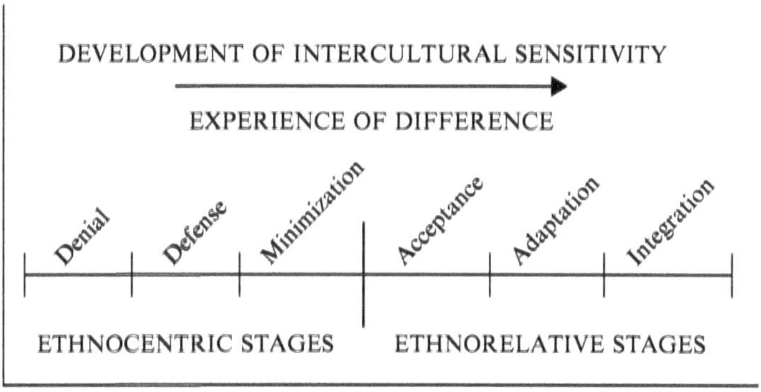

Fig. 1.2 The developmental model of intercultural sensitivity

behaving as the other might behave in a given social situation. For the Bennetts, there are two kinds of adaption: unintentional and intentional. In the first, a person might adapt quite well to one culture, but because the adaptation has not been completely internalized, the adjustment would be incomplete to another culture. Hence, such a person might not be able to guide others in the development of intercultural skills (Bennett & Bennett, 2004, p. 157).

The final stage, which few if any reach, is called integration. The issue here is how to reestablish an identity that encompasses all of their experiences. Their identity becomes marginalized from any one identity, and they are able to transition easily from one cultural context to another. They are able to easily learn a new culture and respond appropriately. Organizations at this stage look at each aspect in its cultural context with little or no concern about the national identity of the company.

Though the DMIS has strength in a set of logical progressions that have a great deal of face validity, that strength is also a weakness. The model does not provide the *deus ex machina* that provides the motivation for the person, or even the organization, to move from one stage to the next. In this aspect, DMIS shares the problem with other stage models (e.g., Freud, Piaget). It is also assumed that movement is unidirectional; that is, it is difficult if not impossible to regress to an earlier stage or at least the rationale for such slippage is not specified.

It has been an additional (Hammer et al., 2003) strength of DMIS that a measure, the Intercultural Development Inventory (IDI), was developed. This measure aims to assess where along the stages a respondent lies. Since the measure uses an interval scale in which statistics that assume a normal distribution are used, the fact that underlying construct is nominal causes an interpretability problem (Stevens, 1946, 1951). Furthermore, application of the IDI often results in values for a particular person indicating multiple stages with perhaps one having the largest index. Although this might vitiate the stage construct, the recent literature on priming of cultural contexts (see Chap. 7, by Glazer, et al.) would suggest that individuals might hold several different stages, each of which can be brought to the fore by appropriate settings or priming (Fig. 1.2).

1.2.4 Ward's ABC Model (Figure from (Ward, 2004))

Ward recognizes that there are multiple theories that attempt to describe the acculturation process. These are capsulated in A (affect), B (behavior), and C (cognitive) theories. So, the model (Fig. 1.3) presented is less a model that would require that the parts fit together in some predictive manner that would allow empirical testing. Rather, she uses the "model" as a template to describe the various theories and their predictions with regard to acculturation. As a heuristic device it is quite compelling and bears serious appraisal.

The ABC model bears a superficial resemblance to Triandis' second equation, described earlier in this chapter. It will be recalled that the equation predicts behavioral intentions as a function of affect, social cognitions (e.g., norms and values), and perceived consequences (which would come about, one would think, as a result of learning). While Triandis did not link those predictors to particular theories, it is quite evident that each one has a rich history. One such history is the extensive literature on emotion, which can be heightened by stress. Hence, the measure of affect would be predicted by such physiological reactions as stress and how the person appraises the threat (Lazarus & Folkman, 1984). Additional stressors are as life changes, appraisal of change particular those in a new culture, and coping strategies. Indeed, Ward has reported several studies looking at such variables in a variety of acculturating groups (Ward, 1996).

A second literature, which relates to Triandis' social cognitions predictor deals with social identity and social cognition. A particularly well-thought example is Stephan and Stephan's Integrated Threat Theory (Stephan & Stephan, 2000). This theory sees as antecedents to discriminatory behavior variables such as prejudice, which in turn are predicted by two kinds of threats (realistic and symbolic), intergroup anxiety, and negative stereotyping. These four mediating variables are themselves predicted by intrapersonal autobiographical aspects such as intergroup conflict, intergroup contact, and so on. Integrated Threat Theory has sparked an extensive literature testing its predictions, in particular those related to the role of anxiety (Stephan, Stephan, & Gudykunst, 1999). This literature is well summarized by Ward (2004) and in several publications by the Stephans (Stephan, Renfro, Esses, Stephan, & Martin, 2005; Stephan & Stephan, 2000).

In regard to the role of anxiety, Stephan et al. (1999) note that a major difference between AUM and ITT is whether it is always negative or sometimes positive in relation to attitudes toward the other. Triandis would seem to come down on the mixed side of the argument since his second equation assumes that affect is experienced as a continuum from bad to good. It may be that, in conformance with much of the early work on anxiety, it acts as a motivator up to a certain point, and then the effects become quickly negative. As a negative, anxiety may act to prevent or disrupt the development of positive behaviors. This was the implication of a study comparing contact and assimilator training where contact alone was found to engender higher levels of anxiety which were correlated with poorer intergroup behaviors (Landis et al., 1985). Anxiety as a felt state resulting from the activation of noradrenergic

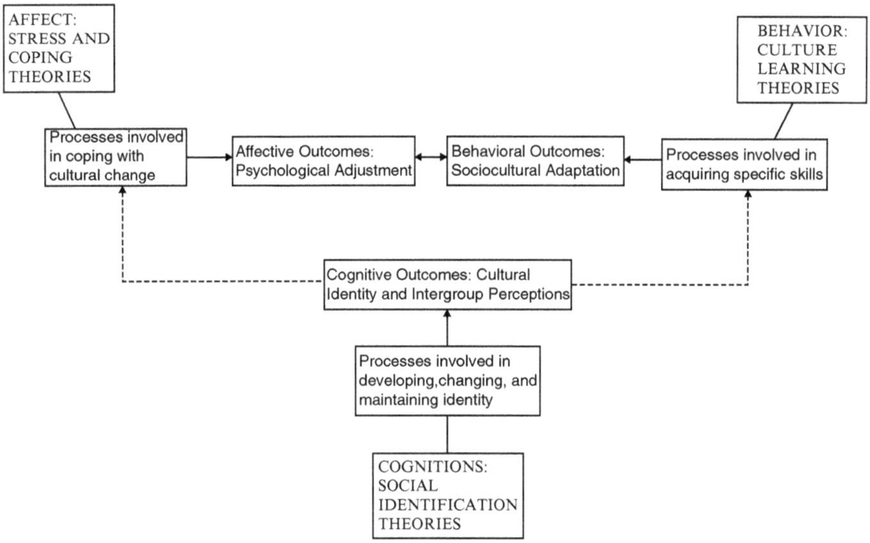

Fig. 1.3 Ward's ABC model of culture contact

mechanisms is most likely present when racial bias is measured implicitly. This hypothesis is lent credence by a study in which a well-known beta-blocker (propranolol) was used to reduce certain proprioceptive awareness (e.g., heart rate) and also to significantly reduce implicit racial bias without effect explicit bias (Terbeck et al., 2012). However, most of these studies were done within the framework of racial bias (e.g., where the targets were African-Americans or Latinos in the United States or South Asians in the United Kingdom). It is unknown whether the same negative attributes of anxiety would apply in less emotionally charged situations.

1.2.5 Berry's Two-Dimensional Acculturation Model (Berry, 2004)

Perhaps the most influential model of acculturation in the past five decades has been his two-dimensional description of the strategies adopted by immigrants (Berry, 1974). This model suggests that immigrants, of whatever stripe, have two choices to make: how much of their home culture to retain and how much of the new culture to adopt. That this model has influenced countless studies, e.g., Berry, Phinney, Sam, and Vedder (2006), is obvious. In 2008, Ward noted that a PsycINFO search produced over 800, certainly now a severe underestimate, studies dealing with the topic (Ward, 2008). To be sure the model has not been without dissenters (Boski, 2008; Chirkov, 2009; Rudmin, 2009). Nevertheless, it has shown itself to be robust enough to spawn numerous empirical studies (Berry et al., 2006) as well as cavils against its measurement processes (Rudmin & Ahmadzadeh, 2001) and conceptual underpinnings (Boski, 2008).

Though the present model envisions a continuum between the endpoints of the two dimensions, most discussion (perhaps engendered by the phrasing of the questions used to mark each facet) has identified just four combinations or strategies: *integration* when both the home and receiving cultures are adopted, *assimilation* when the receiving culture is adopted to the exclusion of the home culture, *separation* when the home culture is retained to the exclusion of the receiving culture, and *marginalization* when neither the home nor the new culture is adopted (Berry, 2004).[4]

The large study of immigrant youth, sometimes referred as the ICSEY study, seemed to show that youth in a variety of countries prefer the integration option (Berry et al., 2006). However, Boski (2008) has criticized these findings on a number of bases and suggested that the high scores for integration, often bumping up against the ceiling, may well be artifacts. These artifacts may be based on a very natural positive affect toward biculturalism and bilingualism as an ideal and not with any actual experience or competence with either phenomenon.

Berry also has related the four acculturation strategies to government policies. As Fig. 1.4 suggests, integration is related to a multicultural policy, while assimilation is related to a melting pot ideology, separation leads to segregation, and, lastly, marginalization is based on an exclusion policy. Bourhis and colleagues proposed a different set of governmental policies (Bourhis, Moise, Perreault, & Senecal, 1997) in their Interactive Acculturation Model: pluralism, civic, assimilation, and ethnist ideologies. Pluralism has three aspects: first, immigrants are expected to adopt the public values of the host country; second, there is respect for the private values of the immigrant; and, finally, the public money can be used to support the immigrants' private values. A civic ideology is in agreement on the first two aspects of pluralism but insists that immigrant cultural groups must be responsible on their own for the maintenance of their values and norms, what Triandis would call their subjective culture. Bourhis pointed out that both pluralism and civic ideologies are related to Berry's multiculturalism policies and, hence, to the integration at the individual level. Assimilation insists that first, immigrant groups must adopt the values of the host culture; second, no public funds are to be spent on these groups to maintain their original culture; and third, the state may intrude on some public domains to help the groups shed their sending cultural values. An ethnist ideology defines who can become a citizen of the host culture and that many immigrants are forever forbidden to attain citizenship status. They may be guest workers, but no matter how long they reside the host culture, they will never become full citizens no matter what the importance of their contributions. There are two variations within this ideology: exclusion and individualism. Exclusion is a dilemma that many countries (who adopt either overtly or covertly this policy) are facing in the modern world. For example, Israel has to deal with providing true and complete citizenship to Arabs who did not leave after the founding of the state in 1948. While Israeli Arabs carry Israeli passports, their day-to-day experience impresses upon them that they are not, and many never become, full citizens (Avishai et al., 2014). This ideology has often

[4] This structure is referred to in several of the chapters in this book.

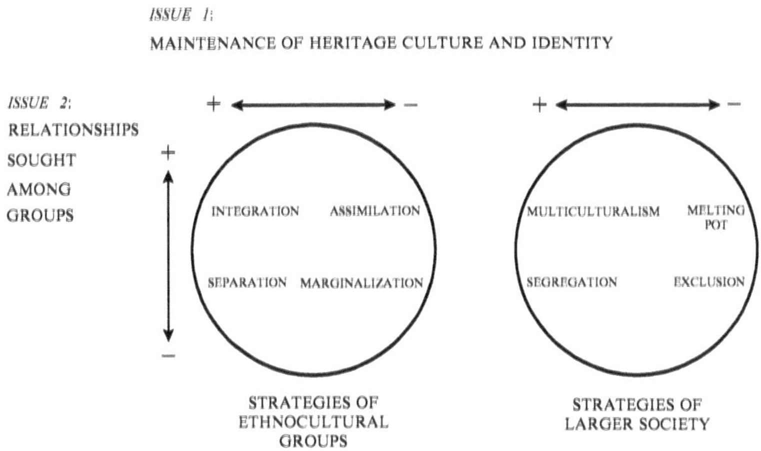

Fig. 1.4 The Berry model of acculturation (from Berry, 2004)

led to a recommended and controversial policy of partition (Habyariman, Humphreys, Posner, & Weinstein, 2008; Muller, 2008). The second variation is a policy of individualism in which cultural differences are not recognized in either the host or sending communities. Rather, immigrants are encouraged to think of themselves as individuals free to choose any aspects or none of either culture (this is similar to constructive marginality in the Bennett scheme, discussed above). In any case, Bourhis's taxonomy demonstrates that the links proposed by Berry between the micro- and the macro levels are not quite as strong as might be supposed. This lack of synchronic relationship between the levels is echoed by van de Vijer, van Hemert, and Poortinga (2008) who noted "…there is no evidence for the assumption that multiculturalism attitudes and policies are related" (p. 101).

One critical aspect of the Bourhis formulation is the insistence on the importance of the host's ideology in forming an acculturation strategy by the migrant. Actually, it is probably the *perception* of the host's ideology that is the critical variable (Noels, Leavitt, & Clement, 2010). When the two dimensions (host versus immigrant community views), and we note whether each combination is consensual, conflictual, or problematic, it turns out that there are only three combinations that result in a nonconflictual interaction (when each group favored an integrative strategy, or when assimilation is favored, and, lastly, when each group prefers the individual strategy). All other possibilities are either problematic or producing conflict (Bourhis et al., 1997). This insistence on the importance of the ideology favored by the host must be included in any reasonable theory of acculturation.

Despite the problematic aspects of Berry's model, it remains the most widely used and accepted theory of acculturation. Therefore, for the purpose of this book, it must be included in any attempt to find an empirical nexus between intercultural relations and brain science.

1.2.6 Navas' RAEM Model (Navas et al., 2005)

Navas and colleagues' Relative Acculturation Extended Model introduced the concept that which acculturation strategy was selected dependent on the domain in which the immigrant was operating. Indeed, depending on the domain, both the host and the immigrant societies might prefer and actually adopt different strategies. Navas et al. (2005) put forth seven different domains as a start to develop a fuller taxonomy of such interactions: religious beliefs, ways of thinking (principles and values), social relationships, family relations, economics (interactions in the economic/business sphere), work, and politics and government. As Navas et al. note:

> …there is no one single acculturation strategy or attitude. The adaptation process is complex (different options can be preferred and adopted at the same time) and relative, since the same strategies are not always used on the same options preferred when the interaction with other cultures takes place in different domains …. (p. 27)

So, domains that are private (e.g., religious beliefs, family relations) are likely to be problematic in relationships between the heritage and host societies. On the other hand, domains that are public (e.g., economics, work, and politics and government) are usually less problematic. This model also differentiates between attitudes (desired strategies) and strategies (the actual behaviors adopted by either society). So, the distinction here is between the ideal strategy to be used in a given domain and the actual strategy implemented. They suggest that the greater the disparity between the ideal and actual, the greater will the frustration and conflict that will ensue. A further suggestion is that the ideal and actual strategies will vary by ethnocultural group and host society. If the immigrant group is made up of individuals whose migration is desired by the host society, there may be more willingness, on the part of the hosts, to accept greater disparity between the two groups. This might well not be true in the case of immigrant groups that are not desired. The skill level of the various groups might also be an important factor in acceptance or rejection. In other words, the process of acculturation "…affects the majority group as much as the minority group…" (Navas et al., 2005, p. 31) (Fig. 1.5).

A significant addition to the study of acculturation was an expansion of the Navas et al. concept by Doucerain and colleagues in a recent paper (Doucerain, Dere, & Ryder, 2013). Those authors start from the obvious axiom that attitudes are not behaviors (Boski, 2008; Ward & Kus, 2012). While attitudes toward multiculturalism may well have some predictive power over behaviors, those actions remain stubbornly multideterminate. They also add the proposition that the heritage/mainstream structure is too gross to capture how people negotiate the cultural divide. Following Navas, they assert that there are likely to be other dimensions that develop because of hybridization as, in any metropolitan context, cultures mix and develop complex new structures. One way such hybridization occurs is through intermarriage. In locales such as Hawaii, where the vast majority of marriages are "mixed," individuals may trace their cultural lineage to 10 or 20 different groups (e.g., Hawaiian, Japanese, Chinese, African-American, Filipino, Portuguese, Puerto

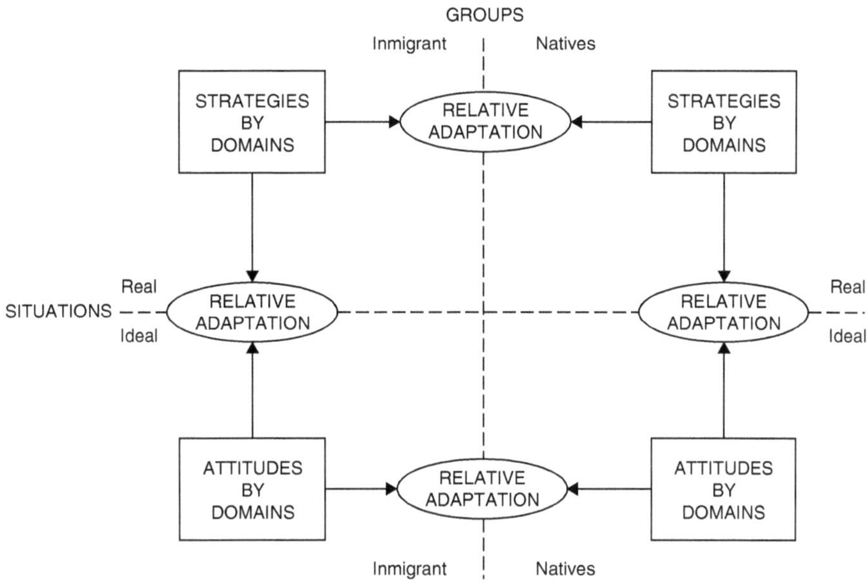

Fig. 1.5 Relative acculturation extended model (RAEM)

Rican, and so on). Doucerain et al. note that such mixing occurs quite regularly and consistently in Montreal, the site of their research.

The third observation made by Doucerain et al. follows closely both Bourhis and Navas and others (Arends-Toth & Vijer, 2004; Bourhis et al., 1997; Navas et al., 2005) and suggests that whatever strategy is developed by the migrant is highly situation specific. This should come as no surprise when one considers the extensive research on priming effects (noted in Chap. 7, by Glazer and her colleagues). For example, individuals who profess either a collectivist or individualist perspective could evince the other cultural syndrome when so primed (Hong, Morris, Chiu, & Benet-Martinez, 2000). Thus, context distinctions such as private/public as well as those listed by Navas, above, become crucial to understanding multicultural behavior.

In order to operationalize the three concerns, Doucerain et al. started from ground zero to see if they could discern and demark the interaction between context, multidimensionality, and domains. In essence, they decided to ask a sample of immigrants to the Montreal area what happened as they interacted during a number of days. Basically, the technique they used was a version of representative design (Brunswik, 1956). Since they couldn't actually follow the respondents around, as did Brunswik, the method of choice was to use a diary covering all the interactions in a particular day. This method is certainly closer to accessing actual behavior as compared to gathering preferences among a restricted set of strategies (Arends-Toth & Vijer, 2004; Boski, 2008; Ward & Kus, 2012). Such a shift to behavior from preferences is certainly to be desired.

The diary, the Cultural Day Reconstruction Method (C-DRM), starts by asking the participants what cultural group they most identify with. They are encouraged to include hyphenated (e.g., Haitian-Canadian) groups to get an indication of hybridization. Then they list all of the episodes that they encountered on that day and for each answer a series of questions (e.g., when the episode began, what the participant was doing when the episode occurred, what language was spoken during the episode, where did the episode occur (e.g., at school, home, work, etc.), with whom did you interact with, and, most important, what cultural group did the participant identify with during the episode). It will be obvious that the diary is a more structured critical incident (Flanagan, 1954). The episodes are then coded by independent coders, and multilevel analyses are performed (Vijer, Hemert, & Poortinga, 2008) because the episodes were nested within individuals. The general finding was that the cultural identification of the participants varied as the context. The authors concluded:

> ...characteristics of the local context of an episode, combined with an assessment of general acculturation attitudes, allowed use to accurately predict momentary cultural affiliation is particularly noteworthy. It underscores the fluid nature of acculturation and suggests that a systematic portrayal of the local context can, at least, partially account for variability in affiliation. (p. 607)

The models we have described up to this point are what might be termed "middle range theories." They take a few (usually very few) antecedent variables and look to assess their relationship to intercultural behaviors; however, those behaviors are defined. While this is in the tradition of Occam's razor, the amount of variance accounted for is often very small even if it is statistically significant. We feel it is probably better to have an overarching model in the forefront that can be systematically separated into sub-models, each of which can then be tested, with the results finally combined into a test of the total model. The section deals with such a process.

1.2.7 Landis-Brislin-Brandt-Bhagat-Bhawuk Intercultural Behavior Model (Landis & Bhawuk, 2004)

Brislin, Landis, and Brandt in 1983 proposed an earlier version of a model of intercultural behavior that is depicted in the figure below (Brislin, Landis, & Brandt, 1983). In 2004, Landis and Bhawuk separated the model into five sub-models and argued that each could be tested independently of the total structure. For convenience, we have adopted those sub-models and for each we suggest the regions of interest (ROIs) that might well be innervated for each cognitive variable. To be sure, the specific relationships have yet to be tested with appropriate brain mapping (e.g., fMRI) technology. Nevertheless, there is sufficient evidence from the neuroscience literature to suggest that the relationships we suggest are, at least, provocative and worthy of investigation.

In the 2004 chapter, Landis and Bhawuk included detailed descriptions of each of the models. We are going to take the liberty of quoting those descriptions at length rather than developing new arguments that will cover the same ground. It will be understood that, even without specific citation, all of the descriptions are from the 2004 chapter (pp. 458–463).

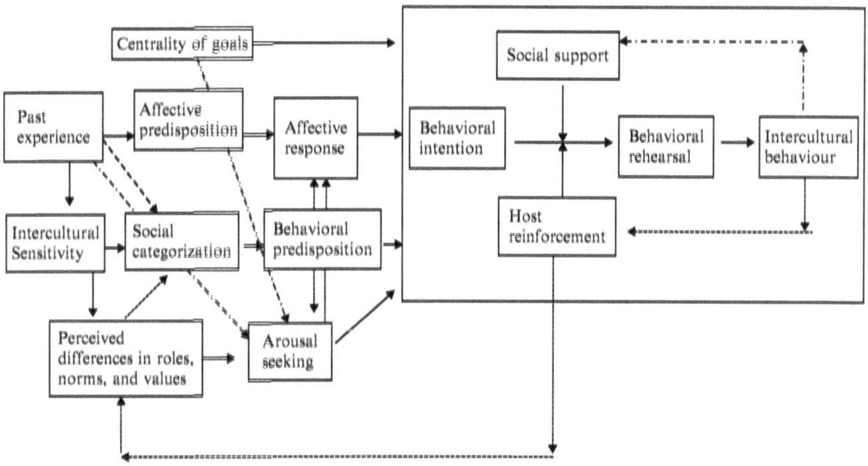

Fig. 1.6 A model of the intercultural behavior process

Before we launch into the sub-models, we include a few comments on the status of brain science (Fig. 1.6).

1.3 Status of Brain Models

The differences and similarities found between cultures in behavior, brain function, and structure were once used to further the nature-nurture debate, e.g., McCrae et al. (2000). However, this dichotomy has been replaced by the epigenetic approach, which is becoming the prevalent model used to explain these culturally bound biobehavioral traits (Kitayama & Tompson, 2010). Epigenetics is the biological subdiscipline that looks at the effects of genes (e.g., behavior) and the environmental variables (e.g., stress) that modify these genetic functions (Zhang & Meaney, 2010). Within this model, genes are not seen as silent determiners of our biological condition but as being responsive to the environment throughout one's lifetime. This approach provides novel ways to investigate the biological structures we find between cultures and the changes we see throughout the acculturation process. The data presented in the chapters in this book will provide the foundation on which epigenetic research can further understand the impact of culture on the biological human (Fig. 1.7).

1.3.1 Model 1

This model is derived from attitude research, where it is quite well established that behavioral intention is the best predictor of a behavior (Triandis, 1977). It is important that intercultural behaviors be measured directly and not through self-report

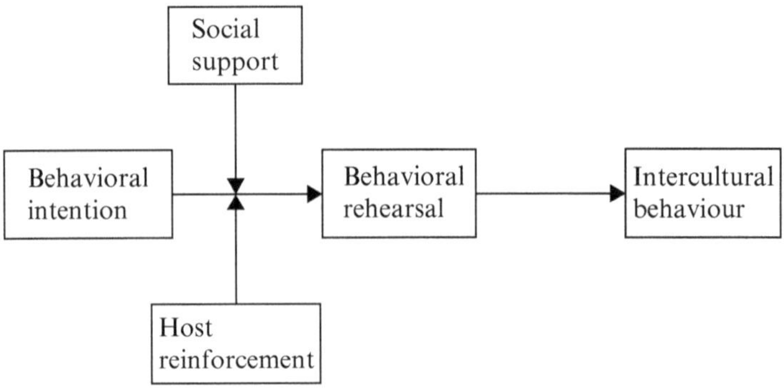

Fig. 1.7 Model 1: Moderators of intercultural behaviors

questionnaires, which are necessarily affected by memory and demand characteristics. Research personnel can measure such behaviors using host ratings or actual observations. Another technique can be adapted from personnel psychology in which employees are rated by themselves, their peers, their subordinates, and their superiors (the so-called 360° degree approach). It is also desirable to assess intercultural behaviors at various periods during the sojourn. Due care, of course, should be taken to assure to have adequate psychometric properties.

Social support has a number of meanings. A skein of mechanisms provided by the home culture organization is one such meaning. Personal support by an accompanying spouse or other family members is another. In any case, this is a poorly researched variable. There is some research pointing to the importance to the expatriate sojourner of the well-being of the accompanying spouse, but it is preliminary, and the findings to date are not strong. Also unstudied is the impact of cross-cultural marriages on successful intercultural behaviors.

The model also includes another moderator variable: host reinforcement. Most theories suggest that what the host does and how often he or she does it is important to the sojourner, but this aspect has been difficult to measure. Ward and Rana-Deuba found that the quality rather than the quantity of the host's behavior is a predictor of total mood disturbance in the sojourner (Ward & Rana-Deuba, 2000).

We propose that behavioral rehearsal is necessary in the intercultural context because people are acquiring new behaviors from another culture, and acquisition of such behaviors will necessarily follow social learning theory (Bandura, 1977). The acquisition of these new behaviors would be moderated by social support as well as host reinforcement. This model can be tested for a number of intercultural behaviors, from learning foreign languages to learning gestures and body language. This model will be the first building block that we will use in the subsequent models (Fig. 1.8) (Table 1.1).

Table 1.1 Regions of interest for model 1 (Fig. 1.7)

Cognitive variable	Region of interest	Biological function
Behavior rehearsal	Caudate	Neuroimaging research has suggested that the ROI is associated with behavioral rehearsal (Grahn, Parkinson, & Owen, 2009)
Host reinforcement	Caudate, mPFC, anterior insula (AI), nucleus accumbens	Studies (e.g., neuroimaging) have suggested that variables related to behavioral intention produce activation in these regions. These ROIs are areas critical for processing emotional stimuli. Further, the Caudate (Grahn et al., 2009) and the mPFC (Shimamura, 2000) are regions implicated in executive functioning. The nucleus accumbens is involved in processing reward states (Sabatinelli, Lang, Bradley, Costa, & Versace, 2007)
Social support	Nucleus accumbens, ventral tegmental area (VTA)	Neuroimaging studies have shown that variables associated with social support activate these ROIs. These ROIs are known to be involved in the processing of reward states (Hikosaka, Bromberg-Martin, Hong, & Matsumoto, 2008; Sabatinelli et al., 2007)
Behavioral intention	Dorsal anterior cingulate cortex (dACC)	Neuroimaging and studies of neurodegenerative (e.g., Parkinson's disease) and neuropsychiatric (e.g., Tourette syndrome) disorders have implicated these regions in the phenomenon of behavioral intention. The dACC (Bush, Lum, & Posner, 2000) and DLPFC (Duncan & Owen, 2000) are ROIs involved in decision-making processes. The premotor cortex (Roland, Larsen, N. Lassen, & Skinhoj, 1980) and basal ganglia (Mizuguchi et al., 2013) are tasked with planning movement and action selection, respectively. The cerebellum is a key region in motor control. The primary somatosensory area processes the sense of touch, and the cuneus is involved in the processing of visual information. See also Ouden, Frith, Firth, and Blakemore (2005)
	Dorsolateral prefrontal cortex (DLPFC), premotor cortex, primary somatosensory area, cuneus, cerebellum, basal ganglia, precuneus/ posterior cingulate cortex, mPFC	

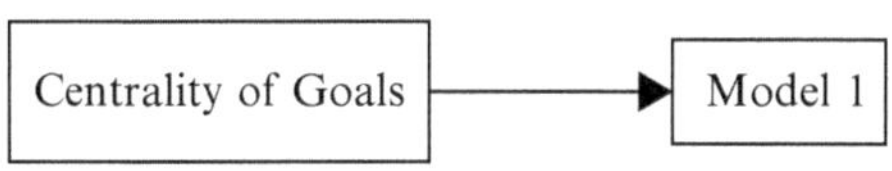

Fig. 1.8 Model 2: Centrality of goals and intercultural behaviors

1.3.2 *Model 2*

Most intercultural interactions have a functional component. A manager needs to run an international organization or project; a student needs to earn a degree abroad; a volunteer needs to carry out a development project; a doctor or nurse needs to provide healthcare service; a peacekeeping force needs to maintain peace; an immigrant needs to adjust to a new home far away from where he or she grew up; and so forth. Therefore, in all intercultural interactions, tasks take a central stage, and centrality of goal is likely to take a central stage, and centrality of goals will have a direct impact on behavioral intentions and, ultimately, on intercultural behaviors.

When change agents are specified in theoretical models common in the field of intercultural research, such changes seems to be based on drive-reduction theory, which was popular in psychology in the 1940s and 1950s (Brown, 1953; Harlow, 1953; Spence, 1951). It seems to be an article of faith that people will act to avoid or reduce an uncomfortable state (such as might occur when entering an unfamiliar culture). However, Harlow (1953) demonstrated rather conclusively that monkeys, at least, will often seek to increase drives in order to arrive at a desirable result. Deciding to travel and live in another culture, if only for a short time, in incomprehensible under drive-reduction type theories, but people do it just as they climb Mount Everest or do other highly anxiety producing activities. We suggest here that the importance of the goal will override any long- or short-term increase in anxiety. In contrast when the goal is not important, then there may well be a withdrawing from the intercultural behavior (Fig. 1.9) (Table 1.2).

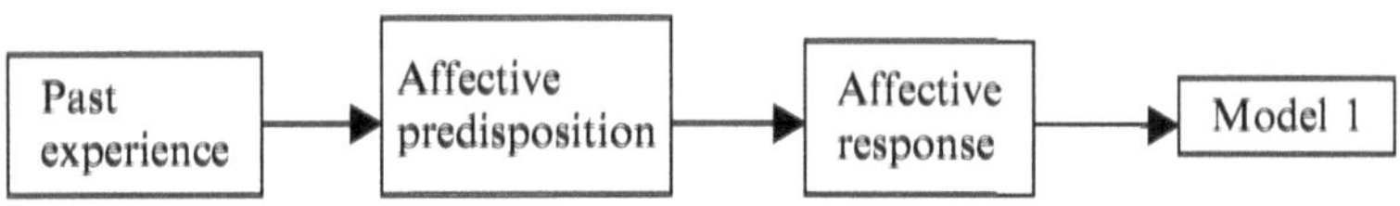

Fig. 1.9 Model 3: Affect and intercultural behavior

Table 1.2 Regions of interest for model 2 (Fig. 1.8)

Cognitive variable	Regions of interest	Biological function
Centrality of goals (figure/ground)	Prefrontal cortex, inferotemporal cortex, right inferior frontal gyrus, medial frontal parietal, caudate, right temporal-parietal junction	Neuroimaging research and neurology case studies have implicated the importance of these regions in motivation. The prefrontal cortex (Shimamura, 2000) and caudate are ROIs involved in executive functioning and decision-making processes (Grahn et al., 2009). The right temporal-parietal junction (Decety & Lamm, 2007) is implicated in moral decision-making and in making distinctions between self-other. The right inferior frontal gyrus (Aron, Robbins, & Poldrack, 2004) is a region controlling behavioral inhibition. The inferotemporal cortex (Denys et al., 2004) is involved in object recognition and processing the visual field
Motivation		

1.3.3 Model 3

The dependent variable, intercultural behavior, is necessarily multidimensional in character. It must consist of both sociocultural (Ward & Kennedy, 1999) and psychological facets. The two dimensions while correlated may have different antecedents. The former may be partially predicted by cultural distance between heritage and receiving cultures, and the latter may be more related to internal techniques for handling unfamiliar social stimuli. These techniques for the sociocultural dimension may be better predicted by past experiences with members from the new culture.

Bennett and Castiglioni place great stress on affect as a predictor, component, and director of intercultural behaviors (Bennett & Castiglioni, 2004). We present a second model in which affect appears in two forms: as a predisposition to be emotionally labile or "affective predisposition" and "affective response" or the display of actual emotional states. Additionally, the Stephans (Stephan & Stephan, 1985, 1992; Stephan et al., 1999) have pointed to intercultural interaction as being inherently anxiety producing. The type of affect that Bennett and Castiglioni describe which is a positive affect rooted in body awareness and which also directed behavior in an automatic fashion bears some similarity to successive orders of perception described by Ezra Krendle and Duane McRuer in modeling landing aircraft on a carrier (Krendle & McRuer, 1960). This affect may be more closely related to arousal seeking (Model 5) than to the negative emotion described by the Stephans. As such, the two affects may represent the opposite poles of the same emotional state (Fig. 1.10) (Table 1.3).

Table 1.3 Regions of interest for model 3 (Fig. 1.9)

Cognitive variable	Region of interest	Biological function
Affective response	Amygdala	Electrical stimulation studies have implicated the amygdala as having a critical role in the affective response to the environment (Lanteaume et al., 2007)
Affective predisposition	Amygdala	fMRI and PET scan studies have implicated the amygdala in the formation of the emotional components of autobiographical memories. The amygdala is a key region in the formation of memories related to emotional events (Parkinson, Robbins, & Everitt, 2000)
Past experience Autobiographical memory	Prefrontal, medial and lateral temporal cortex, cingulate cortex, temporoparietal cortex (TPJ), cerebellum	Numerous neuroimaging studies have associated the following ROIs in the formation of autobiographical memories. The temporal cortex, cingulate cortex (Squire, Stark, & Clark, 2004), and cerebellum (Timmann & Daum, 2007) are key ROIs involved in storing memory
		The prefrontal cortex and cingulate cortex are involved in executive functioning. The temporal-parietal cortex is implicated in the self-other distinction (Decety & Lamm, 2007)

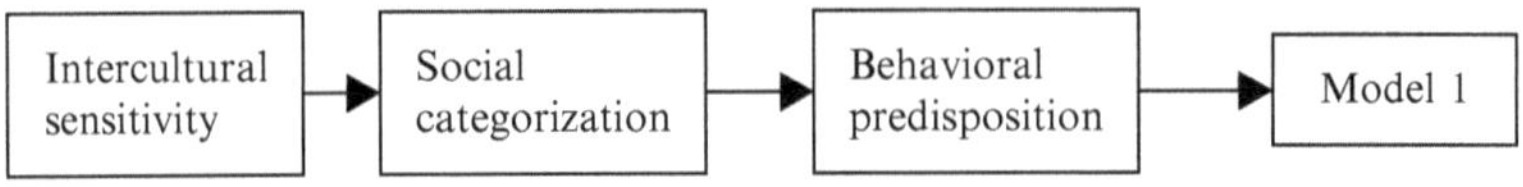

Fig. 1.10 Model 4: Intercultural sensitivity and intercultural behaviors

1.3.4 Model 4

Model 4 links intercultural sensitivity, behavioral predisposition, and intercultural behaviors (as represented by Model 1). From the earliest interest in training for international assignments, researchers have searched for individual differences that could predict success in such situations (Jasinskaja-Lahti & Leibkind, 2000; Kealey, 1996; Mak & Tran, 2001). These efforts have generally produced weak effects if they produced any effects at all (Mendenhall et al., 2004).

One possible reason for the rather disappointing results may be a level of analysis issue. The behaviors being examined tend to be rather specific, whereas the individual difference measures are far more diffuse and can apply to many situations. Recall our discussion earlier on the importance of context in assessing acculturative strategies. Another reason is apparent if we grant the general accuracy of the Landis et al. parent model, above. There are a number of intervening variables between the personal characteristics that the person brings to the situation and the actual intercultural behavior. It should not be surprising, therefore, that the direct effect is modest.

We should recognize that there are many intercultural sensitivity measures (of varying reliability and validity) that can help test this model (Paige, 2004; Pedersen, Neighbors, Larimer, & Lee, 2011; Stuart, 2009). Much work has been done on the development and validation of the Intercultural Development Inventory (Hammer, 2011; Hammer et al., 2003). There is tentative evidence that the IDI does predict some aspects of intercultural behavior. If these results prove to be robust, it may be desirable to modify the model to include a direct path between intercultural sensitivity and intercultural behavior.

Many attempts at reducing prejudice have focused on increasing the permeability of social categories (e.g., Brewer & Gaertner, 2001; Gaertner & Dovidio, 2000; Shachar & Amir, 1996). These studies have mostly assumed that the affect toward the in-group is quite different than that directed toward the out-group (Allport, 1954; Brewer, 1999). It is that difference that reinforces the strength of the boundaries. Accordingly, many studies have worked on changing the cognitions and affects associated with the out-groups so that they are seen as similar to the in-group (Bagci, Kumashiro, Smith, Blumberg, & Rutland, 2014). These studies have worked generally when the two groups have been in a dominance-submission relationship (e.g., European-American versus African-American, Pettigrew & Troop, 2006). They have rarely, if ever, been applied to expatriate managers or other sojourners. Model 4 gives some prominence to the role of social categories (Fig. 1.11) (Table 1.4).

Table 1.4 Regions of interest for model 4 (Fig. 1.10)

Cognitive variable	Regions of interest	Biological function
Intercultural sensitivity Attribution	Temporoparietal junction, medial prefrontal cortex	Neuroimaging studies have implicated the temporoparietal junction (Decety & Lamm, 2007) as having a critical role in the self-other distinction and the mPFC (Shimamura, 2000) as being implicated in executive functioning
Social categorization	TPJ, posterior cingulate cortex	fMRI studies have implicated that these ROIs are implicated in the self-other distinction (Decety & Lamm, 2007)
Behavior predisposition	Caudate, anterior insula, nucleus accumbens	Neuroimaging studies have suggested that variables related to behavior predisposition produce activation in these regions. The caudate (Aron, Fisher, Mashek, Strong, & Brown, 2005) and the anterior insula (Phan, Wager, Taylor, & Liberzon, 2002) are ROIs important in emotion processing. The nucleus accumbens is involved in processing reward states

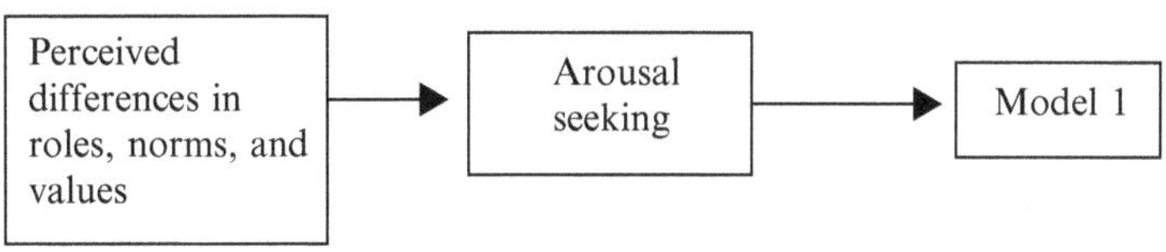

Fig. 1.11 Model 5: Subjective culture differences and intercultural behaviors

1.3.5 Model 5

Model 5 proposes that the greater the perceived differences in subjective culture (Triandis, 1972, 1976), the greater will be the affective reaction. We suggest that at some optimum level of difference, the individual will seek greater information, even at the risk of increasing levels of stress. At a point above this level, the arousal seeking will decrease, and the person will seek to return to earlier homeostatic levels, as suggested by Kim (Kim, 2004). The trajectory we propose bears similarity to the Yerkes-Dodson curve relating motivation to performance (Spence, 1951). For simplicity's sake, we have not included the effects of cognitive-perceptual sets or styles (Riding & Rayner, 2000) on arousal seeking.

We have, however, included a new predictor to arousal seeking: wayfinding (references in Table 1.5). This trait, which involves the ability to develop cognitive maps, allows the person to travel from point to point in space. The neuroscience research on wayfinding is extensive and compelling. Our reasoning is that having a well-developing wayfinding ability would reduce the level of anxiety when in a new environment and hence lead to an increased willingness to seek new experiences even at the expense of a higher level of arousal.

Table 1.5 Regions of interest for model 5 (Fig. 1.11)

Cognitive variable	Regions of interest	Biological function
Perceived differences in roles, norms, values (self-definition)	mPFC (both ventral and dorsal), amygdala	fMRI studies have implicated the mPFC in the self-other distinction (Decety & Lamm, 2007). Neuroimaging studies have implicated the amygdala in the formation of the emotional components of autobiographical memories (Parkinson et al., 2000)
Arousal seeking	Pleasure network, mPFC, lateral hypothalamus	fMRI studies have suggested that the mPFC is involved in executive functioning (Shimamura, 2000). The lateral hypothalamus is involved in pleasure seeking and the processing of emotional stimuli (Berridge & Kringelbach, 2008)
Wayfinding	Posterior hippocampus, caudate	Many MRI studies have implicated these areas in people finding their ways in the physical environment. It is predicted that wayfinding is a predictor of arousal seeking (Chrastil, 2013; Maguire et al., 1998, 2000; Maguire, Woollett, & Spiers, 2006; Wegman et al., 2014)

1.4 Plan of This Book

The chapters in this book were arranged in a manner meant to guide you through the burgeoning field of cultural neuroscience and its potential application to intercultural relations. The first four chapters offer reviews that detail the cultural neuroscience of specific psychological phenomena. For example, Shihui Han offers the lead chapter, and it discusses a common topic we have all experienced, namely, empathy for others' in pain. Na and Chan's Chap. 3 breaks down the topic of cognitive differences in reasoning styles between cultures. Coderre thoroughly reviews the neuroscience of bilingualism. Franklin et al. provide a detailed review of the ability to gain information (e.g., emotion) from looking into another person's eyes and how that ability can be disrupted by cultural factors.

The next two chapters are methodological in nature and provide unique perspectives in studying cultural neuroscience. Chen et al. review their work in studying the neural changes that occur during acculturation in a longitudinal manner. Glazer et al. suggest that instead of using priming of cultural syndromes to investigate cultural differences, researchers should utilize the relational models of Fiske and colleagues.

The final three chapters look at how experience and self-relevant memories are major components of cultural differences. Doole et al. review the differences in Westerners and Easterners is the result of repeated experiences that ultimately change the neurological structure of the brain. Huff et al. provide a detailed analysis of how differences in the focus of attention have an impact on social and self-relevant memory between cultures. Abraham reviews how cultural factors can impact personal relevance of events and eventually helps define the line between reality and fiction.

References

Albert, R. (1983). The intercultural sensitizer or culture assimilator: A cognitive approach. In D. Landis & R. Brislin (Eds.), *Handbook of intercultural training* (pp. 186–217). Elmsford, NY: Pergamon.

Allport, G. W. (1954). *The nature of prejudice*. Reading, MA: Addison-Wesley.

Ambady, N., & Bharucha, J. (2009). Culture and the brain. *Current Directions in Psychological Science, 18*(6), 342–345.

Arends-Toth, J., & van de Vijer, F. (2004). Domains and dimensions in acculturation: Implicit theories of Turkish-Dutch. *International Journal of Intercultural Relations, 28*(1), 19–35.

Aron, A., Fisher, H., Mashek, D., Strong, G., & Brown, L. (2005). Reward, motivation, and emotion systems associated with early-stage intense romantic love. *Journal of Neurophysiology, 94*, 327–337.

Aron, A., Robbins, T., & Poldrack, R. (2004). Inhibition and right inferior frontal cortex. *Trends in Cognitive Science, 8*, 170–177.

Avishai, B., Dayan, D., Hussein, F., Illouz, E., Khoury, B., Margalit, E., et al. (2014). Israel and Palestine: Where do we go from here. *Harpers, 329*(1972), 29–43.

Bagci, S., Kumashiro, M., Smith, P., Blumberg, H., & Rutland, A. (2014). Cross-ethnic friendships: Are they really rare? Evidence from secondary schools around London. *International Journal of Intercultural Relations, 41*(July), 125–137.

Bandura, A. (1977). *Social learning theory*. Englewood Cliffs, NJ: Prentice-Hall.

Benedict, R. (1934). *Pattern of culture*. Boston, MA: Houghton Mifflin.

Bennett, M. (1993). Towards ethnorelativism: A developmental model of intercultural sensitivity. In R. Paige (Ed.), *Education for the intercultural experience* (2nd ed., pp. 21–71). Yarmouth, ME: Intercultural Press.

Bennett, J., & Bennett, M. (2004). Developing intercultural sensitivity: An integrative approach to global and domestic diversity. In D. Landis, J. Bennett, & M. Bennett (Eds.), *Handbook of intercultural training* (3rd ed., Vol. 147–163). Thousand Oaks, CA: Sage.

Bennett, M., & Castiglioni, I. (2004). Embodied ethnocentrism and the feeling of culture: A key to training for intercultural competence. In D. Landis, J. Bennett, & M. Bennet (Eds.), *Handbook of intercultural training* (3rd ed., pp. 249–265). Thousand Oak, CA: Sage.

Berridge, K., & Kringelbach, M. (2008). Affective neuroscience of pleasure: Reward in humans and animals. *Psychobiology, 28*, 207–218.

Berry, J. (1974). Psychological aspects of cultural pluralism. *Topics in Culture Learning, 2*, 17–22.

Berry, J. (2004). Fundamental psychological processes in intercultural relations. In D. Landis, J. Bennett, & M. Bennett (Eds.), *Handbook of intercultural training* (3rd ed., pp. 166–184). Thousand Oaks, CA: Sage.

Berry, J., Phinney, J., Sam, D., & Vedder, P. (2006). *Immigrant youth in cultural transition: Acculturation, identity, and adaptation across national contexts*. Mahwah, NJ: Erlbaum.

Bhawuk, D., & Triandis, H. (1996). The role of culture theory in the study of cultural and intercultural training. In D. Landis & R. Bhagat (Eds.), *Handbook of intercultural training* (2nd ed., pp. 17–34). Thousand Oaks, CA: Sage.

Boas, F. (1911). *The mind of primitive man*. New York, NY: Macmillan.

Boski, P. (2008). Five meanings of integration in acculturation research. *International Journal of Intercultural Relations, 33*(2), 142–153.

Bourhis, R., Moise, L., Perreault, S., & Senecal, S. (1997). Towards an interactive acculturation model: A social psychological approach. *International Journal of Psychology, 32*(6), 369–386. doi:10.1080/002075997400629.

Brewer, M. (1999). The psychology prejudice: Ingroup love or outgroup hate. *Journal of Social Issues, 55*(3), 429–444.

Brewer, M., & Gaertner, S. (2001). Toward the reduction of prejudice: Intergroup contact and social categorization. In R. Brown & S. Gaertner (Eds.), *Blackwell handbook of social psychology: Intergroup processes* (pp. 451–474). Oxford, UK: Blackwell.

Brislin, R., Landis, D., & Brandt, M. (1983). Conceptualizations of intercultural behavior and training. In D. Landis & R. Brislin (Eds.), *Handbook of intercultural training* (Vol. 1, pp. 1–35). Elmsford, NY: Pergamon.
Brown, J. (1953). *Problems presented by the concept of acquired drives Nebraska Symposium on Motivation* (pp. 1–21). Lincoln, NE: Nebraska University Press.
Brunswik, E. (1956). *Perception and the representative design of psychological experiments*. Berkeley, CA: University of California Press.
Bush, G., Lum, P., & Posner, M. (2000). Cognitive and emotional influences in anterior cingulate cortex. *Trends in Cognitive Science, 4*, 215–222.
Carr, N. (2010). *The shallows*. New York, NY: Norton.
Chirkov, V. (2009). Critical psychology of acculturation: Who we study and how do study it when we investigate acculturation. *International Journal of Intercultural Relations, 33*(2), 94–105.
Chrastil, E. (2013). Neural evidence supports a novel framework for spatial navigation. *Psychonomic Bulletin and Review, 20*(2), 208–227.
Cikara, M., & Bavel, J. V. (2014). The neuroscience of intergroup relations: An integrative review. *Perspectives on Psychological Science, 9*, 245–267.
Cushner, K., & Landis, D. (1996). The intercultural sensitizer. In D. Landis & R. Bhagat (Eds.), *Handbook of intercultural training* (2nd ed., pp. 185–202). Thousand Oaks, CA: Sage.
Decety, J., & Lamm, C. (2007). The role of the right temporoparietal junction in social interaction: How low-level computational processes contribute to meta-cognition. *The Neuroscientist, 13*, 580–593.
den Ouden, H., Frith, U., Firth, C., & Blakemore, S. (2005). Thinking about intentions. *NeuroImage, 28*, 787–796.
Denys, K., Vanduffel, W., Fize, D., Nellsson, K., Peuskens, H., Essen, D. V., et al. (2004). The processing of visual shape in the cerebral cortex of human and nonhuman primates: A functional magnetic resonance imaging study. *The Journal of Neuroscience, 24*, 2551–2563.
Doucerain, M., Dere, J., & Ryder, A. (2013). Travels in hyperdiversity: Multiculturalism and the contextual assessment of acculturation. *International Journal of Intercultural Relations, 37*(6), 686–699.
Duncan, J., & Owen, A. (2000). Common regions of the human frontal lobe recruited by diverse cognitive demands. *Trends in Neurosciences, 23*, 475–483.
Feldman, R., & Goodrich, J. (1999). The Edwin Smith surgical papyrus. *Child's Nervous System, 15*, 281–284.
Flanagan, J. (1954). The critical incident technique. *Psychological Bulletin, 51*, 327–358.
Gaertner, S., & Dovidio, J. (2000). *Reducing intergroup bias: The common ingroup identity model*. Philadelphia, PA: Taylor and Francis.
Gao, G., & Gudykunst, W. (1990). Uncertainty, anxiety, and adaptation. *International Journal of Intercultural Relations, 14*, 301–317.
Grahn, J., Parkinson, J., & Owen, A. (2009). The role of the basal ganglia in learning and memory: Neuropsychological studies. *Behavioural Brain Research, 199*, 53–60.
Gudykunst, W. (1988). Uncertainty and anxiety. In Y. Kim & W. Gudykunst (Eds.), *Theories in intercultural communication* (pp. 123–156). Newbury Park, CA: Sage.
Gudykunst, W. (1995). Anxiety/uncertainty management theory: Current status. In R. Wiseman (Ed.), *Intercultural communication theory* (pp. 1–58). Thousand Oaks, CA: Sage.
Gudykunst, W., Guzley, R., & Hammer, M. (1996). Designing intercultural training. In D. Landis & R. Bhagat (Eds.), *Handbook of intercultural training* (2nd ed., pp. 61–80). Thousand Oaks, CA: Sage.
Gudykunst, W., & Sudweeks, S. (1992). Applying a theory of intercultural adaptation. In W. Gudykunst & Y. Kim (Eds.), *Reading on communicating with strangers* (pp. 358–368). New York, NY: McGraw-Hill.
Habyariman, J., Humphreys, M., Posner, D., & Weinstein, J. (2008). Better institutions, not partition. *Foreign Affairs, 87*(4), 138–141.
Hammer, M. (2011). Additional cross-cultural validity testing of the Intercultural Development Inventory. *International Journal of Intercultural Relations, 35*(4), 474–487.

Hammer, M., Bennett, M., & Wiseman, R. (2003). Measuring intercultural sensitivity: The Intercultural Development Inventory. *International Journal of Intercultural Relations, 27*(4), 421–443.
Harlow, H. (1953). *Motivation as a factor in new responses Nebraska Symposium on Motivation*. Lincoln, NE: Nebraska University Press.
Hikosaka, O., Bromberg-Martin, E., Hong, S., & Matsumoto, M. (2008). New insights on the subcortical representation of reward. *Current Opinion in Neurobiology, 18*, 203–208.
Hofstede, G. (1999). *Culture and organizations: Software of the mind*. New York, NY: McGraw-Hill.
Hong, Y., Morris, M., Chiu, C., & Benet-Martinez, V. (2000). Multicultural minds: A dynamic constructivist approach to culture and cognition. *American Psychologist, 55*(7), 709–720.
Jasinskaja-Lahti, I., & Leibkind, K. (2000). Predictors of the actual degree of acculturation of Russian-speaking immigrant adolescents in Finland. *International Journal of Intercultural Relations, 24*(4), 503–518.
Kealey, D. (1996). The challenge of international personnel selection. In D. Landis & R. Bhagat (Eds.), *Handbook of intercultural training* (2nd ed., pp. 81–105). Thousand Oaks, CA: Sage.
Kidder, L. (1977). The inadvertent creation of a neocolonial culture: A study of western sojourners in India. *International Journal of Intercultural Relations, 1*(1), 48–60.
Kim, Y. (2004). Long-term cross-cultural adaptation: Training implications of an integrative theory. In D. Landis, J. Bennett, & M. Bennett (Eds.), *Handbook of intercultural training* (3rd ed., pp. 317–362). Thousand Oaks, CA: Sage.
Kim, U., Triandis, H., Kagitcibasi, C., Choi, S., & Yoon, G. (1994). *Individualism and collectivism: Theory, method, and applications*. Thousand Oaks, CA: Sage.
Kitayama, S., & Tompson, S. (2010). Envisioning the future of cultural neuroscience. *Asian Journal of Social Science, 13*, 92–101.
Klineberg, O. (1980). Historical perspectives: Cross-cultural psychology before 1980. In H. Triandis (Ed.), *Handbook of cross-cultural psychology* (Vol. I). Boston, MA: Allyn and Bacon.
Kluckholn, F., & Strodtbeck, F. (1961). *Variations in value orientation*. New York, NY: Row, Peterson.
Krendle, E., & McRuer, D. (1960). A servomechanisms approach to skill development. *Journal of The Franklin Institute, 269*, 24–42.
Landis, D., & Bhawuk, D. (2004). Synthesizing theory building and practice in intercultural training. In D. Landis, J. Bennett, & M. Bennett (Eds.), *Handbook of intercultural training* (3rd ed., pp. 453–468). Thousand Oaks, CA: Sage.
Landis, D., Brislin, R., & Hulgus, J. (1985). Attribution training versus contact in acculturative training: A laboratory study. *Journal of Applied Social Psychology, 15*, 466–482.
Landis, D., Day, H., McGrew, P., Miller, A., & Thomas, J. (1976). Can a black culture assimilator increase racial understanding? *Journal of Social Issues, 32*, 169–184.
Landis, D., McGrew, P., & Triandis, H. (1975). Behavioral intentions and norms of urban school teachers. In G. Verman & C. Bagley (Eds.), *Race and education across cultures* (pp. 117–144). London: Heinemann.
Landis, D., Triandis, H., & Adamopoulous, J. (1978). Habit and behavioral intentions as predictors of social behaviors. *Journal of Social Psychology, 108*, 227–237.
Lanteaume, L., Khalfa, S., Regis, J., Marquis, P., Chauvel, P., & Bartolomei, F. (2007). Emotion induction after direct intracerebral stimulation of human amygdala. *Cerebral Cortex, 17*, 1307–1313.
Lazarus, R., & Folkman, S. (1984). *Stress, coping, and appraisal*. New York, NY: Springer.
Legrenzi, P., & Umilta, C. (2011). *Neuromania: On the limits of brain science*. Oxford, UK: Oxford University Press.
Lewin, K. (1946). Action research and minority problems. *Journal of Social Issues, 2*(4), 34–46.
Lieberman, M., Berkman, E., & Wager, T. (2009). Correlations in social neuroscience aren't voodoo: Commentary on Vul, et al. *Perspectives on Psychological Science, 4*, 299–307.
Maguire, E., Burgess, N., Donnett, J., Frackowiak, R., Frith, C., & O'Keefe, J. (1998). Knowing where and getting there: A human navigation network. *Science, 280*, 921–924.

Maguire, E., Gadian, D., Johsrude, I., Good, C., Ashburner, J., Frackowiak, R., et al. (2000). Navigation-related structural change in the hippocampi of taxi drivers. *PNAS, 97*(8), 4398–4404.
Maguire, E., Woollett, K., & Spiers, H. (2006). London taxi drivers and bus drivers: A structural MRI and neuropsychological analysis. *Hippocampus, 16*, 1091–1101.
Mak, A., & Tran, C. (2001). Big five personality and cultural relocation factors in Vietnamese Australian students' intercultural social self-efficacy. *International Journal of Intercultural Relations, 25*, 181–201.
Mayberg, N. H., Lozano, A., Voon, V., McNeely, H., Seminowicz, D., Hamani, C., et al. (2005). Deep brain stimulation for treatment of resistant depression. *Neuron, 45*, 651–660.
McCrae, R., Costa, P., Ostendorf, F., Hrebickova, M., Avia, M., Sanz, J., et al. (2000). Nature over nurture: Temperament, personality, and life span development. *Journal of Personality and Social Psychology, 78*, 173–186.
Mendenhall, M., Stahl, G., Ehnert, I., Oddou, G., Osland, J., & Kuhlmann, T. (2004). Evaluation studies of cross-cultural training programs: A review of the literature from 1988 to 2000. In D. Landis, J. Bennett, & M. Bennett (Eds.), *Handbook of intercultural training* (3rd ed., pp. 129–144). Thousand Oaks, CA: Sage.
Mizuguchi, N., Nakata, H., Hayashi, T., Sakamoto, M., Uchida, Y., & Kanosue, K. (2013). Brain activity during motor imagery of an action with an object: A functional magnetic imaging study. *Neuroscience Research, 76*, 150–155.
Muller, J. (2008). Us and them. *Foreign Affairs, 87*(2), 18–35.
Navas, M., Garcia, M., Sanchez, J., Rohas, A., Pumares, P., & Fernandez, J. (2005). Relative acculturation extended model (RARM): New contributions with regard to the study of acculturation. *International Journal of Intercultural Relations, 29*(1), 21–37.
Noels, K., Leavitt, P., & Clement, R. (2010). "To see ourselves as others see us": On the implications of reflected appraisals for ethnic identity and discrimination. *Journal of Social Issues, 66*(4), 740–758.
OECD. (2010). *Education at a glance: OECD indicators*. Paris: OECD.
Osgood, C., May, W., & Miron, M. (1975). *Cross-cultural universals of affective word meaning*. Urbana, IL: University of Illinois Press.
Osgood, C., Suci, G., & Tannenbaum, P. (1957). *The measurement of meaning*. Urbana, IL: University of Illinois Press.
Paige, R. (2004). Instrumentation in intercultural training. In D. Landis, J. Bennett, & M. Bennett (Eds.), *Handbook of intercultural training* (3rd ed., pp. 85–128). Thousand Oaks, CA: Sage.
Parkinson, J., Robbins, T., & Everitt, B. (2000). Dissociable roles of the central and basolateral amygdala in appetitive emotional learning. *European Journal Neuroscience, 12*, 405–413.
Pedersen, E., Neighbors, C., Larimer, M., & Lee, C. (2011). Measuring sojourner adjustment among American students. *International Journal of Intercultural Relations, 35*(6), 881–889.
Penfield, W. (1958). Some mechanisms of consciousness discovered during electrical stimulation of the brain. *Proceedings of the National Academy of Sciences of the United States of America, 44*, 51–66.
Pettigrew, T. F., & Troop, L. R. (2006). A metanalytic test of intergroup contact theory. *Journal of Personality and Social Psychology, 90*(3), 751–783. doi:10.1037/0022-3514.90.3.751.
Phan, K., Wager, T., Taylor, S., & Liberzon, I. (2002). Functional neuroanatomy of emotion: A meta-analysis of emotion activation studies in PET and fMRI. *NeuroImage, 16*, 331–348.
Racine, E., Bar-Ilan, O., & Illes, J. (2006). Brain imaging: A decade of coverage in the print media. *Science Communication, 28*, 122–143.
Randolf, G., Landis, D., & Tzeng, O. (1977). The effects of time and practice upon culture assimilator training. *International Journal of Intercultural Relations, 1*, 105–119.
Riding, R., & Rayner, S. (Eds.). (2000). *International perspectives on individual differences* (Cognitive styles, Vol. 1). Stamford, CT: Ablex.
Roland, P., Larsen, B., Lassen, N., & Skinhoj, E. (1980). Supplementary motor area and other cortical areas in organization of voluntary movements in man. *Journal of Neurophysiology, 43*, 118–136.

Rudmin, F. (2009). Constructs, measurements, and models of acculturation and acculturative stress. *International Journal of Intercultural Relations, 33*(2), 106–123.

Rudmin, F., & Ahmadzadeh, V. (2001). Psychometric critique of acculturation psychology: The case of Iranian migrants in Norway. *Scandinavian Journal of Psychology, 42*, 41–56.

Sabatinelli, D., Lang, P., Bradley, M., Costa, V., & Versace, F. (2007). Pleasure rather than salience activates human nucleus accumbens and medial prefrontal cortex. *Journal of Neurophysiology, 98*, 1374–1379.

Shachar, H., & Amir, T. Y. (1996). Training teachers and students for intercultural cooperation in Israel. In D. Landis & R. Bhagat (Eds.), *Handbook of intercultural training* (2nd ed., pp. 400–413). Thousand Oaks, CA: Sage.

Shimamura, A. (2000). The role the prefrontal cortex in dynamic filtering. *Psychobiology, 28*, 207–218.

Spence, K. (1951). Theoretical interpretations of learning. In S. Stevens (Ed.), *Handbook of experimental psychology* (pp. 690–729). New York, NY: Wiley.

Squire, L., Stark, C., & Clark, R. (2004). The medial temporal lobe. *Annual Review of Neuroscience, 27*, 279–306.

Stephan, W., Renfro, C., Esses, V., Stephan, C., & Martin, T. (2005). The effects of feeling threatened on attitudes toward immigrants. *International Journal of Intercultural Relations, 29*(1), 1–19.

Stephan, W., & Stephan, C. (1985). Intergroup anxiety. *Journal of Social Issues, 41*, 157–176.

Stephan, C., & Stephan, W. (1992). Reducing intercultural anxiety through intercultural contact. *International Journal of Intercultural Relations, 16*(1), 89–106.

Stephan, C., & Stephan, W. (2000). An integrated threat theory of prejudice. In S. Oskamp (Ed.), *Reducing prejudice and discrimination* (pp. 23–46). Hillsdale, NJ: Lawrence Erlbaum.

Stephan, W., Stephan, C., & Gudykunst, W. (1999). Anxiety in intergroup relations: A comparison of anxiety/uncertainty management they and integrated threat theory. *International Journal of Intercultural Relations, 23*(4), 613–628.

Stevens, S. (1946). On the theory of scales of measurement. *Science, 103*, 670–680.

Stevens, S. (1951). Mathematics, measurement, and psychophysics. In S. S. Stevens (Ed.), *Handbook of experimental psychology* (pp. 1–49). New York, NY: John Wiley & Sons.

Stuart, D. (2009). Assessment instruments for the global workforce. In M. Moodian (Ed.), *Contemporary leadership and intercultural competence* (pp. 175–190). Thousand Oaks, CA: Sage.

Terbeck, S., Kahane, G., McTavish, S., Savulescu, J., Cowen, P., & Hewstone, M. (2012). Propranolol reduces implicit negative racial bias. *Psychopharmacology, 222*, 419–424.

Timmann, D., & Daum, I. (2007). Cerebellar contributions to cognitive functions: A progress report after two decades of research. *Cerebellum, 6*, 159–162.

Triandis, H. (1972). *The analysis of subjective culture*. New York, NY: Willey-Interscience.

Triandis, H. (Ed.). (1976). *Variations in black and white perceptions of the social environment*. Urbana, IL: University of Illinois Press.

Triandis, H. (1977). *Interpersonal behavior*. Monterey, CA: Brooks/Cole.

United Nations. (2013). *Migrants around the world*. Retrieved August 25, 2014, from www.imigration.org

van de Vijer, F., van Hemert, F., & Poortinga, Y. (Eds.). (2008). *Multilevel analysis of individuals and cultures*. New York, NY: Lawrence Erlbaum Associates.

Vul, E., Harris, C., Winkielman, P., & Pashler, H. (2009). Puzzling how correlations in fMRI studies of emotion, personality, and social cognition. *Perspectives on Psychological Science, 4*, 274–290.

Ward, C. (1996). Acculturation. In D. Landis & R. Bhagat (Eds.), *Handbook of intercultural training* (2nd ed., pp. 124–147). Thousand Oaks, CA: Sage.

Ward, C. (2004). Psychological theories of culture contact and their implications for intercultural training and intervention. In D. Landis, J. Bennett, & M. Bennett (Eds.), *Handbook of intercultural training* (3rd ed., pp. 185–216). Thousand Oaks, CA: Sage.

Ward, C. (2008). Thinking outside the Berry boxes: New perspectives on identity, acculturation, and intercultural relations. *International Journal of Intercultural Relations, 32*(2), 105–114.

Ward, C., & Kennedy, A. (1999). The measurement of sociocultural adjustment. *International Journal of Intercultural Relations, 23*(4), 659–678.
Ward, C., & Kus, L. (2012). Back to and beyond Berry's basics: The conceptualization, operationalization, and classification of acculturation. *International Journal of Intercultural Relations, 36*(4), 472–485.
Ward, C., & Rana-Deuba, A. (2000). Home and host culture influences on sojourner adjustment. *International Journal of Intercultural Relations, 24*(3), 291–306.
Wegman, J., Fonteljn, H., van Ekert, J., Tyborowska, A., Jansen, C., & Janzen, G. (2014). Gray and white matter correlates of navigational ability in humans. *Human Brain Mapping, 35*(6), 2561–2572.
Weldon, D., Carston, D., Rissman, A., Slobodin, L., & Triandis, H. (1975). A laboratory test of effects of culture assimilator training. *Journal of Personality and Social Psychology, 21*, 300–310.
Zhang, T., & Meaney, M. (2010). Epigenetics and the environmental regulation of the genome and its function. *Annual Review of Psychology, 61*, 19.11–19.28.

Chapter 2
Intergroup Relationship and Empathy for Others' Pain: A Social Neuroscience Approach

Shihui Han

Abstract Han reviews the neuroimaging evidence for the brain regions involved in empathy for pain. The implicated regions, the anterior cingulate and anterior insula, overlap with those involved in firsthand pain experiences. However, several factors, including sociocultural variables, can influence empathy toward others' pain states.

Han discusses the evidence of racial bias in feelings of empathy for pain states. Han shows that racial bias in empathy of pain can potentially produce real-world effects like differences in medical treatment between racial groups. Several brain regions are implicated including the anterior cingulate, the supplementary motor cortex, the anterior insula, and the medial prefrontal cortex.

Han concludes by discussing evidence that intercultural experiences can decrease racial bias of empathy of pain. Living in a country with an other-race majority can decrease the racial bias shown for empathy of pain and alter the neural responses to seeing pictures of pain expressions. Han suggests that future research should investigate how educational opportunities can be offered to eliminate racial bias in empathy toward others in pain.

2.1 Empathy for Pain and its Neural Correlates

Imagine that you are watching a friend who is cutting a cucumber into pieces to make a salad. She accidently cuts one of her fingers and shows a pain expression. What would you feel and what would you do in such a situation? You quickly understand that she is feeling pain and may immediately try to find a Band-Aid to cover her finger and console her. Such life experiences illustrate well that we have an ability called *empathy* that engages complicated psychological processes and has

S. Han (✉)
Department of Psychology, PKU-IDG/McGovern Institute for Brain Research,
Peking University, Room 1614, Wang Kezhen Building, No. 52, Haidian Road,
Haidian District, Beijing 100871, China
e-mail: shan@pku.edu.cn

J.E. Warnick, D. Landis (eds.), *Neuroscience in Intercultural Contexts*,
International and Cultural Psychology, DOI 10.1007/978-1-4939-2260-4_2

been associated with prosocial behavior. Psychologists define empathy in many different ways. For example, according to Berger (1987), empathy refers to "the capacity to know emotionally what another is experiencing from within the frame of reference of that other person, the capacity to sample the feelings of another or to put one's self in another's shoes." de Waal (2008) defined empathy as "the capacity to (a) be affected by and share the emotional state of another, (b) assess the reasons for the other's state, and (c) identify with the other, adopting his or her perspective." Regardless of subtle differences in the way psychologists define empathy, there are two common components among different definitions of empathy, that is, to understand and share emotional states of others.

How are the psychological processes involved in empathy mediated by the human brain? This issue is critical for understanding of the neurobiological mechanisms of empathy. It is also pivotal for understanding of the human prosocial nature and has important social implications. Empathy for pain provides a good model to study the neural correlates of empathy because most of us have vivid experiences of feeling others' pain. Neural substrates of empathy for pain have been addressed extensively by recent brain imaging studies (see Fan, Duncan, de Greck, & Northoff, 2011; Lamm, Decety, & Singer, 2011 for review). A common paradigm used in the brain imaging research is to record neural responses to perceived painful or non-painful stimuli applied to others. An early functional magnetic resonance imaging (fMRI) study scanned female subjects while they received painful vs. non-painful stimuli or while they were informed by a visual symbol that their partners were receiving painful vs. non-painful stimuli (Singer et al., 2004). It was found that knowing others in pain activated brain regions such as the anterior cingulate (ACC) and anterior insula and these activations overlapped with those engaged in the first-hand pain experience. The following fMRI studies recorded brain activity in response to static images of body parts (hand or foot) receiving painful vs. non-painful stimulations (Gu & Han, 2007; Jackson, Meltzoff, & Decety, 2005) or painful vs. non-painful facial expressions (Han et al., 2009; Saarela et al., 2007). These studies also found increased activity in the ACC, insula, and somatosensory cortex (SII) in response to perceived pain in others. Moreover, the magnitude of neural activities in specific brain regions (e.g., ACC, Jackson et al., 2005) positively correlated with subjective feelings of the intensity of others' pain. The findings indicate that the neural activity in the pain matrix that mediates the firsthand pain experience can differentiate between painful and non-painful stimuli applied to others and is associated with one's own subjective feelings of others' pain and thus provide evidence for shared neural representation of one's own pain and others' pain.

The neural activity underlying empathy for pain may occur quite early during perception of others in pain. Fan and Han (2008) conducted the first event-related potential (ERP) study that examined the time course of empathy for pain. They recorded ERPs from healthy adults while they perceived pictures of hands that were in painful (e.g., being cut by a scissor) or non-painful (holding a scissor) situations and had to judge whether or not models in the pictures were feeling painful. It was found that early neural activity underlying differentiation between painful and non-painful stimuli occurred over the frontal lobe at 140 ms after sensory stimulation.

Painful stimuli elicited a positive shift of the ERP amplitudes compared to non-painful stimuli. A long-latency positive activity over the central–parietal regions also showed increased amplitude to painful vs. non-painful stimuli after 380 ms. Moreover, the mean ERP amplitudes at 140–180 ms were correlated with subjective reports of the degree of perceived pain of others and of self-unpleasantness. Similar results were replicated in the following ERP research (Decety, Yang, & Cheng, 2010; Han, Fan, & Mao, 2008; Li & Han, 2010). Mu, Fan, Mao, and Han (2008) also reported evidence that non-phase-locked neural oscillations are also involved in empathic responses. They showed that, relative to perceiving non-painful stimuli, perceiving painful stimuli applied to others' body parts induced increased theta (3–8 Hz) event-related synchronization (ERS) at 200–500 ms but decreased alpha (9–14 Hz) event-related desynchronization (ERD) at 200–400 ms. In addition, subjective ratings of perceived pain and self-unpleasantness positively correlated with theta band ERS but negatively correlated with alpha band ERD related to painful stimuli, suggesting that theta and alpha oscillations are, respectively, involved in emotional sharing and regulation during empathy for pain.

Taken together, the previous brain imaging studies uncovered neural correlates of empathy for pain by showing that both blood oxygen level-dependent (BOLD) signals and scalp electrical activities can differentiate between perceived painful vs. non-painful stimuli applied to others and painful vs. neutral expressions. In addition, the neural activity elicited by perceived pain in others is associated with subjective feeling of others' pain intensity and of one's own unpleasant feelings. Therefore, from the neuroscience perspective, the brain imaging findings demonstrate that the human brain can understand and share others' painful feelings. Moreover, the insular activity in response to other's suffering predicted how frequently individuals helped the others by enduring physical pain themselves to reduce the other's pain (Hein, Silani, Preuschoff, Batson, & Singer, 2010). Empathy-related activity in the inferior frontal and secondary somatosensory cortices also predicted the amount of monetary donation given to a real charitable organization (Ma, Wang, & Han, 2011). Therefore, the neural correlates of empathy for others' pain may be linked to prosocial behaviors toward other individuals.

However, the neural activity underlying empathy for pain is not invariant. There has been evidence that the neural activity related to empathy for pain is influenced by task demand (Fan & Han, 2008; Gu & Han, 2007), prior knowledge about painful stimuli (Fan & Han, 2008; Gu & Han, 2007), personal experience (Cheng et al., 2007), attitude (Singer et al., 2006), etc. For instance, distracting attention from painful stimuli applied to others or decreasing the reality of perceived painful stimuli reduced empathic neural responses to others' pain (Fan & Han, 2008; Gu & Han, 2007). Personal experiences such as being exposed to painful stimuli frequently (Cheng et al., 2007) and negative attitude toward a target person (Singer et al., 2006) also weakened empathic neural responses. More recently, there has been increasing behavioral and brain imaging evidence that racial social group relationship strongly modulates empathy for others' pain. These findings have important social significance given that empathy provides a proximate mechanism of prosocial behavior (de Waal, 2008). Therefore, this chapter will focus on the variation of empathic neural responses to others' pain as a function of racial group relationship. These brain imaging findings not only uncover the psychological and neurobiological mechanisms underlying racial bias in empathy for pain but also provide clues for how to reduce the racial bias in empathy. Related social significance of these brain imaging findings is also discussed.

2.2 Racial Bias in Empathy and Empathic Neural Responses

A human does not live alone. To be affiliated to a social group makes it possible for a person to accomplish tasks that he cannot do by himself and to get social support from others. Thus, a social group provides a basis for an individual to survive in a human society. Race is "a dynamic set of historically derived and institutionalized ideas and practices that sorts people into ethnic groups according to perceived physical and behavioral human characteristics" (Moya & Markus, 2011). It is common that people quickly categorize others, based on external attributes, such as skin tone and facial and body shapes, into "White," "Black" or "Asian," "Caucasian." Although it is debated whether there are racial differences in psychological tendencies and behavior and whether such differences are biologically determined, race as a sociocultural construction does produce social groups characterized by different values, power, and social status. In addition, racial group membership defines coalitions and alliances during evolution (Cosmides, Tooby, & Kurzban, 2003), and the concept of race "emerges when groups are perceived (a) to pose a threat (political, economic, or cultural) to each other's world view or way of life; and/or (b) to justify the denigration and exploitation (past, current, or future) of, and prejudice toward, other groups" (Moya & Markus, 2011). Thus, race gives a way to quickly categorize others as in-group or out-group members.

In human history, racially defined social groups often fought against each other to compete for natural resources. Imagine that two racial groups are fighting against each other. While a person watches an in-group member being hurt by out-group members during fighting and showing painful expression, he or she may empathize the in-group member's pain and help the victim, given that empathetic emotion evokes altruistic motivation to benefit the person for whom empathy is felt (Batson, 1987, 1991) and provides a proximate mechanism of prosocial behavior in response to another's pain (de Waal, 2008). However, if viewing an out-group member being hurt during fighting similarly induces empathy that provokes altruistic behavior, this would prevent one from further fighting against out-group members. Obviously, this is not the case we see in human history. A soldier usually keeps on fighting against out-group members even when he makes them feel painful and show painful expressions. Therefore, the human brain must evolve a mechanism to bias empathy for pain of in-group rather than out-group members so as to switch between different behaviors toward others' suffering during social group interactions such as fighting.

As race is often used to categorize people into racial in-group members (same-race individuals) and out-group members (other-race individuals), early behavioral studies tested racial bias in empathy and related social consequences. Johnson and colleagues (2002) first asked Caucasian students to read a passage involving a Black or a White young man who was charged with a criminal act. The participants were induced to feel high empathy (by imagine how the defendant feels while reading the passage) or low empathy (by trying to be objective while reading the passage) for the defendant. Participants then had to answer five questions on a 7-point Likert-type scale that assessed their feelings of sympathy, compassion, warmth, soft-heartedness, and how moved they were. The responses to these five questions were averaged to yield a self-reported empathy score for each participant. Finally, participants were asked to answer two questions on a 7-point Likert scale that assessed their perceptions of the appropriate punishment for the defendant and the attributions regarding the defendant's actions. Johnson et al. reported two findings. First, Caucasian students reported greater feelings of empathy for the White defendant compared to the Black defendant. Second, Caucasian students assigned more lenient sentences to the White defendant relative to the Black defendant. These results suggest a tendency in Caucasian students for enhanced empathy for racial in-group compared to out-group members, and such racial bias in empathy may lead to different social behaviors such as judicial decision making toward racial in-group/out-group members.

Drwecki and colleagues (2011) further investigated racial bias in pain treatment decisions and empathy. They showed college students (all White) and nursing professionals (31 White out of 40) with videos of real Black and White patients' genuine facial expressions of pain. They then asked participants to make pain treatment recommendations using a 4-item treatment questionnaire (e.g., how much "pain medication" and "physical therapy" they would prescribe for each patient). They also measured their empathic reactions to each patient using the Empathic Concern Scale (Batson et al., 1977, 1988; Batson, Early, & Salvarani, 1977). Drwecki et al. found that participants exhibited significant pro-White pain treatment biases by assigning enhanced pain treatment to White than Black patients. Participants also reported higher mean levels of empathy for White patients than Black patients. Moreover, pro-White empathy biases were highly predictive of pro-White pain treatment biases. However, asking participants to imagine how patients' pain affected patients' lives significantly reduced pain treatment bias in comparison to controls.

While these behavioral observations suggest the existence of racial bias in empathy, the neural correlates of racial bias in empathy remain unclear. In particular, given that empathy for pain engages multiple brain regions associated with sensory, affective, and cognitive processes, it is critical to understand which part of the pain matrix involved in empathy for pain is modulated by racial intergroup relationship. We performed the first fMRI study to investigate the neural basis of racial bias in empathy for pain (Xu, Zuo, Wang, & Han, 2009). This study scanned both Chinese and Caucasian healthy college students in Beijing, China, who were matched in age. Caucasian students were from American and European countries. Participants watched video clips showing faces of six Chinese and six Caucasian models. Each clip lasted 3 s and depicted a face with neutral expressions that received either painful (needle penetration) or non-painful (Q-tip touch) stimulation applied to the left or right cheeks. After each video clip, participants were asked to judge whether or not the model was feeling pain by pressing a button. BOLD signals were recorded to examine whether empathic neural responses were modulated by racial group membership between individuals. To examine whether participants showed explicit racial bias in empathy, after scanning, the participants were shown half of the video clips again and had to rate the pain intensity felt by the model and the unpleasantness felt by the onlooker.

Both Chinese and Caucasians reported greater rating scores of pain intensity and self-unpleasantness for painful than non-painful stimulations, but the differential rating scores (painful vs. non-painful stimuli) of pain intensity and self-unpleasantness did not differ between racial in-group and out-group members. Thus, neither Chinese nor Caucasian participants showed explicit racial bias in empathy for others' pain. fMRI results first showed that, relative to watching non-painful stimulation, watching painful stimulation applied to others significantly activated the ACC/supplementary motor cortex and the inferior frontal/insular cortex in both racial groups. Moreover, we found that the ACC/supplementary motor activation was significantly stronger for racial in-group members than for racial out-group members (Fig. 2.1). Post hoc analysis further confirmed that watching painful vs. non-painful stimulations activated the ACC/supplementary motor cortex when the stimulations were applied to racial in-group faces but not when applied to

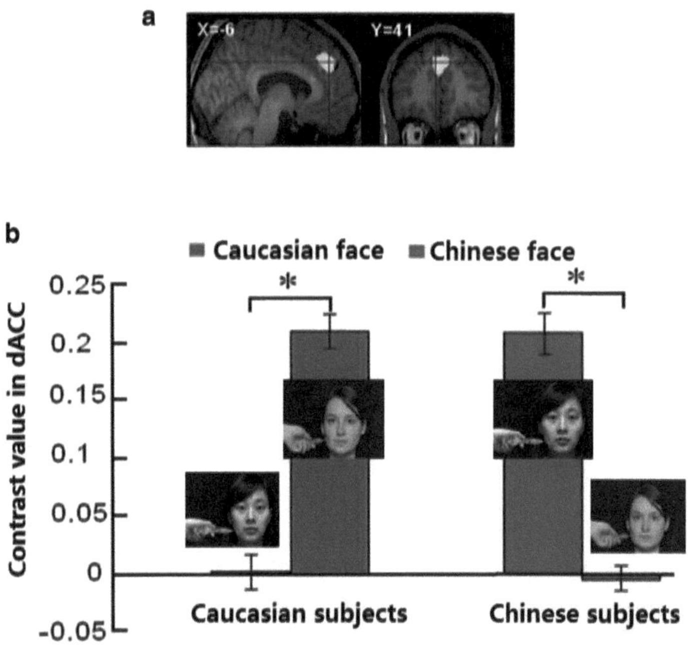

Fig. 2.1 Illustration of racial bias in empathic neural responses. (**a**) Viewing needle penetration vs. Q-tip touch to neutral faces significantly activated the ACC (from Fig. 2 in Han et al., 2009). (**b**) The activity in the ACC to painful vs. non-painful stimuli was stronger to racial in-group than out-group members (modified from Figure 1 in Xu et al., 2009)

racial out-group faces, providing fMRI evidence for racial bias in empathic neural responses. In addition, the racial bias in the ACC/supplementary motor activity to perceived pain did not differ between Chinese and Caucasian participants, indicating similar racial bias in empathy in both racial groups.

Recent meta-analysis studies have assigned distinct functions to different parts of the neural circuit involved in empathy for pain. The core network involved in empathy for pain consists of the ACC and anterior insula in which activations are independent of stimuli perceived (e.g., pictures of body parts in painful situations or abstract visual information about others' pain, Lamm et al., 2010). Fan et al. (2011) suggest that the ACC/supplementary motor cortex is recruited more frequently in the cognitive–evaluative form of empathy whereas the right anterior insula is engaged in the affective–perceptual form of empathy only and the left anterior insula is active in both forms of empathy. Thus, it may be speculated that the racial bias in empathic neural responses in the ACC implicates enhanced evaluation of racial in-group members' affective states. However, such neural empathic bias toward racial in-group members did not necessarily result in different conscious subjective ratings of others' pain intensity and induced self-unpleasantness related to racial in-group and out-group members.

Does perceived pain in racial in-group and out-group members affect neural activity in other brain regions? Using transcranial magnetic stimulation (TMS), a following research examined whether watching video clips depicting needle penetrating or Q-tip touching hands of stranger Black or White models modulates the excitability of sensorimotor regions in White-Caucasian and Black-African participants (Avenanti, Sirigu, & Aglioti, 2010). It was found that observing the pain of racial in-group models inhibited the onlookers' sensorimotor activity in both Black and White individuals. However, observing the pain of racial out-group models did not affect the onlookers' sensorimotor activity (Avenanti et al., 2010). Moreover, stronger sensorimotor response to in-group relative to out-group models' pain was observed in those subjects who showed greater negative attitude toward racial out-group members (i.e., who scored higher on the race implicit-association test). The same group also recorded BOLD signals from White and Black subjects during watching video clips depicting White and Black hands being either painfully penetrated by a syringe or being touched by a Q-tip (Azevedo et al., 2013). The activity in the bilateral anterior insula was greater for the pain experienced by same-race compared to that of other-race models. Greater implicit racial bias also predicted increased activity within the left anterior insula during the observation of own-race pain relative to other-race pain. These findings suggest stronger vicarious mapping of the pain of individuals culturally marked as in-group compared to out-group members, and the racial bias in empathic neural responses was linked to the difference of subjective attitudes toward racial in-group and out-group members.

Racial bias in empathic neural responses was also observed in the medial prefrontal cortex. Mathur, Harada, Lipke, and Chiao (2010) scanned African-American and Caucasian-American individuals while they perceived naturalistic visual scenes depicting African-American or Caucasian-American individuals in painful (e.g., in the midst of a natural disaster) or neutral (e.g., attending an outdoor picnic) situations. They found that African-American individuals recruit the medial prefrontal cortex specifically when observing racial in-group members who were suffering. Moreover, individuals who showed greater medial prefrontal activity to pain expressed by racial in-group relative to out-group members reported stronger altruistic motivation for racial in-group members. The same research group also reported that Korean participants showed greater activity in the left temporoparietal junction in response to perceived emotional pain from Koreans compared to Caucasian-Americans (Cheon et al., 2011).

Taken together, increasing brain imaging evidence indicates that racial intergroup relationship between an observer and a target person significantly influences the neural activity in multiple brain regions associated with perceived pain. The multiple levels of neural mechanisms involved in cognitive evaluation, affective sharing, and sensorimotor resonance are more sensitive to perceived pain in racial in-group than out-group members. The racial bias in empathic neural responses was confirmed in different ethnic groups, indicating a universal effect of racial intergroup relationship on empathy.

2.3 Psychological Manipulations Reduce Racial Bias in Empathic Neural Responses

While the aforementioned brain imaging findings demonstrate the existence of racial bias in empathic neural responses, the psychological mechanisms underlying racial bias in empathy for pain remain undefined. In addition, it is unknown whether the racial in-group bias in empathic neural responses can be reduced by psychological manipulations. Discovering methods to reduce racial bias in empathic neural responses may further our understanding of the neurocognitive mechanisms underlying racial bias in empathy.

We explored psychological manipulations that may reduce racial bias in empathic neural responses by recording ERPs to Asian and Caucasian faces with pain or neutral expressions from Chinese healthy adults (Sheng & Han, 2012). We hypothesized that an other-race face may be perceived as a symbol of a group rather than of an individual because other-race faces are perceived as more psychologically similar to each other relative to same-race faces (Valentine & Endo, 1992; Vizioli, Rousselet, & Caldara, 2010). Weakened individuation processing of other-race faces may lead an observer to perceive a racial out-group member without any reference to the individual's personal situation and result in decreased empathy. If this hypothesis is correct, one would predict that psychological manipulations that enhance individuation processing of other-race faces should increase empathy for other-race individual's pain. Thus, we tried two manipulations to test whether increasing attention to an individual's painful feelings and including other-race individuals in one's own social group (both lead to individuated processing of perceived faces) reduce racial bias in empathic neural responses to pain expression.

In Experiment 1, Chinese participants were asked to perform a race judgment task that required them to categorize perceived faces in terms of Asian vs. Caucasian but ignore facial expressions. These faces were evaluated by two independent groups of Chinese and Caucasian participants to ensure that the emotional faces used in the study were indeed perceived as painful rather than as portraying any other emotions. In addition, subjective feelings of pain intensity, racial identity, and facial attractiveness were matched for Asian and Caucasian faces. ERP results showed that, relative to neutral expressions, pain expressions increased neural responses at 128–188 ms (P2) after stimulus onset over the frontal/central brain regions. Source estimation suggested that the frontal P2 component might have a source in the ACC. Moreover, the differential P2 amplitudes to pain vs. neutral expressions were positively correlated with subjective ratings of self-unpleasantness induced by perceived pain in others and with subjective ratings of the empathic concern subscale that measured empathic traits. These results indicate that the neural activity in the P2 window is associated with empathy for others' pain. Most important, the empathic neural response in the P2 time window was significantly stronger for same-race faces than for other-race faces. Post hoc analysis further confirmed that the P2 empathic response was evident for same-race faces but not for other-race faces (Fig. 2.2).

The key question addressed in Experiment 2 was whether increased attention to painful feeling of each individual face would reduce racial bias in the P2 empathic responses by increasing the P2 amplitude to Caucasian faces with painful expression. Thus, besides the race judgment task, Chinese participants were also asked to perform a pain judgment task that required them to identify whether each Asian or Caucasian face was feeling painful. It was found that, in the race judgment task, the P2 amplitude was greater to pain than neutral expressions and this effect was evident for Asian faces but not for Caucasian faces. During pain judgments, however, pain vs. neutral expressions elicited a larger P2 amplitude, and the enhanced P2 amplitudes to pain vs. neutral expressions were observed for both Asian and Caucasian faces and did not differ significantly between Asian and Caucasian faces. This suggests that top-down attention to each individual's emotional state significantly reduced racial bias in empathic neural responses by increasing empathy for other-race faces. Moreover, the increased neural responses to pain vs. neutral expressions of Caucasian faces during pain vs. race judgments were positively correlated with the participants' ability of perspective-taking. Thus, it seems that the increased empathy for other-race individuals' pain by top-down attention was stronger in those who were better in taking others' perspective.

Experiment 3 employed minimal group manipulations to examine whether embracing other-race individuals in one's own group can reduce racial bias. Participants were informed that they would be assigned to the blue or green team for a competitive game, and both the fellow team and opponent team consisted of half Asians and half Caucasians. Before EEG recording, participants had to perform learning tasks so as to remember fellow team and opponent team members. If in-group relationships increase empathy for other-race individuals of the fellow team, one would expect increased empathy-related neural activity to Caucasian faces of the fellow team, and thus, the racial bias in empathic neural responses would be reduced for Caucasian faces of the fellow team compared to Caucasian faces of the opponent team. Indeed, it was found that the P2 amplitudes were increased by

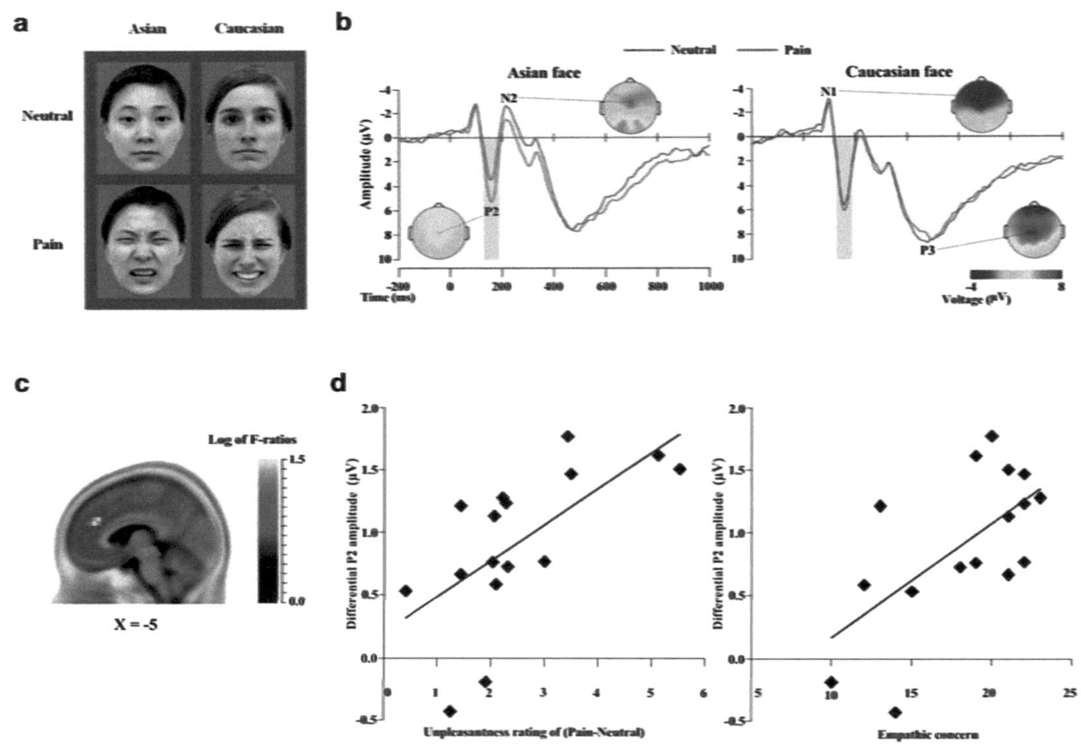

Fig. 2.2 Illustration of the stimuli and results in Sheng and Han (2012). (**a**) Chinese participants were presented with Asian and Caucasian faces with pain or neutral expressions. (**b**) ERPs to Asian and Caucasian faces during race judgments. The P2 was of larger amplitude to pain compared to neutral expressions, but this effect was evident for Asian but not Caucasian faces. (**c**) Source estimation suggested that the P2 might arise from the dACC. (**d**) Ratings of subjective feelings and empathic concern ability predicated the magnitude of empathic neural responses (differential P2 amplitude to pain vs. neutral expressions)

painful vs. neutral expressions of Asian faces but not of Caucasian faces from the opponent team. In contrast, the P2 amplitudes were significantly increased by painful vs. neutral expressions of both Asian and Caucasian faces from the fellow team. Thus, the manipulation of intergroup relationship mainly enhanced the empathic neural responses in the P2 time window to Caucasian faces from the fellow team, and consequently, the racial bias in empathic neural responses was reduced.

In all the three experiments, participants were asked to rate the intensity of pain expression and self-unpleasantness associated with the facial expressions of Asian and Caucasian faces. None of the rating scores was different between Asian and Caucasian faces though participants reported greater pain intensity and distressed feelings when watching painful compared to neutral expressions of both Asian and Caucasian faces. Therefore, while self-reports did not exhibit explicit racial bias in empathy, the neural activity shows clear evidence for implicit racial bias in empathy. However, the racial bias in empathic neural responses is not inevitable. Tasks that facilitate individuation processes of others can significantly enhance empathic neural responses to other-race individuals and result in reduction of the racial bias in empathic neural responses.

2.4 Cultural Experiences Reduce Racial Bias in Empathic Neural Responses

Sheng and Han (2012) showed evidence that psychological manipulations that enhanced individuated processing of persons reduced racial bias in empathic neural responses by increasing empathic neural responses to other-race individuals. These findings leave an open question of whether real-life experiences such as living in a society where other-race individuals consist the majority may also reduce the racial bias in empathic neural responses. This is possible because daily experiences require dealing with each individual of other-race population and thus enhance the individuated processing of other-race people.

This hypothesis was tested in a recent fMRI study that scanned Chinese adults who were either born in or immigrated to the Western countries at an early age and thus had ample experiences with Caucasian individuals (Zuo & Han, 2013). Participants were presented with video clips of Asian or Caucasian models who received painful or non-painful stimulations, similar to those used in Xu et al. (2009). Life experiences of interacting with individual Caucasians may enhance individuated processing of Caucasians in general. If this is true, one would expect that Chinese participants have similar empathic neural responses to pain stimulation applied to Asian and Caucasian models. Indeed, Zuo and Han found that viewing painful vs. non-painful stimuli applied to both Asian and Caucasian models significantly activated the ACC, anterior insula, inferior frontal cortex, and somatosensory cortex in their Chinese participants. In addition, painful vs. non-painful stimuli applied to both Asian and Caucasian models induced overlapping activations in these brain regions (Fig. 2.3). Direct comparison between brain activations

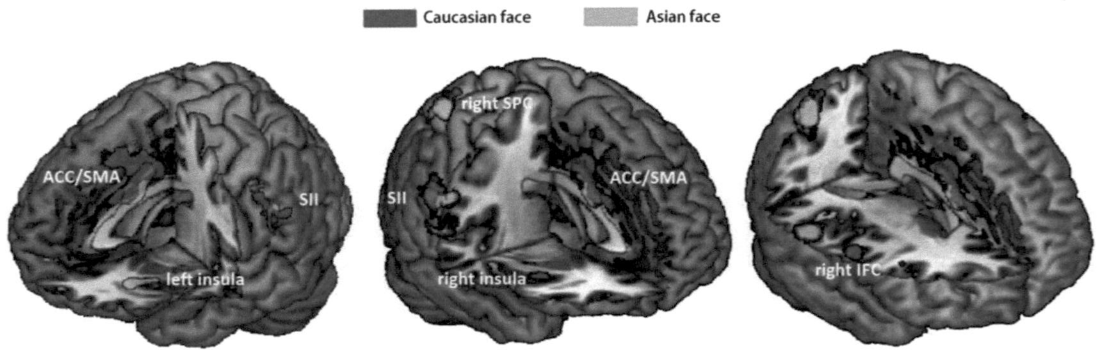

Fig. 2.3 Brain activations to painful vs. non-painful stimuli applied to Asian and Caucasian faces in Zuo and Han (2013). Note that these activations were partially overlapped in the ACC, supplementary motor cortex, somatosensory cortex, and insula

elicited by painful stimuli applied to Asian and Caucasian models did not differ significantly, suggesting comparable empathic neural responses to racial in-group and out-group members. Thus, it appears that cultural experiences with racial out-group members may increase the neural responses to the suffering of other-race individuals. Therefore, both manipulations of cognitive strategies and intergroup relationship in laboratory and real-life experiences can significantly reduce racial bias in empathic neural responses.

2.5 Molecular Mechanisms of Racial Bias in Empathic Neural Responses

The racial bias in empathy for pain reflects the effect of social intergroup relationship on how we understand and share others' painful feelings. However, it remains unclear how social influences on empathic neural responses are mediated by neurobiological factors. Oxytocin is a neuropeptide of nine amino acids that is produced in the hypothalamus and functions as both a hormone and neurotransmitter. Oxytocin receptors are expressed by neurons in many parts of the brain and spinal cord. Recent behavioral studies have shown increasing evidence that oxytocin plays a key role in the maintenance of social groups and development of trust in in-group members (De Dreu, 2012 for a review). For example, intranasally administered oxytocin versus placebo promoted trust or cooperation with in-group members but not with out-group members (De Dreu et al., 2010; De Dreu, Greer, Van Kleef, Shalvi, & Handgraaf, 2011). Oxytocin also enhances the behavioral index of emotional empathy in response to both positive and negative stimuli (Hurlemann et al., 2010) and improves performance on inference of others' emotion (Domes, Heinrichs, Michel, Berger, & Herpertz, 2007). These behavioral findings suggest that oxytocin may improve empathic neural responses specifically to racial in-group members rather than function as a general facilitator of empathy.

Sheng, Liu, Zhou, Zhou, and Han (2013) tested this hypothesis by recording ERPs to Asian and Caucasian faces with pain or neutral expressions from healthy Chinese male adults. Using a randomized, double-blind, within-subject, placebo-controlled design, this study focused on oxytocin effect on the P2 empathic neural responses to pain expressions of racial in-group and out-group faces. If oxytocin plays a role in the racial bias in empathy, the in-group bias in the neural activity in the P2 time window observed in Sheng and Han (2012) should be increased by OT compared to placebo treatment. Sheng et al. (2013) first replicated their previous finding in the placebo condition that the fronto-central P2 amplitude was greater to pain vs. neutral expressions racial in-group members but not of racial out-group members. Oxytocin treatment did not influence the P2 amplitude to pain or neutral expressions of Caucasian faces but significantly increased the P2 amplitude to pain expression of Asian faces. This effect consequently induced greater racial bias in empathic neural responses after oxytocin compared to placebo treatments. Sheng et al. also measured the participants' implicit attitudes toward racial in-group and out-group faces after oxytocin and placebo treatments using the implicit-association test (Greenwald, McGhee, & Schwartz, 1998) to assess whether OT affects the association between the racial bias in empathic neural responses and implicit racial attitudes. It was found that the racial bias in the empathic neural responses in the P2 time window was significantly associated with the racial bias in the implicit racial attitudes in the oxytocin condition but not in the placebo condition.

These findings suggest a molecular mechanism of racial bias in empathic neural responses. It appears that oxytocin does not function as a general facilitator of empathy. Instead, oxytocin improves empathic neural responses specifically to racial in-group members. In addition, it is likely that oxytocin modifies empathic neural responses to racial in-group members by enhancing the association between the implicit positive attitude toward racial in-group members and the racial bias in empathic neural responses. It seems that neither social relationship nor biological factors work alone to affect human empathy for the suffering of others. The final outcome of empathy for others' pain is determined by the interaction between social and biological factors.

2.6 Conclusion

Because bias in empathy is related to both within-group altruism and between-group conflict (Galinsky, Glin, & Maddux, 2011) and race is one of the factors that are most frequently used to categorize people into different social groups, it is highly important to uncover the social and neurobiological mechanisms underlying racial bias in empathic neural response and to discover methods to reduce neural activity related to racial bias in empathy for pain. The empirical brain imaging findings reviewed in this chapter provides solid evidence for racial bias in brain activity involved in empathy for pain. Empathy is an ability that facilitates social bonding not only in humans but in other primates as well (de Waal, 2008), and thus, there may be a long evolutionary history of this ability. Racial bias in empathy may also evolve with a long history during human evolution and function essentially to mediate racial in-group favoritism. On the other hand, racial bias in empathy may foster ignorance of painful feeling of racial out-group members and, in turn, aggravate tension between racial groups. Fortunately, current sociocultural world views do not encourage racial bias, and we have shown brain imaging evidence that racial bias in empathy is not inevitable. Future research should further explore how educational interventions may influence and weaken racial bias in empathy.

Future research should also address another important question raised by the previous brain imaging studies of racial bias in empathy, that is, are similar neural mechanisms engaged in the effects of racial intergroup relationship and other types of intergroup relationship on neural activity in response to others' pain? Sheng and Han (2012) found that minimal group manipulation in laboratory only influenced empathic neural response to perceived pain of racial out-group members but did not affect that to perceived pain of racial in-group members. Hein et al. (2010) found that minimal group manipulation in the laboratory modulated the left insular activity, being stronger when participants saw high versus low pain in the in-group member as compared with high versus low pain in the out-group member. This is apparently different from the effect of racial intergroup relationship on empathic neural responses in the ACC (Sheng & Han, 2012; Xu et al., 2009). The racial intergroup relationship may be different from the minimal group relationship in that the former is powerful for creating stable social categorizations of a large population whereas the latter is useful for social categorizations of a small population and can vary easily. This possibly gives rise to distinct neural substrates underlying the effect of the racial intergroup relationship and the minimal group relationship on empathy for others' pain and may be clarified in future research.

Acknowledgments This work was supported by the National Natural Science Foundation of China (Project 30910103901, 91024032, 81161120539) and the National Basic Research Program of China (973 Program 2010CB833903).

References

Avenanti, A., Sirigu, A., & Aglioti, S. M. (2010). Racial bias reduces empathic sensorimotor resonance with other-race pain. *Current Biology, 20*, 1018–1022.

Azevedo, R. T., Macaluso, E., Avenanti, A., Santangelo, V., Cazzato, V., & Aglioti, S. M. (2013). Their pain is not our pain: Brain and autonomic correlates of empathic resonance with the pain of same and different race individuals. *Human Brain Mapping, 34*, 3168.

Batson, C. D. (1987). Prosocial motivation: Is it ever truly altruistic? In L. Berkowitz (Ed.), *Advances in experimental social psychology* (Vol. 20, pp. 65–122). New York, NY: Academic.

Batson, C. D. (1991). *The altruism question: Toward a social-psychological answer*. Hillsdale, NJ: Lawrence Erlbaum.

Batson, C. D., Dyck, J. L., Brandt, J. R., Batson, J. G., Powell, A. L., McMaster, M. R., et al. (1988). 5 studies testing 2 new egoistic alternatives to the empathy altruism hypothesis. *Journal of Personality and Social Psychology, 55*, 52–77.

Batson, C. D., Early, S., & Salvarani, G. (1977). Perspective taking: Imagining how another feels versus imagining how you would feel. *Personality and Social Psychological Bulletin, 23*, 751–758.

Berger, D. M. (1987). *Clinical empathy*. Northvale, NJ: Jason Aronson.

Cheng, Y., Lin, C. P., Liu, H. L., Hsu, Y. Y., Lim, K. E., Hung, D., et al. (2007). Expertise modulates the perception of pain in others. *Current Biology, 17*, 1708–17013.

Cheon, B. K., Im, D. M., Harada, T., Kim, J. S., Mathur, V. A., Scimeca, J. M., et al. (2011). Cultural influences on neural basis of intergroup empathy. *NeuroImage, 57*, 642–650.

Cosmides, L., Tooby, J., & Kurzban, R. (2003). Perceptions of race. *Trends in Cognitive Sciences, 7*, 173–179.

De Dreu, C. K. W. (2012). Oxytocin modulates cooperation within and competition between groups: An integrative review and research agenda. *Hormones and Behavior, 61*, 419–428.

De Dreu, C. K. W., Greer, L. L., Handgraaf, M. J. J., Shalvi, S., Van Kleef, G. A., Baas, M., et al. (2010). The neuropeptide oxytocin regulates parochial altruism in intergroup conflict among humans. *Science, 328*, 1408–1411.

De Dreu, C. K. W., Greer, L. L., Van Kleef, G. A., Shalvi, S., & Handgraaf, M. J. J. (2011). Oxytocin promotes human ethnocentrism. *Proceedings of the National Academy of Sciences of the United States of America, 108*, 1262–1266.

de Waal, F. B. M. (2008). Putting the altruism back into altruism: The evolution of empathy. *Annual Review of Psychology, 59*, 279–300.
Decety, J., Yang, C., & Cheng, Y. (2010). Physicians down-regulate their pain empathy response: An event-related brain potential study. *NeuroImage, 50*, 1676–1682.
Domes, G., Heinrichs, M., Michel, A., Berger, C., & Herpertz, S. (2007). Oxytocin improves "mind-reading" in humans. *Biological Psychiatry, 61*, 731–733.
Drwecki, B. B., Moore, C. F., Ward, S. E., & Prkachin, K. M. (2011). Reducing racial disparities in pain treatment: The role of empathy and perspective-taking. *Pain, 152*, 1001–1006.
Fan, Y., Duncan, N. W., de Greck, M., & Northoff, G. (2011). Is there a core neural network in empathy? An fMRI based quantitative meta-analysis. *Neuroscience and Biobehavior Reveiw, 35*, 903–911.
Fan, Y., & Han, S. (2008). Temporal dynamic of neural mechanisms involved in empathy for pain: An event-related brain potential study. *Neuropsychologia, 46*, 160–173.
Galinsky, A. D., Glin, D., & Maddux, W. W. (2011). Using both your head and your heart: The role of perspective taking and empathy in resolving social conflict. In J. P. Forgas, A. W. Kruglanski, & K. D. Williams (Eds.), *The psychology of social conflict and aggression* (pp. 103–118). New York, NY: Psychology Press.
Greenwald, A. G., McGhee, D. E., & Schwartz, J. L. K. (1998). Measuring individual differences in implicit cognition: The implicit association test. *Journal of Personality and Social Psychology, 74*, 1464–1480.
Gu, X., & Han, S. (2007). Attention and reality constraints on the neutral processes of empathy for pain. *NeuroImage, 36*, 256–267.
Han, S., Fan, Y., Xu, X., Qin, J., Wu, B., Wang, X., et al. (2009). Empathic neural responses to others' pain are modulated by emotional contexts. *Human Brain Mapping, 30*, 3227–3237.
Han, S., Fan, Y., & Mao, L. (2008). Gender difference in empathy for pain: An electrophysiological investigation. *Brain Research, 1192*, 85–93.
Hein, G., Silani, G., Preuschoff, K., Batson, C. D., & Singer, T. (2010). Neural responses to ingroup and outgroup members' suffering predict individual differences in costly helping. *Neuron, 68*, 149–160.
Hurlemann, R., Patin, A., Onur, O., Cohen, M., Baumgartner, T., Metzler, S., et al. (2010). Oxytocin enhances amygdala-dependent, socially reinforced learning and emotional empathy in humans. *Journal of Neuroscience, 30*, 4999–5007.
Jackson, P. L., Meltzoff, A. N., & Decety, J. (2005). How do we perceive the pain of others? A window into the neural processes involved in empathy. *NeuroImage, 24*, 771–779.
Johnson, J. D., Simmons, C. H., Jordan, A., MacLean, L., Taddei, J., & Thomas, D. (2002). Rodney King and O. J. revisited: The impact of race and defendant empathy induction on judicial decisions. *Journal of Applied Social Psychology, 32*, 1208–1223.
Lamm, C., Decety, J., & Singer, T. (2011). Meta-analytic evidence for common and distinct neural networks associated with directly experienced pain and empathy for pain. *NeuroImage, 54*, 2492–2502.
Li, W., & Han, S. (2010). Perspective taking modulates event related potentials to perceived pain. *Neuroscience Letter, 469*, 328–332.
Ma, Y., Wang, C., & Han, S. (2011). Neural responses to perceived pain in others predict real-life monetary donations in different socioeconomic contexts. *NeuroImage, 57*, 1273–1280.
Mathur, V. A., Harada, T., Lipke, T., & Chiao, J. Y. (2010). Neural basis of extraordinary empathy and altruistic motivation. *NeuroImage, 51*, 1468–1475.
Moya, P., & Markus, H. R. (2011). Doing race: A conceptual overview. In H. R. Markus & P. Moya (Eds.), *Doing race: 21 essays for the 21st century* (pp. 1–102). New York, NY: Norton.
Mu, Y., Fan, Y., Mao, L., & Han, S. (2008). Event-related theta and alpha oscillations mediate empathy for pain. *Brain Research, 1234*, 128–136.
Saarela, M. V., Hlushchuk, Y., Williams, A. C., Schurmann, M., Kalso, E., & Hari, R. (2007). The compassionate brain: Humans detect intensity of pain from another's face. *Cerebral Cortex, 17*, 230–237.

Sheng, F., & Han, S. (2012). Manipulations of cognitive strategies and intergroup relationships reduce the racial bias in empathic neural responses. *NeuroImage, 61*, 786–797.
Sheng, F., Liu, Y., Zhou, B., Zhou, W., & Han, S. (2013). Oxytocin modulates the racial bias in neural responses to others' suffering. *Biological Psychology, 92*, 380–386.
Singer, T., Seymour, B., O'Doherty, J., Kaube, H., Dolan, R. J., & Frith, C. D. (2004). Empathy for pain involves the affective but not sensory components of pain. *Science, 303*, 1157–1162.
Singer, T., Seymour, B., O'Doherty, J. P., Stephan, K. E., Dolan, R. J., & Frith, C. D. (2006). Empathic neural responses are modulated by the perceived fairness of others. *Nature, 439*, 466–469.
Valentine, T., & Endo, M. (1992). Toward an exemplar model of face processing: The effects of race and distinctiveness. *Quarterly Journal of Experimental Psychology: Human Experimental Psychology, 44*(A), 161–204.
Vizioli, L., Rousselet, G. A., & Caldara, R. (2010). Neural repetition suppression to identity is abolished by other-race faces. *Proceedings of the National Academy of Sciences of the United States of America, 107*, 20081–20086.
Xu, X., Zuo, X., Wang, X., & Han, S. (2009). Do you feel my pain? Racial group membership modulates empathic neural responses. *Journal of Neuroscience, 29*, 8525–8529.
Zuo, X., & Han, S. (2013). Cultural experiences reduce racial bias in neural responses to others' suffering. *Culture and Brain, 1*, 34.

Chapter 3
Culture, Cognition, and Intercultural Relations

Jinkyung Na and Micaela Y. Chan

Abstract Na and Chan examine the cultural variations in reasoning style. They highlight the well-documented differences in cognition that paints Easterners as being holistic processors and Westerners being analytic processors. Na and Chan assess this overarching construct by reviewing the cultural differences of attention, attribution, and motivation.

Attention has shown cultural differences where Easterners are more relational and Westerners are more focused. Across various attention tasks, Easterners attend widely to a scene including contextual cues while Westerners are more concerned with focal elements. The neuroimaging evidence for these differences indicates cultural differences in frontoparietal activation for attention tasks.

Cultural differences in attribution show that Easterners use relational reasoning with making attributes about behavior while Westerners focus more on the central figures. Na and Chan detail a study using event-related potential on a lexical decision task that suggests differences in attribution-based neural activity between cultures. Additional neuroimaging studies of phenomena similar to attribution are also discussed.

Easterners have been shown to believe that broad social contexts operate to make choices while Westerners believe a choice is an act of self-expression. Na and Chan detail neuroimaging studies that investigate cognitive dissonance and choice justification to examine the cultural differences. These studies show a wide variety of neural responses that underlie cultural differences in cognitive dissonance.

Na and Chan conclude by discussing how the understanding of cultural differences in reasoning style could be used in our multicultural world.

J. Na (✉)
School of Behavioral and Brain Sciences, University of Texas at Dallas,
800 W Campbell Rd, Richardson, TX 75080-3021, USA
e-mail: jinkyung.na@utdallas.edu

M.Y. Chan
School of Behavioral and Brain Sciences, University of Texas at Dallas,
800 W Campbell Rd, Richardson, TX 75080-3021, USA

Center for Vital Longevity, School of Behavioral and Brain Sciences,
1600 Viceroy Dr. Suite 800, Dallas, TX 75235, USA
e-mail: mchan@utdallas.edu

J.E. Warnick, D. Landis (eds.), *Neuroscience in Intercultural Contexts*,
International and Cultural Psychology, DOI 10.1007/978-1-4939-2260-4_3

On July 5, 2002, a tragic accident occurred in South Korea. Two 14-year-old Korean girls were struck and killed by an American armored vehicle (weighing approximately 57 t) as they walked along a narrow country road on their way to a birthday party. This accident sparked one of the biggest anti-American demonstrations among South Koreans that the country has seen in recent years. Given that South Koreans normally consider the United States as their closest ally, this anti-American sentiment was somewhat striking. Although many reasons were partially responsible for this rare anti-American protest among South Koreans (e.g., such as issues regarding jurisdictional authority of the Korean court on American soldiers), a critical contributor was the stark contrast between how the United States and Korea differed in their approach in dealing with the accident.

Of particular importance is that Americans mainly focused on the fair and impartial nature of legal processes, whereas Koreans additionally emphasized relational implications. For example, some Koreans asked for a sincere apology from the US President George W. Bush, because an apology from the US president is a symbolic way of reassuring the close relations between Korea and the United States. In other words, Koreans were much more relational in their approach than Americans who wanted to concentrate on focal issues. We believe that the difference observed here is closely linked to cultural differences in cognition (Markus & Kitayama, 1991; Nisbett, Peng, Choi, & Norenzayan, 2001). It is well established that Westerners, such as Americans, tend to narrowly attend to focal objects, information, and people in their reasoning; instead, East Asians, such as Koreans, tend to be contextual and relational in their reasoning. As shown vividly in the foregoing example, cultural differences in cognition could have important implications for intercultural relations. Thus, the present chapter reviews cultural differences in cognition and discusses them under the rubric of intercultural relations.

3.1 Culture, Reasoning, and Brain

Although interests in cultural variations among psychologists can be traced back to the very beginning of modern psychology (e.g., James, 1890/1950), empirical research on culture and mind was inspired by two highly influential reviews published around the early 1990s (Markus & Kitayama, 1991; Triandis, 1989). Since then, cultural differences in various psychological processes have been demonstrated (see Henrich, Heine, & Norenzayan, 2010, for a recent review). One of the most influential findings among them is that the way people perceive and think about the world systematically varies depending on their cultural backgrounds (Nisbett & Masuda, 2003; Nisbett et al., 2001). Much empirical evidence for this premise has been shown between Easterners (in particular East Asians) and Westerners (in particular Americans) in various domains ranging from attention (Masuda & Nisbett, 2001), attribution (Choi, Nisbett, & Norenzayan, 1999), to logical reasoning (Norenzayan, Smith, Kim, & Nisbett, 2002). Such cultural variations between East and West can be summarized into one critical difference. Namely,

Easterners are much more holistic (i.e., processing the entire contexts) and relational in their reasoning compared to Westerners who are more analytic (i.e., processing primarily objects) and focused. Thus, the Eastern mode of reasoning is referred to as holistic reasoning, whereas the Western mode of reasoning is referred to as analytic reasoning (Nisbett et al., 2001).

Many researchers believe that these cultural differences in cognition arise because Easterners repeatedly participate in various cultural practices that encourage them to see relations between people and objects, whereas Westerners are habitually encouraged to detach objects and people from the context. For example, Jaein, a Korean, may be socialized to believe that his behavior is fundamentally driven by surrounding contexts and people, whereas David, an American, may be culturally trained to believe that his behavior is mainly driven by his own personal attributes (e.g., personality or attitude). This type of repeated engagement in cultural practices can have significant impact on their brains as do other types of sustained experiences (Hanakawa, Honda, Okada, Fukuyama, & Shibasaki, 2003; Tang et al., 2006). A classic demonstration of this sort was shown by the famed study on the hippocampus of London cab drivers (Maguire et al., 2000). The hippocampus, a seahorse-shaped structure in the limbic system, plays a critical role in spatial memory and navigation (O'Keefe & Nadel, 1978). Maguire and colleagues (2000) found that, compared with matched control subjects, London cab drivers showed substantial enlargement in the posterior part of the hippocampus, as well as sizable reduction in the anterior part. Importantly, the observed structural differences in the hippocampus were significantly correlated with the years of experience as a cab driver. Thus, the result suggested that extensive engagement in spatial navigation (i.e., driving a cab in a highly complex city like London) could significantly alter one's brain.

By the same logic, cultural experiences can also make functional as well as structural changes in the brain. In fact, an emerging literature on cultural neuroscience has accumulated much empirical evidence showing the effect of culture on brain structures and functions (see Kitayama & Uskul, 2011, for a review). Building on this literature, we will review the neural evidence showing cultural differences and discuss it with respect to intercultural relations. In doing so, we will first introduce cultural differences in reasoning across various domains including basic processes such as attention and a more applied area such as choice, then show how these cultural differences are *embrained*, and finally address their implications for intercultural relations.

3.2 Attention

3.2.1 Cultural Differences

Attention is a cognitive domain that clearly demonstrates the cultural differences in which Easterners are relational whereas Westerners are focused (Kitayama, Duffy, Kawamura, & Larsen, 2003; Masuda & Nisbett, 2001). Previous research has shown that Easterners tend to attend to the entire perceptual field including backgrounds as well as focal objects, whereas Westerners tend to pay close attention to focal objects while relatively ignoring backgrounds. For example, Masuda and Nisbett (2001) showed Japanese and American students a series of animations of underwater scenes. Each of these animations contained one or more focal fish (larger, bright, and faster-moving fish) as well as background objects such as plants, rocks, bubbles, and other small animals. After watching them, Japanese and American participants were simply asked to report what they saw. The researchers found that Japanese participants were more likely to mention background objects, whereas American participants were more likely to mention the focal fish in their first sentence. Thus, this study empirically showed the predicted cultural differences in attention to focal and contextual objects.

Essentially, the same differences in visual perception were more elegantly shown by Kitayama and colleagues (Kitayama et al., 2003) using what is called the framed-line task (Fig. 3.1). In the task, participants are shown a square with a line drawn inside it. After studying the original square and line, a new empty square, which is either larger or smaller than the original one, is given to participants. Their job is to draw a line on the new square that has the same length as the original line. The same length, however, is defined in two different ways. In the absolute condition, participants are

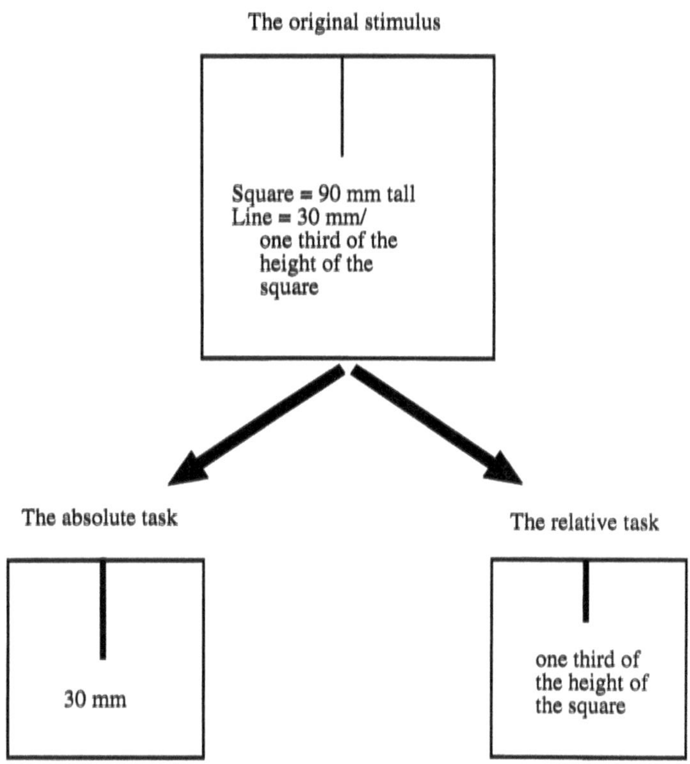

Fig. 3.1 Example of the framed-line task (adapted from Kitayama et al., 2003, *Psychological Science*)

asked to draw a line on the new square that has exactly the same numerical values in millimeters. In contrast, the relative condition requires participants to draw a line that has the same proportion relative to the surrounding frame. In other words, those who can easily detach an object (i.e., the line) from the context (i.e., the surrounding frame) would have advantage in the absolute condition, whereas those who habitually see the relations among parts have advantage in the relative condition. In line with the proposition that East Asians are more relational and Americans are more focused in attention, East Asians were better in the relative condition and Americans were better in the absolute condition than their counterparts.

These differences across cultures in attention are not limited to visual perception. In particular, an attention bias in interpersonal communications has been well noted. For example, Westerners tend to attend primarily to focal information, namely, verbal contents, whereas Easterners tend to pay closer attention to contextual and nonverbal cues such as vocal tone (e.g., Barnlund, 1989). This observation is empirically confirmed in recent studies by Ishii and colleagues (Ishii, Reyes, & Kitayama, 2003; Kitayama & Ishii, 2002). In their studies, participants were presented with utterance in their native language and instructed to either judge word meaning as positive or negative while ignoring vocal tone ("meaning judgment") or judge vocal tone as pleasant or unpleasant while ignoring word meaning ("tone judgment"). For the half of the utterances, verbal content was congruous with vocal tone (congruous trials: positive meaning in pleasant tone and negative meaning in unpleasant tone). For the remaining half, however, verbal content is incongruous with vocal tone (incongruous trials: positive meaning in unpleasant tone and negative meaning in pleasant tone). For example, "grateful" pronounced in pleasant tone or "dislike" pronounced in unpleasant tone is designated as congruous trials, whereas "grateful" pronounced in unpleasant tone or "dislike" pronounced in pleasant tone is designated as incongruous trials. In other words, the task is a type of Stroop task that has been used in previous cross-cultural studies (Oyserman, Sorensen, Reber, & Chen, 2009; Sabbagh, Xu, Carlson, Moses, & Lee, 2006). They found that the interference in reaction time (incongruous trials–congruous trials) was stronger in the tone judgment than in the meaning judgment among Americans, whereas the opposite was the case among Asians (Japanese and Filipinos). That is, for Americans, it was relatively easier to ignore vocal tone than to ignore word meaning. But for Asians, it was relatively easier to ignore word meaning than to ignore vocal tone. The results strongly suggest that Americans primarily attend to verbal contents (i.e., word meaning) whereas Asians to contextual cues (i.e., vocal tone) in interpersonal communications.

Taken together, it has been repeatedly reported that Easterners have a more diffused attention system, whereas Westerners have a more focused attention system. As outlined in the introduction, individuals are socialized to show this cultural difference by habitually engaging in cultural practices that promote culturally representative attention system. Since habitual engagement results in neural changes (Maguire et al., 2000), it is expected that repeated engagement in cultural practices should also result in changes in relevant brain pathways. Thus, in the following section, we review neural evidence that reflects cultural differences in attention.

3.2.2 Neural Evidence

By now, there are quite a number of studies that reported neural evidence on cultural differences in attention. First, Hedden and colleagues used a modified version of the framed-line task (mentioned in the previous section) in a study using functional magnetic resonance imaging (fMRI), a method that could reveal locations of neural functions (Hedden, Ketay, Aron, Markus, & Gabrieli, 2008). In the modified framed-line task, participants viewed a series of stimuli (i.e., a square with a line on it) and judged whether each stimulus matched the preceding one in the relative sense (based on the proportion of a line and a square) or in the absolute sense (based on the length of a line). Interesting cultural differences in neural activations were found in the frontal and parietal brain regions known to be associated with attentional control. When making the relative judgment, the frontoparietal attention network was more activated for Americans than for East Asians. In contrast, when making the absolute judgment, the same brain network was more activated for East Asians than for Americans. In other words, attentional control was greater for the culturally unfamiliar judgment than for the culturally familiar judgment. Furthermore, activation in this brain network correlated with their view of cultural identity.

In addition, studies using event-related potential (ERP), a method of measuring neural signal that is temporally more sensitive than fMRI (measures neural activity in millisecond intervals vs. second intervals by fMRI), also revealed cultural differences in attention. First, Lewis and colleagues (2008) investigated cultural differences in attention with an oddball paradigm. In their study, participants were presented with three different types of stimuli one by one in a random order: standard stimulus (the number 8), target stimulus (the number 6), and oddball stimulus (English words, consonants, and numbers; for example, DOG, TCQ, and 305). Their task is to press the designated key when the target appeared on the computer screen. Three different types of stimuli differed in how frequently they were presented: 76 %, 12 %, and 12 % for the standard, the target, and the oddballs, respectively. Previous research with this type of oddball paradigms has identified two positive ERP components (target P3 and novelty P3). P3 stands for positive deflection peaking around 300 ms after the stimulus presentation. However, although both occurring at similar timing, these two components reflect different types of brain processing. First, the target P3, observed most prominently in posterior regions, occurred in response to a target stimulus. Thus, the strength of the target P3 is believed to indicate how much they focused on the target. In contrast, novelty P3 is observed most prominently in anterior regions and elicited as a response to an oddball. Thus, the strength of the novelty P3 is believed to indicate the amount of attention paid to contextually deviant events. Consistent with previous work on culture and attention, Asian Americans showed stronger novelty P3 than Americans. This suggests that Asian Americans are more sensitive to contextual deviants than Americans. Moreover, the opposite pattern was shown for the target P3 although it was only marginally significant. That is, it can be said that Americans showed a tendency to focus on the target stimulus more than did Asian Americans.

Finally, this type of cultural differences in brain responses was shown to be pronounced more among older adults (Goh et al., 2007). In this study, based on the

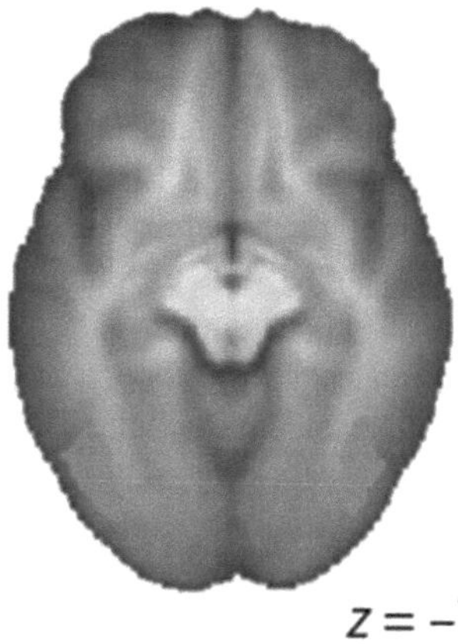

Fig. 3.2 Left and right lateral occipital complex; object processing region of the brain (adapted from Goh et al., 2007, *Cognitive, Affective, & Behavioral Neuroscience*)

phenomenon that brain responses toward repeated stimuli are likely to decrease, participants' adaptation (i.e., lessening of neural response) to objects vs. background contexts was examined. Specifically, young and old participants in the United States and Singapore passively viewed a series of quartets of pictures; on each picture, one central figure was embedded in a particular context (e.g., a lion in a desert). Four different types of quartets were presented in the study: (a) the same object and backgrounds were repeated, (b) the central figure was repeated while backgrounds varied (e.g., a lion in four different backgrounds), (c) the central figure varied while backgrounds were repeated (e.g., four different animals in a desert), and (d) both the central figure and backgrounds varied. By comparing neural responses to these types of quartets, the participants' adaptation to the central figures and backgrounds was measured. The study showed that Singaporean older adults experienced less neural adaptation in the object processing region (lateral occipital complex; see Fig. 3.2), compared to American older adults. In other words, the decrease in neural response to repeated central figures was substantially smaller for old Singaporeans than for old Americans, which indicates that older Singaporeans paid less attention to central figures than older Americans. Moreover, the same effect was not found among younger adults. This suggested that although differences in younger adults across both cultures in neural adaptation may not be prominent, aging may play a role in deepening the differences between cultures due to a longer period of socialization in a particular culture.

In sum, numerous studies have confirmed that behavioral differences in attention between Americans and Asians are closely linked to the corresponding differences in brain responses.

3.2.3 Implications for Intercultural Relations

Given that attention guides subsequent information processing, cultural differences in attention can have important implications for various psychological processes, particularly ones that are relevant to intercultural relations. For example, the tendency for Easterners to pay close attention to the context is also reflected in their communication style, such that they tend to heavily rely on contextual factors when they communicate with others (Holtgraves, 1997; Kashima & Kashima, 1998). In contrast, Westerners are less likely to rely on the context when conveying messages to others (Hall, 1976; Sanchez-Burks et al., 2003). A study by Holtgraves and Yang (1992) demonstrated this differences by showing that Korean language was more indirect than English. Similarly, Ambady, Koo, Lee, and Rosenthal (1996) showed that the way of expressing politeness among Koreans was more influenced by relational or social cues, but in contrast, Americans focused on expressing politeness by the content of the message. These differences in communication style, which corresponds to cultural differences in attention, may cause confusions in intercultural communications since Easterners would express their intentions indirectly using contextual cues, which will likely be ignored by Westerners when interpreting their intentions.

3.3 Attribution

3.3.1 Cultural Differences

Attribution, how one explains or understands others' behaviors, is another domain of cultural differences in reasoning that have important implications for intercultural relations. To the extent that Westerners are analytic in their reasoning (i.e., focusing on central figures), they also tend to focus on information about the person, namely, internal dispositions, in order to explain his or her behavior. However, to the extent that Easterners are relational in their reasoning (i.e., paying attention to the entire contexts), they also tend to utilize contextual information in explaining others' behaviors. Evidence is mounting that Westerners show a strong bias to dispositional attribution (i.e., giving undue weight to one's internal disposition), whereas such bias is much weaker or even absent among Easterners (Choi et al., 1999; Masuda & Kitayama, 2004; Morris & Peng, 1994).

For example, in a classic study by Jones and Harris (1967), participants were asked to infer the essay writer's true attitude toward Castro after reading either "pro-Castro" or "anti-Castro" essay. In the critical condition, participants were told that the position that the writer took in the essay was randomly assigned by the experimenter. In other words, participants would infer that the essay did not reflect the writer's true attitude since their position was given rather than chosen. However, American participants in this study were still influenced by the position of the essay when inferring the essay writer's true attitude even though they were fully aware that the position itself had been randomly assigned. In stark contrast, Choi and Nisbett (1998) showed that East Asians were much more sensitive to situational constraints of behaviors and, thus, less vulnerable to this type of errors. Specifically, Korean participants in their study did not infer the true attitude of the essay writers when the situational constraint (i.e., the assignment of the essay position) was made salient by asking participants themselves to write an essay in the same situation or emphasizing that the writer merely copied the arguments provided by the experimenter. However, American participants were not affected by these additional manipulations and still made inferences about the true attitude of the essay writer based on the assigned position.

Furthermore, a recent study by Na and Kitayama (2011) suggested that spontaneous/subconscious processes are critical to these cultural differences in attribution. In this study, cultural differences in spontaneous trait inference (a tendency to spontaneously infer the corresponding trait from a behavior; Uleman, Saribay, & Gonzalez, 2008) between European Americans and Asian Americans were examined. Participants were asked to perform a memory test on faces paired with two short statements of behaviors. Both behavior statements that were paired with the same face implied the same trait. For example, "She held the handrail on the escalator like the instruction advised" and "She tested her smoke detector's battery before going to bed" were paired with one face, and both statements implied the trait "cautious." Thus, if trait inference or dispositional attribution is indeed spontaneous and automatic, participants would form an association between a face and the implied trait (e.g., she is cautious) during the memory test. After studying the pairs (before testing their memory), participants completed a lexical decision task, a task that measures how quickly a stimuli is being classified. The lexical decision task was administered as a filler task, when in fact it was designed to assess the degree to which participants had spontaneously inferred the corresponding traits (e.g., cautious) from behavior statements. On each trial of the lexical decision task, one of the original faces used in the memory test was first presented as a fixation point. The face was followed by either a trait that was implied by the behavior statements previously paired with the face, a trait that was irrelevant to those behaviors, or a pseudo-word. Participants were instructed to report whether the stimulus was an English word or not as quickly and as accurately as possible by pressing one of two designated computer keys.

The results showed that, for European Americans, lexical decision was significantly faster for the implied traits than for the irrelevant traits. That is, during the memory test, a trait was spontaneously inferred from a behavior and then the inferred trait was linked to the stimuli face. Therefore, the face—acting as a fixation point—automatically activated the trait during the lexical decision task, thereby facilitating the performance of lexical decision for the implied trait relative to the irrelevant trait. However, for Asian Americans, lexical decision was no faster for the implied traits than for the irrelevant traits, suggesting that they did not make any spontaneous trait inference. Importantly, when given explicit instructions to form an impression on the face stimuli, Asian Americans were able to show the same effect as European Americans (i.e., faster RT for implied traits than for irrelevant traits). Taken together, the results indicated that upon exposure to another's behavior, European Americans automatically infer a corresponding trait from a behavior and ascribe it to the actor. However, Asian Americans tend to make such inference only when it is really necessary.

Taken together, behavioral evidence has repeatedly shown cultural differences in attribution such that Americans are highly biased toward dispositional attribution whereas such bias is much weaker among Asians.

3.3.2 Neural Evidence

Cultural differences in attribution have also been demonstrated with neural measures. Since behavioral data reviewed above clearly showed cultural differences in spontaneous trait inferences, Na and Kitayama (2011) further reasoned that this behavioral outcome would have corresponding neural underpinning. Therefore, a similar cross-cultural difference might be obtained with neural indicators. Based on this notion, they conducted another study similar to their behavioral study (see Sect. 3.1). In this study, participants also studied pairs of face and statement of trait-implying behavior and then performed a lexical decision task. A notable difference in this version of the lexical decision task is that the targets were either traits implied by the stimulus behaviors, antonyms of these traits (instead of irrelevant traits used in the behavioral version of this study), or pseudo-words. During the lexical decision task, ERP was used to measure stimulus-locked electrical activities of the brain. The ERP component of interest is the N400, a negative deflection peaking approximately 400 ms after stimulus presentation. The N400 is typically observed in posterior electrodes, which is thought to index detection of semantic incongruity (Kutas & Hillyard, 1980).

The behavioral version of the study showed that during the memory phase of the study, European Americans spontaneously inferred a trait corresponding to each behavior statement and associated the trait to the stimulus face. When presented during the lexical decision task, the face automatically activated the inferred trait. Thus, if the activation of the inferred trait was followed by presentation of its antonym, a strong N400 component may be expected. In contrast, Asian Americans do not infer any traits during the memory phase of the study. Hence, there will be no N400 component even when the antonym of a trait linked to a priming face is presented during the subsequent lexical decision task.

The results confirmed this prediction. Figure 3.3 shows the time course of ERPs at the posterior central (Pz) scalp location. For European Americans, a clear N400 component was identified when the targets were antonyms of the traits implied by the relevant behaviors (but not when the targets were the implied traits). The N400 component can be expected only if the face induces a trait implied by his or her previous behaviors. Hence, the pattern observed here is a clear indication that European Americans spontaneously inferred a trait of a person based on his or her behaviors during the memory phase of the study. For Asian Americans, however, there was no such incongruity effect. This lends further support to the hypothesis that Asian Americans do not engage in spontaneous trait inference when asked to memorize trait-implying behaviors.

Similar observation was also made in an fMRI study investigating the theory of mind in American and Japanese children (Kobayashi, Glover, & Temple, 2007). The theory of mind is the ability to predict behaviors of others based on the inferred underlying intentions of others. In this study, American and Japanese children were presented with both stories and cartoons that either (1) required them to use the theory of mind in order to infer others' minds (ToM condition) or (2) did not require

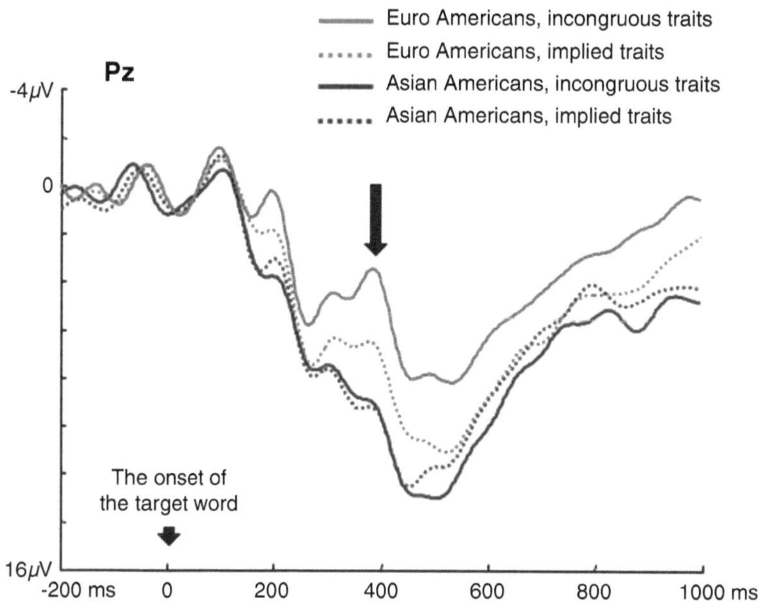

Fig. 3.3 Grand-averaged ERPs at Pz in the implied trait condition (*dotted lines*) and in the incongruous trait (antonym) condition (*solid lines*) among European Americans (*blue lines*) and Asian Americans (*red lines*). Note that negative deflections of ERPs are shown in the upward direction on the *y*-axis (adapted from Na & Kitayama, 2011, *Psychological Science*)

understanding the mind of others (non-ToM condition). Compared to the non-ToM condition, the ToM condition was associated with the activation of brain regions such as the temporoparietal junction (TPJ) and medial prefrontal cortex (mPFC), which are both related to inference about another's internal disposition (Mitchell, Banaji, & Macrae, 2005; Mitchell, Macrae, & Banaji, 2005). More importantly, interesting cultural differences were also observed in the activation of TPJ. That is, ToM-specific activation in the TPJ was stronger for American children than for Japanese children, which suggested that American children might engage in inferences about others' internal disposition more than Japanese children. Then, the results are highly consistent with previous research in culture and attribution.

Taken together, an emerging literature is beginning to identify neural pathways underlying cultural differences in attribution.

3.3.3 Implications for Intercultural Relations

It is almost impossible, if not impossible, to live without interacting with individuals from other cultural background given the increasingly global nature in business and other industries alike. Thus, it is needless to say how important it is to properly understand and explain others' behaviors. Therefore, substantial cultural differences in attribution should have important implications for intercultural relations. First, compared to those who narrowly focus on internal dispositions in explaining others' behaviors (e.g., Americans), those who take into account other contextual information are likely to have very different causal beliefs. Such differences may end up with various types of misunderstanding (e.g., a conflict illustrated in the introduction).

Second, a recent study showed that cultural differences in attribution are closely linked to corresponding differences in interpersonal liking (Na, Choi, & Sul, 2013). In this study, Korean and American participants were asked to form an impression of two people and evaluate them along various dimensions. An interesting twist is that one of the two target people showed the East Asians' style of attribution, whereas the other showed the American's style of attribution. For example, one of them considered more information before making final attribution than the other since the former weighs contextual factors as well as dispositional factors. The results showed that Koreans favored the one who considered more information whereas Americans favored the other who emphasized internal disposition. In other words, culture influences not only the way people make attribution but also the way they judge others based on their attribution style. This suggests that what should be taken seriously in intercultural relations is not just how we make attribution. Rather, we also need to think about how our attribution style would be perceived by our partners in intercultural relations.

3.4 Motivation and Choice

3.4.1 Cultural Differences

Cultural differences in cognition are also evident in more applied domains of reasoning. For example, it is well reflected in the way individuals make choices. On the one hand, Westerners tend to focus on central information, and so their choices tend to be made narrowly based on their personal preferences. On the other hand, Easterners are relational, and thus, broader social contexts play important roles in making choices (Kim & Sherman, 2007). For example, Americans assume that a choice is an act of self-expression whereas these assumptions are much weaker among Koreans (Kim & Drolet, 2003). Additionally, compared to Americans, Asians are more likely to choose brand name products over generic products presumably because of their relational concerns (e.g., signaling social status) (Kim & Drolet, 2009).

These differences in choice have downstream consequences for motivation as well. Kitayama and colleagues (2004) found that Americans are highly motivated to justify their choice when the choice is believed to be made solely based on their personal preferences, whereas Japanese are highly motivated to justify their choice when the choice is believed to have relational implications. In their study, participants were asked to rank 10 CDs according to their own preference. Then, participants were given a choice between their 5th and 6th ranking CD, thinking that those were the only two CDs left in stock. After a 10 min delay, they were told to rank the

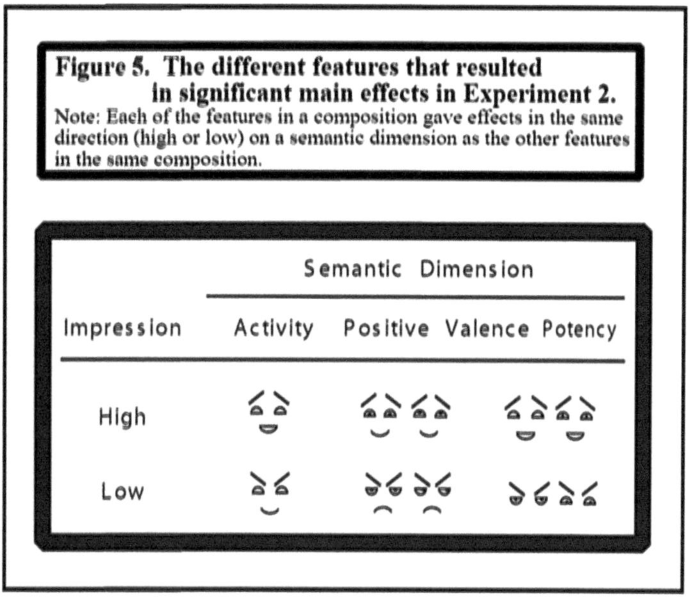

Fig. 3.4 The watching-eyes poster (adapted from Kitayama, Snibbe, Markus, & Suzuki, 2004, *Psychological Science*)

ten CDs again according to their own preference, without the CD cover, allegedly because the sponsor of the study wants to know how rankings may differ based on the CD cover. Interestingly, there were two conditions. In one condition, the choice between the 5th and 6th ranked CDs was made in a completely private situation (private condition). In the other condition, the choice was made in front of a poster including several watching eyes (public condition; see Fig. 3.4). The poster was hung right in front of the participants at their eye level, and so, from their point of view, the eyes on the poster appeared to be watching them. That is, it is logical to believe that the choice made in the private condition only had personal consequences whereas the choice made in the public condition had social consequences. The results showed that participants justified their choice by increasing liking for the chosen CD and decreasing liking for the rejected CD, and yet this justification effect was observed only in the private condition for Americans and only in the public condition for Japanese.

Not only does motivation change attitudes, but it also leads to changes in actual behaviors. In another study (Na & Kitayama, 2010), the participant made a choice in the private condition as well as in the public condition as in the study described above. Specifically, participants were asked to choose one IQ test they would like to take among the three different IQ tests (analytic, fluid, and creative test). That is, participants in the private condition freely chose an IQ test in a completely private situation. In the public condition, participants made a choice in front of the watching-eyes poster (Fig. 3.4). There was also an assignment condition where one IQ test was just assigned to each participant by an experimenter. After making a choice, participants moved to a cubicle and performed the chosen/assigned test for 5 min. They found that Americans performed best in the private condition, and further, the performance in the public condition was no different from the assignment condition. In contrast, Koreans performed best in the public condition, and further, the performance in the private condition was no different from the assignment condition. The results suggested that Americans were highly motivated to perform better on the IQ test in the private choice condition, whereas Koreans were highly motivated in the public condition.

Taken together, these studies suggest that choices can be motivating in all cultures, and yet, the type of choice that is motivating may vary across cultures. Americans become highly motivated when their choice was made in the absence of any potentially imposing social others. In contrast, Asian culture places a greater emphasis on social relations. Therefore, a choice becomes highly motivating when it is made in the presence of social others and, thus, experienced as bearing significance on social aspects of the self.

3.4.2 Neural Evidence

Recent fMRI studies sought to identify neural pathways that underlie cognitive dissonance, which refers to the tendency for someone to alter their preference to match their choices (i.e., choice justification effects), and relevant cultural differences. In one study (Jarcho, Berkman, & Lieberman, 2010), participants first reported their preferences for names and paintings, then made choices between similarly rated names and paintings, and finally rerated the same names and paintings. Their brain was scanned while they were making choices. They found that choice justification effects (i.e., increase in liking for a chosen item and decrease in liking for a rejected item) were associated with activations in the right inferior frontal gyrus and medial frontoparietal regions during the choice. That is, post-choice changes in attitudes could be predicted by brain activities during the choice. However, attitude changes after the choice may be related to brain activities after making choices as well as brain activities during the choice. In a study by Sharot and colleagues (Sharot, De Martino, & Dolan, 2009), participants made choices between hypothetical vacation destinations. The results showed that post-choice changes associated with chosen and rejected items after choice were reflected in corresponding neural activities in the caudate, a nucleus located within the basal ganglia. Neural activations in the caudate have previously been found to be associated with rewards (Knutson, Fong, Adams, Varner, & Hommer, 2001) and imagined positive events (Sharot, Martorella, Delgado, & Phelps, 2007). In other words, changes in psychological preferences induced by their choices (measured with subjective ratings) corresponded to the changes in biological hedonic values.

Although these two studies provide important clues to understand neural correlates of choice justification effects, participants in these studies were Westerners (Americans in Jarcho et al., 2010 and British in Sharot et al., 2009). As we reviewed in the previous section, there has been substantial behavioral evidence showing cultural differences in choice justification between Westerners and Easterners. Therefore, similar cultural differences should exist in neural correlates of choice justification. In this regard, a recent fMRI study on the neural pathways of choice justification by Qin and colleagues (2011) is very informative because they recruited Chinese participants. In their study, participants made choice between two CDs and indicated their preferences for CDs before and after making choices. Most relevant to the present chapter, attitude changes (i.e., increase of preference of chosen items minus decrease of preference of rejected items) were predicted by activations in the right temporal–parietal junction (rTPJ). Given that the rTPJ is associated with perspective taking (Ruby & Decety, 2003), this finding suggested that taking others' perspectives played a critical role in choice justification for Chinese participants in this study. Then, the results are very much in line with behavioral evidence showing cultural differences in choice. As we noted above, Asians are highly motivated by their choice, and thus, after making a choice, they change their attitudes to justify their choice and work harder on the chosen task. However, these motivational consequences only occur when their choice is believed to have social implications (i.e., when the presence of others are primed by the watching-eyes poster). Taken together, behavioral and neural evidences showed that the relational aspects of choice are critical in the motivating effects of choice among Asians. In stark contrast, Americans justify their choice by aligning their preferences with their choice and work harder on a chosen task when the choice is believed to be based on their internal attributes (e.g., personal preferences) without any hint of others' influences. Moreover, the absence of activations in rTPJ for Westerners also suggested that taking others' perspective is not necessary for choice justification for them. Then, these findings strongly suggest that Westerners focus on their own internal attributes when making a choice.

In a similar vein, Kitayama and Park (2014) also investigated the relational aspect of motivation among Asians or lack thereof among Americans. In this study, participants performed the flanker task (Eriksen & Eriksen, 1974), where participants were presented with a set of five letters (HHHHH or SSSSS in consistent trials and SSHSS or HHSHH in conflict trials) and they were instructed to specify the center letter. Participants were then told that their performance would be monitored and converted into points, which they could use to choose one gift for themselves and another gift for their friend after the experiment. The half of the trials were self-blocks during which earned points were used for a gift for themselves, whereas the other half were friend-blocks during which earned points were used for a gift for their friend. Thus, each participant performed for both themselves and their friends. The researchers were interested in whether participants would work harder for the self or for their friend. In particular, they looked at error-related negativity (ERN), an event-related brain potential that is observed within 100 ms after an error. The ERN is known to increase as a function of motivational significance (Gehring et al., 1993; Hajcak, Moser, Yeung, & Simons, 2005). In other words, a stronger ERN will be observed when an individual cares more about an error (more significance). Thus, by comparing ERN in both blocks, which corresponds to the errors in the self- and the friend-blocks, motivational significance between the two blocks could be compared.

They found that European Americans showed significantly larger ERN during self-blocks than during friend-blocks, whereas there was no such difference among Asians. In other words, European Americans were more motivated in self-blocks than in friend-blocks, and thus, an error in self-blocks was perceived as more significant. In stark contrast, Asian participants did not make such distinction. Furthermore, this cultural difference was mediated by interdependent self-construal.

3.4.3 *Implications for Intercultural Relations*

In the previous section, we presented behavioral and neural evidence showing cultural differences in the way individuals make choices. These differences are consistent with the differences in other domains of reasoning. That is, while Easterners tend to be relational when they make choices, this tendency is much weaker among Westerners. This difference is likely to have important implications for intercultural relations.

Choice is an integral part of our life in many ways. First, we make many choices in our daily life, ranging from trivial one (e.g., what to wear today) to more significant ones (e.g., which university to attend). Second, by making choices, we actually decide to invest our limited resources to one course of action at the expense of other equally attractive alternatives. Considering such pervasiveness and significance of choice, we argue that cultural differences in choices can have impact on various domains and intercultural relation is not an exception. The following example included in Markus and Kitayama (Markus & Kitayama, 1991) well illustrates such impact.

> ...... imagine that one has a friend over for lunch and has decided to make a sandwich for him. The conversation might be: "Hey, Tom, what do you want in your sandwich? I have turkey, salami, and cheese." Tom responds, "Oh, I like turkey." Note that the friend is given a choice because the host assumes that friend has a right, if not a duty, to make a choice reflecting his inner attributes, such as preferences or desires. And the friend makes his choice exactly because of the belief in the same assumptions.........What would happen if the friend were a visitor from Japan? A likely response to the question "Hey, Tomio, what do you want?" would be a little moment of bewilderment and then a noncommittal utterance like "I don't know." ....it is the responsibility of the host to be able to "read" the mind of the friend and offer what the host perceives to be the best for the friend. And the duty of the guest, on the other hand, is to receive the favor with grace and to be prepared to return the favor in the near future, if not right at the next moment....... "Hey, Tomio, I made you a turkey sandwich because I remember that last week you said you like turkey more than beef." And Tomio will respond, "Oh, thank you, I really like turkey.".... (p. 229)

In this example, the American host and guest could communicate to each other without any problem because they had the same belief that choice should be made based on one's internal attributes. Likewise, the Japanese host and guest had the same assumption that one should take into account relational concerns when making choices, which prevented them from misunderstanding or even offending each other. However, things would not go this smoothly if one of them is American and the other is Japanese. For example, the Japanese host may feel offended if the

American guest explicitly says that he would like to have a turkey sandwich. Or the American host may feel awfully confused or even angry if the Japanese guest refuses to let the host know his or her preferences by saying "I don't know." As vividly illustrated in this hypothetical example, cultural differences in the way we make choices can affect the very way we understand others' choices, which will guide our subsequent interaction with them.

3.5 Conclusion

Nowadays, all of us are living in a multicultural context in one way or another. For example, it is not uncommon to encounter international students from various cultures around the campus of a typical university in America. Likewise, most of the latest Hollywood movies are released in countries with various cultural backgrounds. Furthermore, with the advent of the Internet, we can be instantly exposed to a variety of cultural contents and important news around the world. Besides, we can interact with people from other cultures even without leaving our own cultures through social network services like Facebook or Twitter. Finally, given the necessary cooperation between business partners across the globe, more and more business teams are composed of personnel with varying cultural backgrounds.

To the extent that opportunities for intercultural relations become increasingly frequent, proper understanding of other cultures is becoming a necessity. Reflecting this growing need for cultural understanding, much effort has been invested into examining cultural differences in various psychological processes (Markus & Kitayama, 2010; Na et al., 2010; Norenzayan, Choi, & Peng, 2007). In this chapter, we focused on cultural differences in reasoning. More specifically, drawing on the recent findings in cultural psychology, we reviewed behavioral and neural evidences showing differences between Westerners and Easterners and discussed their implications for intercultural relations.

Both behavioral and neural data point to the conclusion that Westerners are more focused and analytic, whereas Easterners are more relational and holistic in their reasoning. Westerners tend to detach objects/people from the contexts by paying narrow attention to focal objects or people while largely ignoring the contexts. This tendency in attention corresponds to the way they make attribution. Namely, they make internal dispositional rather than situational attribution (i.e., focusing on internal or personal causes versus external causes of an action). Moreover, Westerners tend to base their choices solely on personal preferences. If they feel that their choice is contaminated by factors other than personal preferences, their motivation toward their choices decreases. In stark contrast, for Easterners, relational factors loom larger for their reasoning. They tend to believe that objects/people are embedded in the contexts, so they pay attention to the entire context including focal objects or people. Essentially, the same tendency is observed in the way they make attribution. That is, Easterners are less vulnerable to dispositional bias (giving undue weight to one's internal disposition). Instead, in order to understand others' behaviors,

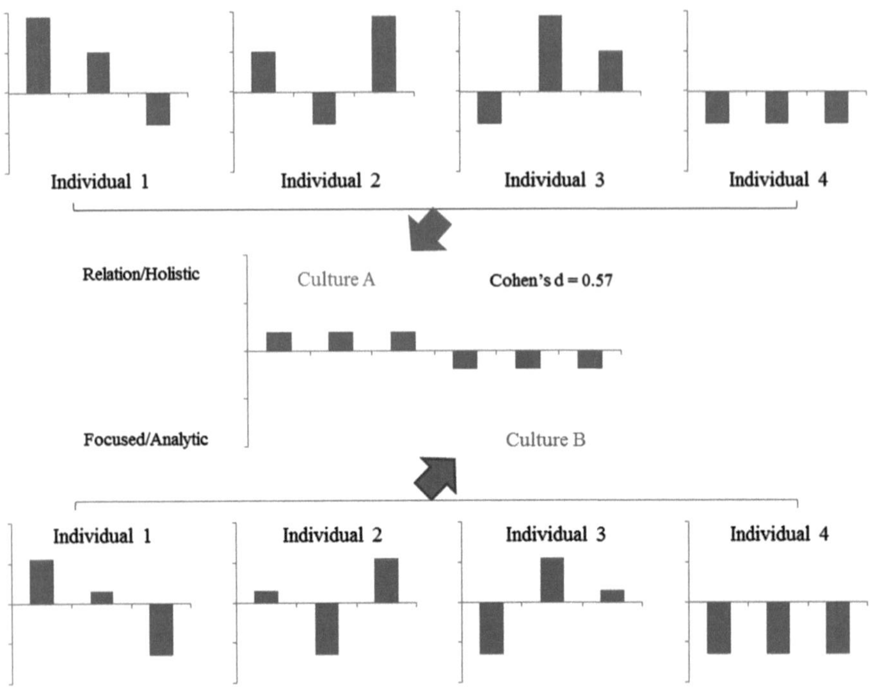

Fig. 3.5 An illustration of independence between individual and cultural differences (adapted from Na et al., 2010, *PNAS*)

Easterners tend to take contextual factors into account. Furthermore, their choices are heavily influenced by relational factors. Therefore, they are more motivated toward the consequences of their choices when they believe that their choice has relational implications.

Obviously, when people are not fully aware of these differences in reasoning, it poses obstacles in their attempt to interact with people from other cultures. Ironically, however, knowing too much about these differences may cause some problems as well. This type of problems occurs when one stereotypically assumes that all the attributes of a cultural mode of reasoning would be observed in every member of the respective culture. For example, after learning about cultural differences in reasoning, Michael may naively believe that his Korean friend, Chulsu, has broader attention, makes situational attribution, and justifies choice only when it has social implications. Likewise, Chulsu may assume that his American friend, Michael, is less likely to consider relational concerns across all aspects of reasoning. These naïve assumptions may be true, but the recent findings in the literature suggest otherwise. Although cultural differences in reasoning are coherent at the group level, substantial individual differences are still observed within a group (Na et al., 2010). As shown in Fig. 3.5, Culture A is relational and holistic in all three aspects of reasoning that was discussed in this chapter (i.e., attention, attribution, or choice), whereas Culture B is focused and analytic across the three domains. Furthermore, the differences between these two cultures were equally strong in all domains (Cohen's $d = 0.0.57$), and thus, differences at the group level can be said to be coherent. However, also shown in Fig. 3.5, individuals among each culture have highly idiosyncratic profiles. That is, Individual 1 in Culture A can be very relational in one domain (let's say attention), but she/he could be not relational at all in another domain (let's say choice). Similarly, Individual 4 in Culture B can be not at all focused in any of the domains, even though Culture B is generally more focused. In sum, coherent differences between cultures may be observed only when idiosyncratic individuals are aggregated at the group level. Supporting this observation, the same study also found that although these measures of reasoning (attention, attribution, and choice) successfully differentiated cultural groups, yet these measures did not correlate among themselves at the individual level. In other words, members of each culture indeed had idiosyncratic profiles. Thus, any given American may be focused and analytic only in some aspects of reasoning but not in all domains. Therefore, if cultural differences in reasoning are stereotypically applied to each individual member of a cultural group, it will pose another obstacle for successful intercultural relations.

So far, we have discussed cultural differences in reasoning with respect to East vs. West differences, mostly focusing on Americans and East Asians. Thus, these implications for intercultural relations may also be restrained accordingly. However, a couple of recent research programs suggest that the scope of cultural differences between focused/analytic and relational/holistic reasoning may not be limited to Americans and East Asians. First, Americans and East Asians are not the only groups whose reasoning styles vary along the dimension of focused/analytic vs. relational/holistic. This dimension has been used to explain the differences between other cultural groups ranging from Eastern vs. Western Europe (Varnum, Grossmann, Katunar, Kitayama, & Nisbett, 2008), Northern Italia vs. Southern Italia (Knight & Nisbett, 2007), to Hokkaido Japanese vs. Mainland Japanese (Kitayama, Ishii, Imada, Takemura, & Ramaswamy, 2006). Moreover, a similar type of differences was also observed between working-class and middle-class individuals in the United States (Kraus, Piff, Mendoza-Denton, Rheinschmidt, & Keltner, 2012). Second, the priming literature has demonstrated that one can be primed to temporally think in an analytic (or holistic) way through various priming techniques (Oyserman & Lee, 2008). For example, repeated exposure to first-person singular pronouns (e.g., I, my, me, mine) or thinking about differences between themselves and their family members was closely linked to focused/analytic reasoning, whereas repeated exposure to first-person plural pronouns (e.g., we, our, us, ourselves) or thinking about similarities between themselves and their family members was closely linked to relational/holistic reasoning. Taken together, these programs of research suggest that the distinction between the analytic and holistic reasoning is not necessarily limited to Americans and East Asians nor does it necessarily constrain to a specific ethnic culture. Nevertheless, future research should explore other cultural groups. Particularly, the Hispanic population would be an interesting group to study as they are one of the fastest growing groups in the United States.

Another interesting domain for future research is investigating various social networking services. These days, many social interactions occur in the online space.

According to Facebook, as of September 2012, 1.01 billion people actively use Facebook each month and 584 million people each day all over the world. Moreover, one's behaviors on Facebook are closely associated with various aspects of their off-line behaviors. For example, one's activities on Facebook (e.g., the usage of the Like function) could predict important personal attributes such as sexual orientation, ethnicity, religion, and political orientation (Kosinski, Stillwell, & Graepel, 2013). To the extent that one's Facebook activities correctly reflect his or her psychological tendencies, cultural differences in reasoning may also be revealed in one's use of Facebook. In fact, Huang and Park (2013) found cultural differences in the profile photos on Facebook between Americans and East Asians. Reflecting their focused attention style, Americans are more likely to focus on the individual's face. However, the profile photos of East Asians tend to de-emphasize the face and include background features, which is consistent with their diffused attention style. Given that cultural differences are evident on Facebook (Na, Kosinski, & Stillwell, 2014), each of us can come across numerous opportunities to interact with people in other cultures and more generally learn about other cultures. Thus, it would be of great interest to examine the effect of these unprecedented opportunities in intercultural relations.

To conclude, cultural differences in reasoning have been extensively documented for the past two decades or so, and these differences are apparently *embrained*. These documentations make it possible to recognize the importance of cultural differences in reasoning. Thus, the field is now beginning to address new and possibly more interesting issues than simply demonstrating cultural differences. As reviewed in the present chapter, these differences in reasoning have important implications for intercultural relations, which are, in turn, important to everyday social life and business development of our increasingly globalized world. Therefore, future research should continue to apply findings from cross-cultural studies in cognition and brain to the domain of intercultural relations.

References

Ambady, N., Koo, J., Lee, F., & Rosenthal, R. (1996). More than words: Linguistic and nonlinguistic politeness in two cultures. *Journal of Personality and Social Psychology, 70*(5), 996–1011.

Barnlund, D. C. (1989). *Communicative styles of Japanese and Americans: Images ad realities*. Belmont, CA: Wadsworth.

Choi, I., & Nisbett, R. E. (1998). Situational salience and cultural differences in the correspondence bias and actor-observer bias. *Personality and Social Psychology Bulletin, 24*(9), 949–960.

Choi, I., Nisbett, R. E., & Norenzayan, A. (1999). Causal attribution across cultures: Variation and universality. *Psychological Bulletin, 125*(1), 47–63.

Eriksen, B. A., & Eriksen, C. W. (1974). Effects of noise letters upon the identification of a target letter in a nonsearch task. *Perception & Psychophysics, 16*, 143–149. doi:10.3758/BF03203267.

Gehring, W. J., Goss, B., Coles, M. G. H., Meyer, D. E., & Donchin, E. (1993). A neural system for error detection and compensation. *Psychological Science, 4*, 385–390. doi:10.1111/j.1467-9280.1993.tb00586.x.

Goh, J. O., Chee, M. W., Tan, J. C., Venkatraman, V., Hebrank, A., Leshikar, E. D., et al. (2007). Age and culture modulate object processing and object-science binding in the ventral visual area. *Cognitive, Affective and Behavioral Neuroscience, 7*(1), 44–52. doi:10.3758/cabn.7.1.44.

Hajcak, G., Moser, J. S., Yeung, N., & Simons, R. F. (2005). On the ERN and the significance of errors. *Psychophysiology, 42*, 151–160. doi:10.1111/j.1469-8986.2005.00270.x.

Hall, E. T. (1976). *Beyond culture*. New York, NY: Anchor Books.

Hanakawa, T., Honda, M., Okada, T., Fukuyama, H., & Shibasaki, H. (2003). Neural correlates underlying mental calculation in abacus experts: A functional magnetic resonance imaging study. *NeuroImage, 19*(2 Pt 1), 296–307.

Hedden, T., Ketay, S., Aron, A., Markus, H. R., & Gabrieli, J. D. E. (2008). Cultural influences on neural substrates of attentional control. *Psychological Science, 19*(1), 12–17.

Henrich, J., Heine, S. J., & Norenzayan, A. (2010). The weirdest people in the world? *Behavioral and Brain Sciences, 33*(2–3), 61–83.

Holtgraves, T. (1997). Styles of language use: Individual and cultural variability in conversational indirectness. *Journal of Personality and Social Psychology, 73*(3), 624–637.

Holtgraves, T., & Yang, J.-n. (1992). Interpersonal underpinnings of request strategies: General principles and differences due to culture and gender. *Journal of Personality and Social Psychology, 62*(2), 246–256.

Huang, C.-M., & Park, D. C. (2013). Cultural influences on Facebook photographs. *International Journal of Psychology, 48*, 334. doi:10.1080/00207594.2011.649285.

Ishii, K., Reyes, J. A., & Kitayama, S. (2003). Spontaneous attention to word content versus emotional tone: Differences among three cultures. *Psychological Science, 14*, 39–46.

James, W. (1890/1950). *Principles of psychology*. New York, NY: Holt.

Jarcho, J., Berkman, E., & Lieberman, M. D. (2010). The neural basis of rationalization: Cognitive dissonance reduction during decision making. *Social Cognitive and Affective Neuroscience, 6*, 460.

Jones, E. E., & Harris, V. A. (1967). The attribution of attitudes. *Journal of Experimental Social Psychology, 3*, 1–24.

Kashima, E. S., & Kashima, Y. (1998). Culture and language: The case of cultural dimensions and personal pronoun use. *Journal of Cross-Cultural Psychology, 29*(3), 461–486.

Kim, H. S., & Drolet, A. (2003). Choice and self-expression: A cultural analysis of variety-seeking. *Journal of Personality and Social Psychology, 85*(2), 373–382.

Kim, H. S., & Drolet, A. (2009). Express your social self: Cultural differences in choice of brand-name versus generic products. *Personality and Social Psychology Bulletin, 35*(12), 1555–1566.

Kim, H. S., & Sherman, D. K. (2007). 'Express Yourself': Culture and the effect of self-expression on choice. *Journal of Personality and Social Psychology, 92*(1), 1–11.

Kitayama, S., Duffy, S., Kawamura, T., & Larsen, J. T. (2003). Perceiving an object and its context in different cultures: A cultural look at New Look. *Psychological Science, 14*(3), 201–206.

Kitayama, S., & Ishii, K. (2002). Word and voice: Spontaneous attention to emotional utterances in two languages. *Cognition & Emotion, 16*(1), 29–59.

Kitayama, S., Ishii, K., Imada, T., Takemura, K., & Ramaswamy, J. (2006). Voluntary settlement and the spirit of independence: Evidence from Japan's 'Northern Frontier'. *Journal of Personality and Social Psychology, 91*(3), 369–384.

Kitayama, S., Snibbe, A. C., Markus, H. R., & Suzuki, T. (2004). Is there any 'Free' choice?: Self and dissonance in two cultures. *Psychological Science, 15*(8), 527–533.

Kitayama, S., & Uskul, A. K. (2011). Culture, mind, and the brain: Current evidence and future directions. *Annual Review of Psychology, 62*, 419–449.

Kitayama, S., & Park, J. (2014). Error-related brain activity reveals self-centric motivation: Culture matters. *Journal of Experimental Psychology: General, 143*(1), 62–70. doi:10.1037/a0031696.

Knight, N., & Nisbett, R. E. (2007). Culture, class and cognition: Evidence from Italy. *Journal of Cognition and Culture, 7*(3), 283–291.

Knutson, B., Fong, G. W., Adams, C. M., Varner, J. L., & Hommer, D. (2001). Dissociation of reward anticipation and outcome with event-related fMRI. *NeuroReport: For Rapid Communication of Neuroscience Research, 12*(17), 3683–3687. doi:10.1097/00001756-200112040-00016.

Kobayashi, C., Glover, G. H., & Temple, E. (2007). Cultural and linguistic effects on neural bases of 'Theory of Mind' in American and Japanese children. *Brain Research, 1164*, 95–107. doi:10.1016/j.brainres.2007.06.022.

Kosinski, M., Stillwell, D., & Graepel, T. (2013). Private traits and attributes are predictable from digital records of human behavior. *Proceedings of the National Academy of Sciences of the United States of America, 110*, 5802. doi:10.1073/pnas.1218772110.

Kraus, M. W., Piff, P. K., Mendoza-Denton, R., Rheinschmidt, M. L., & Keltner, D. (2012). Social class, solipsism, and contextualism: How the rich are different from the poor. *Psychological Review, 119*(3), 546–572. doi:10.1037/a0028756.

Kutas, M., & Hillyard, S. A. (1980). Reading senseless sentences: Brain potentials reflect semantic incongruity. *Science, 207*(4427), 203–205.

Lewis, R. S., Goto, S. G., & Kong, L. L. (2008). Culture and context: East Asian American and European American differences in P3 event-related potentials and self-construal. *Personality and Social Psychology Bulletin, 34*(5), 623–634.

Maguire, E. A., Gadian, D. G., Johnsrude, I. S., Good, C. D., Ashburner, J., Frackowiak, R. S., et al. (2000). Navigation-related structural change in the hippocampi of taxi drivers. *Proceedings of the National Academy of Sciences of the United States of America, 97*(8), 4398–4403.

Markus, H. R., & Kitayama, S. (1991). Culture and the self: Implications for cognition, emotion, and motivation. *Psychological Review, 98*, 224–253.

Markus, H. R., & Kitayama, S. (2010). Cultures and selves: A cycle of mutual constitution. *Perspectives on Psychological Science, 5*(4), 420–430.

Masuda, T., & Kitayama, S. (2004). Perceiver-induced constraint and attitude attribution in Japan and the US: A case for the cultural dependence of the correspondence bias. *Journal of Experimental Social Psychology, 40*(3), 409–416.

Masuda, T., & Nisbett, R. E. (2001). Attending holistically versus analytically: Comparing the context sensitivity of Japanese and Americans. *Journal of Personality and Social Psychology, 81*(5), 922–934.

Mitchell, J. P., Banaji, M. R., & Macrae, C. N. (2005). The link between social cognition and self-referential thought in the medial prefrontal cortex. *Journal of Cognitive Neuroscience, 17*(8), 1306–1315.

Mitchell, J. P., Macrae, C. N., & Banaji, M. R. (2005). Forming impressions of people versus inanimate objects: Social-cognitive processing in the medial prefrontal cortex. *NeuroImage, 26*, 251–257.

Morris, M. W., & Peng, K. (1994). Culture and cause: American and Chinese attributions for social and physical events. *Journal of Personality and Social Psychology, 67*(6), 949–971.

Na, J., Choi, I., & Sul, S. (2013). I like you because you think in the "right" way: Culture and ideal thinking. *Social Cognition, 31*(3), 390–404.

Na, J., Grossmann, I., Varnum, M. E. W., Kitayama, S., Gonzalez, R., & Nisbett, R. E. (2010). Cultural differences are not always reducible to individual differences. *Proceedings of the National Academy of Sciences of the United States of America, 107*(14), 6192–6197.

Na, J., & Kitayama, S. (2010). *Will people work hard on a task they choose? Both culture and social-eyes priming matter*. Unpublished Manuscript. University of Michigan.

Na, J., & Kitayama, S. (2011). Spontaneous trait inference is culture-specific: Behavioral and neural evidence. *Psychological Science, 22*(8), 1025–1032. doi:10.1177/0956797611414727.

Na, J., Kosinski, M., & Stillwell, D. J. (2015). When a new tool is introduced in different cultural contexts: Individualism-collectivism and social networks on Facebook. *Journal of Cross-Cultural Psychology, 46*, 355–370. doi:10.1177/0022022114563932.

Nisbett, R. E., & Masuda, T. (2003). Culture and point of view. *Proceedings of the National Academy of Sciences of the United States of America, 100*, 11163–11175.

Nisbett, R. E., Peng, K., Choi, I., & Norenzayan, A. (2001). Culture and systems of thought: Holistic versus analytic cognition. *Psychological Review, 108*(2), 291–310.

Norenzayan, A., Choi, I., & Peng, K. (2007). Perception and cognition. In S. Kitayama & D. Cohen (Eds.), *Handbook of cultural psychology* (pp. 569–594). New York, NY: Guilford Press.

Norenzayan, A., Smith, E. E., Kim, B. J., & Nisbett, R. E. (2002). Cultural preferences for formal versus intuitive reasoning. *Cognitive Science, 26*, 653–684.

O'Keefe, J., & Nadel, L. (1978). *The hippocampus as a cognitive map*. Oxford: Oxford University Press.

Oyserman, D., & Lee, S. W. S. (2008). Does culture influence what and how we think? Effects of priming individualism and collectivism. *Psychological Bulletin, 134*(2), 311–342.

Oyserman, D., Sorensen, N., Reber, R., & Chen, S. X. (2009). Connecting and separating mind-sets: Culture as situated cognition. *Journal of Personality and Social Psychology, 97*(2), 217–235. doi:10.1037/a0015850.

Qin, J., Kimel, S., Kitayama, S., Wang, X., Yang, X., & Han, S. (2011). How choice modifies preference: Neural correlates of choice justification. *NeuroImage, 55*(1), 240–246. doi:10.1016/j.neuroimage.2010.11.076.

Ruby, P., & Decety, J. (2003). What you believe versus what you think they believe: A neuroimaging study of conceptual perspective-taking. *European Journal of Neuroscience, 17*(11), 2475–2480. doi:10.1046/j.1460-9568.2003.02673.x.

Sabbagh, M. A., Xu, F., Carlson, S. M., Moses, L. J., & Lee, K. (2006). The development of executive functioning and theory of mind: A comparison of Chinese and U.S. preschoolers. *Psychological Science, 17*(1), 74–81. doi:10.1111/j.1467-9280.2005.01667.x.

Sanchez-Burks, J., Lee, F., Choi, I., Nisbett, R., Zhao, S., & Koo, J. (2003). Conversing across cultures: East-West communication styles in work and nonwork contexts. *Journal of Personality and Social Psychology, 85*(2), 363–372.

Sharot, T., De Martino, B., & Dolan, R. J. (2009). How choice reveals and shapes expected hedonic outcome. *The Journal of Neuroscience, 29*(12), 3760–3765. doi:10.1523/jneurosci.4972-08.2009.

Sharot, T., Martorella, E. A., Delgado, M. R., & Phelps, E. A. (2007). How personal experience modulates the neural circuitry of memories of September 11. *PNAS, 104*(1), 389–394. doi:10.1073/pnas.0609230103.

Tang, Y., Zhang, W., Chen, K., Feng, S., Ji, Y., Shen, J., et al. (2006). Arithmetic processing in the brain shaped by cultures. *Proceedings of the National Academy of Sciences of the United States of America, 103*(28), 10775–10780.

Triandis, H. C. (1989). The self and social behavior in differing cultural contexts. *Psychological Review, 96*, 269–289.

Uleman, J. S., Saribay, S. A., & Gonzalez, C. M. (2008). Spontaneous inferences, implicit impressions, and implicit theories. *Annual Review of Psychology, 59*, 329–360.

Varnum, M. E. W., Grossmann, I., Katunar, D., Kitayama, S., & Nisbett, R. E. (2008). Holism in a European cultural context: Differences in cognitive style between Central and East Europeans and Westerners. *Journal of Cognition and Culture, 8*(3), 321–333.

Chapter 4
The Neuroscience of Bilingualism: Cross-Linguistic Influences and Cognitive Effects

Emily L. Coderre

Abstract Coderre reviews the cognitive and neural effects of bilingualism, beginning with an overview of how and where language is processed in the monolingual brain, and then extending this to multiple languages in the bilingual brain. Cross-linguistic effects of bilingualism are specifically discussed, including how a bilingual's two languages can interact with each other during production and comprehension, and how these interactions can lead to facilitation or interference in processing. Coderre also discusses the cognitive advantages and disadvantages of bilingualism. In particular, bilinguals have a delay in lexical processing speed but have an increase in the efficiency of cognitive control.

Coderre concludes by discussing how the neuroscience of bilingualism could impact intercultural relations by improving second-language education. Examples are provided to demonstrate how previous findings could be used to change language instruction and how neuroscience could be used to insure appropriate learning goals are being met.

Language is a uniquely human capacity, arguably the greatest ability mankind has ever developed. Modern humans use language constantly and in many different contexts; it is so highly practiced, in fact, that it becomes automatic and is often taken for granted. Yet language is vastly complicated, consisting of complex interacting processes like acoustic recognition of speech, production of fluent and meaningful speech, visual word recognition, and parsing complicated grammar. More impressive is the range of different languages that have emerged throughout the course of human history and the dynamic way in which they are constantly

E.L. Coderre (✉)
Cognitive Neurology/Neuropsychology, Department of Neurology, Johns Hopkins University School of Medicine, 1629 Thames Street, Suite 350, Baltimore, MD 21231, USA
e-mail: ecoderr1@jhmi.edu

J.E. Warnick, D. Landis (eds.), *Neuroscience in Intercultural Contexts*,
International and Cultural Psychology, DOI 10.1007/978-1-4939-2260-4_4

evolving, merging, and sometimes fading. The various languages of the world share vastly different structures, appearances, and sounds, creating a multifaceted lens through which to view the world.

The majority of the world today speaks two or more languages, a trend reflected in the increasing interest within psychology in the neuropsychological effects of bilingualism. The maintenance and control of multiple languages place huge demands on the cognitive system and subsequently affect many aspects of development. Bilingualism research aims to understand not only how the brain deals with the presence and interaction of multiple languages but also the cognitive and developmental effects of such experience.

This chapter provides a broad overview of the cognitive effects of bilingualism. To provide a basic framework, Section 4.1 begins with a brief introduction to language processing in monolinguals, with specific focus on the neuroimaging indices of language processing and how specific linguistic characteristics can influence the structure and function of these processes. Section 4.2 discusses bilingual language processing, reviewing evidence that is bringing us closer to being able to answer questions such as: How and where are two languages represented in the brain? How are two languages accessed? Can bilinguals operate in a "monolingual mode"? How are multiple languages controlled such that bilinguals can successfully avoid cross-language speech errors and prevent one language from intruding during the processing of the other? How do two languages interact with each other, and do interactions still persist when languages are vastly different in their linguistic properties? Section 4.3 discusses the cognitive effects of managing multiple languages in the brain, reviewing evidence that bilinguals experience a delay in lexical processing speed but also an increase in the efficiency of cognitive control abilities and possibly a cognitive protection against the detrimental effects of aging. Finally, Section 4.4 relates this research on bilingualism to the broader contexts of intercultural relations.

To first define some terminology of language research, "semantics" refers to the representations of meaning in language: for example, the mental image that comes to mind when one thinks of a chair. "Orthography" refers here to the written form of language, such as visual words and how they look: for example, Chinese characters and English letters have very different orthographies. "Phonology" refers to the sounds and pronunciations of language (whole words or single letters). "Syntax" refers to grammatical rules and structures. The "lexicon" is referred to as the store of language-specific words and their links to meaning; "lexical access" refers to the process of accessing semantic meaning given the input of a visual or spoken word, and vice versa. "Production" refers to speaking, whereas "comprehension" refers to reading or hearing language. The term "writing system" refers here more generally to the genre of a language's symbolic system, such as alphabetic (in which each symbol represents a letter, as in English), syllabic (in which each symbol represents a syllable, as in Japanese kana), or logographic (in which each symbol represents an entire word or concept, as in Chinese). Finally, "script" refers to the specific symbols within a language; for example, Japanese has two scripts: the syllabic kana and the logographic kanji.

4.1 Monolingual Language Processing: Neuroimaging Correlates and Language-Specific Effects on Production, Comprehension, and Neural Representation

In language research, the two main neuroimaging techniques employed are electroencephalography (EEG) and functional magnetic resonance imaging (fMRI). EEG records electrical activation from the surface of the scalp. When recording EEG during a particular cognitive task, the neural response can be time locked to the onset of a certain stimulus, such as the final word in a sentence, and an event-related potential (ERP) is measured which reflects the cognitive activity in response to that event or stimulus. EEG has excellent temporal resolution, being able to capture neural function on a millisecond time scale, but relatively poor spatial resolution due to the distortion of the electrical signal by the scalp. fMRI uses the blood-oxygenation-level-dependent (BOLD) signal as a proxy for neural activity, under the assumption that neural firing in a particular brain area recruits more oxygen and creates a larger BOLD signal. fMRI has relatively accurate spatial resolution, being able to localize brain activity within a few millimeters, but its temporal resolution is on the order of seconds due to the lag of the hemodynamic response. Other neuroimaging techniques such as positron emission tomography (PET) and magnetoencephalography (MEG) provide similar temporal and spatial information of cognitive function. Together, these techniques have led to spectacular advancements in our understanding of the human brain.

Monolingual language studies using EEG have identified a series of ERPs reflecting specific steps along the time course of linguistic processing (Sereno, Rayner, & Posner, 1998). Word recognition, for example, first elicits a positive peak at approximately 100 ms known as the P1, which is believed to reflect perceptual and attentional processes (e.g., Luck, Heinze, Mangun, & Hillyard, 1990; Mangun, Buonocore, Girelli, & Jha, 1998), although some have reported linguistic influences at the P1 component (Segalowitz & Zheng, 2009; Sereno et al., 1998). Following the P1 is a negative peak at approximately 170 ms, known as the N1 or N170. This component distinguishes between words and symbol strings and is therefore thought to index orthographic processing (Appelbaum, Liotti, Perez, Fox, & Woldorff, 2009; Bentin, Mouchetant-Rostaing, Giard, Echallier, & Pernier, 1999; Grossi, Savill, Thomas, & Thierry, 2010; Maurer, Brem, Bucher, & Brandeis, 2005; Ruz & Nobre, 2008). Higher-level linguistic processes such as semantic retrieval and integration are indexed by the N400, which is elicited approximately 300–600 ms after the presentation of a semantically incongruous word in a sentence (Kutas & Hillyard, 1980; see Lau, Phillips, & Poeppel, 2008 for a review). Finally, a later P600 (a positive wave from approximately 600–1,000 ms) is elicited by syntactic anomalies and thus indexes grammatical processing (e.g., Friederici & Meyer, 2004).

While EEG studies primarily concern the timing of linguistic processes, fMRI is employed to address the spatial representations of language in the brain. In monolinguals, the various aspects of language (reading, listening to speech, semantic and syntactic processing) tend to generate activation in a left-lateralized (although see Section 4.1.2) frontotemporal network of brain areas (e.g., Binder et al., 1997; Ferstl, Neumann, Bogler, & Von Cramon, 2008; Gitelman, Nobre, Sonty, Parrish, &

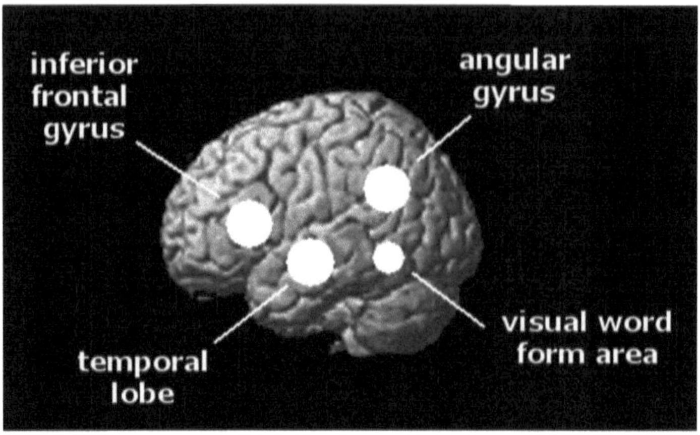

Fig. 4.1 Major areas of the left-hemisphere language network

Mesulam, 2005; Price, 2010; Richardson, Seghier, Leff, Thomas, & Price, 2011; Vogel et al., 2013). This network includes the left inferior frontal gyrus (LIFG, commonly associated with Broca's area) and left angular gyrus, associated with semantic retrieval; the left fusiform gyrus, associated with written orthography; and the left temporal lobe, associated with speech processing (see Fig. 4.1). In neuroimaging studies of word recognition, the region of interest is usually the visual word form area (VWFA), localized to the left fusiform gyrus. This region is sensitive to the processing of letter strings (words and pseudowords) compared to non-orthographic symbols and is thought to integrate the letters of a word together into a "visual word form" while being relatively insensitive to other perceptual variations such as font, case, and size (Cohen et al., 2000; McCandliss, Cohen, & Dehaene, 2003).

4.1.1 *Effects of Language-Specific Characteristics*

The various languages of the world vary widely in the sounds and symbols they use, in how sounds are mapped on to symbols, and in grammatical rules for combining words into sentences. Perhaps unsurprisingly, differences in these language-specific characteristics can influence how a particular language is processed in the brain.

4.1.1.1 Visual Word Recognition

Effects of Orthographic Depth

In the ERP literature, an N170, which indexes orthographic processing, is consistently reported across different languages and writing systems, including Chinese (Lin et al., 2011), French, and Arabic (Simon, Bernard, Lalonde, & Rebaï,

2006; Simon, Petit, Bernard, & Rebaï, 2007), and Hebrew (Bar-Kochva, 2011). However, language-specific script differences, specifically those relating to orthographic depth, have been reported at the N170. Orthographic depth refers to the consistency of the mappings from orthography to phonology. Shallow orthographies, such as Italian and Finnish, have consistent grapheme-phoneme conversion rules such that only one sound is generally associated with one letter. In contrast, deep orthographies, such as English, French, and Arabic, contain common irregularities. Bar-Kochva (2011) investigated the dual scripts of Hebrew, which contain different orthographic depths, reporting that the shallow Hebrew script generated a larger N170 than the deep orthographic script. Comparing French and Arabic, Simon et al. (2006) reported a larger and more left-lateralized N170 for the shallower French orthography compared to the relatively deeper orthography of Arabic, in which N170 amplitude was attenuated and the component was more bilateral. Interestingly, while shallow orthographies seem to generate larger N170 amplitudes than deep orthographies, no differences have been reported in the latency of the effect: orthographic distinction occurs at the N170 peak across languages regardless of other linguistic differences. This suggests that orthographic processing occurs at a relatively similar time course for all languages.

Script differences at later ERP components on the linguistic processing timeline are less widely reported, although Simon et al. (2006) report differences at an N320 component, thought to index grapheme-to-phoneme conversion. In their comparison of French and Arabic, an N320 was present for French but not for Arabic, indicating a heavier reliance on spelling-to-sound conversion in the more regular French orthography. Therefore, differences have been reported between orthographies at early ERP components reflecting lexical processing. However, there have been few studies explicitly addressing differences between languages along the full spectrum of linguistic processing levels, so a more thorough exploration of the influences of orthography is needed.

The Role of Phonology

Throughout the last few decades, much research in neurolinguistics has focused on how alphabetic and logographic languages are processed and how they differ. One heated debate centers around whether lexical access in visual word recognition requires activation of phonology. In alphabetic writing systems such as English, letters correspond directly onto sounds; therefore, the transition from orthography to semantics is thought to require activation of phonology (e.g., Frost, 1998; see review in Perfetti, Liu, & Tan, 2005). In logographic writing systems like Chinese, however, the role of phonology is more ambiguous: rather than letters that map onto sounds, every character has a specific pronunciation. Many characters can share the same pronunciation, creating a high number of homophones. Because phonology is less reliable in Chinese, some researchers have theorized that Chinese character recognition proceeds using a "direct-access" route directly from orthography to semantics, bypassing phonology altogether (Saalbach & Stern, 2004; Taft & van Graan, 1998). In contrast, other evidence suggest that phonology is activated in Chinese word recognition, even in the absence of lexical activation (Chua, 1999; Guo, Peng, & Liu, 2005; Liu, Perfetti, & Hart, 2003; Perfetti et al., 2005; Saalbach & Stern, 2004; Spinks, Liu, Perfetti, & Tan, 2000; Tan, Laird, Li, & Fox, 2005; Xu, Pollatsek, & Potter, 1999).

In an attempt to assimilate this contradictory literature, the lexical constituency model of word recognition (Perfetti et al., 2005) proposes that phonology is activated in all languages, but script can modulate the degree to which phonology contributes to lexical access. For example, Tan and Perfetti (1997), using a phonologically mediated priming paradigm in Chinese, have demonstrated that the phonological mediation effect in Chinese is affected by homophone density: the more homophones attributed to a particular Chinese character, the smaller the mediation priming effect. They proposed that when a certain character has a large number of homophones, phonology has such a distributed spread of activation that it does not significantly contribute to semantic access. However, when a character has fewer homophones, phonological activation contributes more significantly to semantic access, making it a more central step in orthographic recognition. Therefore in a language with many homophones like Chinese, lexical access is effectively a direct link between orthography and semantics (supporting the direct-access hypothesis) because phonological activation is not helpful. In languages with fewer homophones like English, lexical access is phonologically mediated because phonological activation is more pronounced.

4.1.1.2 Phonological Processing

Another major difference between languages, besides the way they look (orthography), is the way they sound (phonology). One of the biggest differences across languages is that of tonal versus nontonal languages. For example, Chinese is a tonal language, in which each character is associated not just with a specific syllable but also with one of four possible tones of pronunciation. To illustrate, in Mandarin the syllable "ma" can be spoken as ma1 (high, level tone), ma2 (low rising tone), ma3 (falling then rising), and ma4 (high, falling tone), all of which are represented by different characters and have different meanings. Tone therefore contains important lexical information; a similar importance of prosodic information is not present in alphabetic languages like English.

In a meta-analysis of phonological processing in reading, Tan et al. (2005) found differences in the phonological processing of Chinese characters compared to English words. Specifically, Chinese reading consisted of a different neural network than that for English alphabetic written words: Chinese phonology, which requires direct retrieval of phonological information, recruited the left middle frontal gyrus, whereas alphabetic phonology recruited left temporoparietal regions for assembled phonology and grapheme-phoneme conversions. Studies of tone perception have also reported differences between tonal and nontonal languages in how this auditory information is processed (e.g., Gandour et al., 2000; Klein, Milner, Zatorre, Zhao, & Nikelski, 1999). For example, Klein et al. (1999) performed a PET study investigating how native Mandarin speakers and native English speakers discriminated pitch patterns in Chinese words. They found that only Chinese speakers showed brain activation in left frontal, parietal, and parieto-occipital regions, which may reflect the fact that the tones contained linguistic meaning. In contrast, the English speakers showed activation in the right inferior frontal cortex, indicating a more right-sided activation for pitch processing. Thus, experience with a tonal language can change the way the brain processes both grapheme-to-phoneme conversion and auditory signals.

4.1.1.3 Language Production Speed

Even in language production, the time needed to produce a word can vary depending on a number of linguistic characteristics. Comparing picture-naming times in seven languages (English, German, Spanish, Italian, Bulgarian, Hungarian, and Mandarin Chinese), Bates et al. (2003) found that factors such as word length, syllable structure, compounding, initial frication, and degree of word order flexibility all affected naming times across languages. Thus, the very orthographic, phonological, and syntactic characteristics that make a particular language unique can also contribute to differences at many levels of linguistic processing.

4.1.2 Language Organization in the Brain: fMRI Evidence of Orthographic Effects

In early fMRI studies of language processing, the majority of work was done with Western or European participants and was therefore all based on speakers of alphabetic languages. As more research has emerged, it is becoming increasingly clear that differences in writing systems, as well as more subtle differences within writing systems, affect how language is processed and organized in the brain.

Within alphabetic languages, orthographic depth primarily affects the recruitment of brain areas involved in phonological activation. For example, Paulesu et al. (2000) found that Italian, a shallow language, showed more activation of areas related to phoneme processing compared to English, a deep language, which showed more activation of word retrieval. Workman, Brookman, Mayer, Rees, and Bellin (2000) demonstrated that Welsh speakers showed greater left-hemisphere processing than English speakers and proposed that Welsh's shallower script recruited more phonological processing, localized in the left hemisphere, rather than ideographic processing of words, localized in the right.

With a growing interest in script differences, fMRI studies of language processing discovered larger effects of script and writing system on language organization in the brain (e.g., Bick, Goelman, & Frost, 2011; Bolger, Perfetti, & Schneider, 2005; Coderre, Filippi, Newhouse, & Dumas, 2008; Nelson, Liu, Fiez, & Perfetti, 2009; Perfetti et al., 2007; Sakurai et al., 2000; Tan, Spinks, Eden, Perfetti, & Siok,

2005; Tan et al., 2001). As more research with logographic languages such as Chinese and Japanese kanji was performed, researchers discovered that the patterns of language organization differed significantly from the previously reported literature on alphabetic languages. While there are some general overlaps in language-processing areas for alphabetic and logographic writing systems, such as the left superior temporal gyrus and left inferior frontal gyrus (Bolger et al., 2005; Tan et al., 2005), there are also notable disparities. In a meta-analysis of alphabetic versus logographic writing systems, Bolger et al. (2005) observed system-specific differences in the posterior superior temporal gyrus, left anterior dorsal frontal region, and right occipitotemporal cortex. They attributed these findings to language-specific differences in phonological processing, the integration of phonology and orthography, and orthographic processing, respectively. Even more notable are differences in language lateralization between writing systems. Specifically, because logographic languages like Chinese are more spatial and require more visuospatial mechanisms in reading, they tend to activate the right hemisphere, which is more involved in spatial processing, to a greater degree than alphabetic languages. This leads to a bilateral language network for logographic languages, compared to a left-lateralized network for alphabetic languages (Perfetti et al., 2007; Siok, Spinks, Jin, & Tan, 2009; Tan et al., 2001; see Bolger et al., 2005 and Tan et al., 2005 for reviews).

Other differences in neural connectivity have also been reported between writing systems. For example, Kawabata Duncan et al. (2014) compared functional connectivity, a measure of how different regions of the brain “talk” to each other during a task, in response to reading Japanese kanji or kana. Similar areas were activated for both scripts, but kanji showed more interhemispheric connectivity than kana, specifically in connectivity of the right and left ventral occipitotemporal cortex. This was interpreted as a greater reliance on integrating higher-order visual stimuli processed in the right hemisphere with language and semantics processed in the left. Interhemispheric connectivity strength was also higher for those more proficient in reading kanji, suggesting increased efficiency of neural communication. In contrast to kanji, kana showed stronger intrahemispheric connectivity from the frontal cortex to the parietal cortex, potentially indicating more reliance on phonological assembly.

4.2 Bilingual Language Processing: The Cognitive and Neural Effects of Having More Than One Language in the Brain

The previous section has illustrated that the specific characteristics of a language can have a significant impact on how and where that language is processed, with specific respect to the electrophysiological and neural indices of processing. With all these complexities of language processing in monolinguals, the case of bilingualism becomes even more interesting. Bilinguals must handle two languages in one brain, creating ample opportunities for cross-linguistic transfer and influence.

This section addresses how multiple languages are represented, accessed, and controlled in the bilingual brain, as well as how the various linguistic characteristics of different languages can interact.

Before discussing the literature addressing these topics, a brief introduction to the terminology of bilingualism research is required. The term "bilingual" itself can be controversial, as it comprises a multitude of dynamic linguistic and cognitive factors. Bilinguals, for the sake of this chapter, are individuals who have achieved a reasonable level of proficiency in two languages and who use both languages on a frequent or daily basis. Language "proficiency," referred to here as fluency or language skill, can be assessed with subjective or objective measures like questionnaires or vocabulary tests. However, this is also a fluid concept, as proficiency can change over years, months, or weeks, depending on factors like immersion or immigration (Hansen, 2001; Linck, Kroll, & Sunderman, 2009; Tokowicz, Michael, & Kroll, 2004). For the purposes of this discussion, the language with the higher subjective proficiency is considered a bilingual's "dominant" language. The "first language" (L1) is considered the native language, which was learned first; the "second language" (L2) is the later-acquired, "nonnative" language. "Early" bilinguals are referred to here as individuals who learned both languages from birth or early childhood, whereas "late" bilinguals acquired their L2 after approximately age 7 (the precise cutoff between early and late bilinguals also differs among researchers). Age of acquisition is referred to as the age of first substantial exposure to the L2. Finally, "balanced" bilinguals are defined here as individuals who have learned two languages from birth and are equally proficient in both, whereas "unbalanced" bilinguals are more dominant in one language than the other. These concepts are necessarily simplified: as Luk and Bialystok (2013) have argued, bilingualism is not a categorical variable; language proficiency and usage can change dynamically in a short time span and over the lifetime. These factors can also become confounded with language experience: for example, bilinguals may become more proficient in their nonnative language after immigration to a foreign country.

4.2.1 How (and Where) Are Two Languages Represented? fMRI Evidence of the Neural Organization of L1 and L2

When considering bilingualism, it seems inevitable that having two different language representations alters the organization of the language network in the brain, especially when two languages are integrated in the brain from birth. A wealth of fMRI research has investigated how bilingual language representation differs from that of monolinguals and how the L1 and L2 are differentially represented (see reviews in Abutalebi, 2008; Abutalebi & Green, 2007; Indefrey, 2006; Stowe & Sabourin, 2005; van Heuven & Dijkstra, 2010). Language processing generally elicits more extensive neural activation in bilinguals than in monolinguals (Kovelman, Baker, & Petitto, 2008; Kovelman, Shalinsky, Berens, & Petitto, 2008; Parker Jones et al., 2011). For example, Parker Jones et al. (2011) reported that during naming and reading tasks, bilinguals demonstrated extensive activation for both the L1 and L2 in regions of the left frontal and temporoparietal cortex: regions that, in monolinguals, are associated with native language processing and/or control of interference. Using a grammaticality judgment task, Kovelman, Baker, et al. (2008) observed that although both monolinguals and bilinguals activated the LIFG, activation in this area was stronger and more extensive for bilinguals when performing in the L2. Using the same task, Kovelman, Shalinsky, et al. (2008) reported that bilinguals recruited additional working memory and attention areas more bilaterally than monolinguals. This work illustrates that language processing is more cognitively demanding in bilingualism, especially in an L2.

An ongoing debate in the field of neurobilingualism concerns whether the L1 and L2 have common or distinct neural representations. On the one hand, some research has reported common spatial activation for both languages (e.g., Briellmann et al., 2004; Consonni et al., 2013; Hasegawa, Carpenter, & Just, 2002; Hernandez, Dapretto, Mazziotta, & Bookheimer, 2001; Hernandez, Martinez, & Kohnert, 2000; Illes et al., 1999; Mahendra, Plante, Magloire, Milman, & Trouard, 2003; Rüschemeyer, Zysset, & Friederici, 2006; Vingerhoets et al., 2003). For example, an extensive review by Abutalebi and Green (2007) concluded that the representations of a second language largely converge with those of the native language. Zhang et al. (2014) performing a functional connectivity study of reading in L1 (Chinese) and L2 (English) recently reported similar patterns of neural connectivity during reading in both languages. Common representations have been reported for both early and late bilinguals (e.g., Consonni et al., 2013; Hernandez et al., 2000; Illes et al., 1999), suggesting a subordinate role of age of acquisition in determining neural language representations. Alternatively, Mahendra et al. (2003) reported that similar regions of activation were elicited for both early and late bilinguals, but the extent of activation was greater for late bilinguals, suggesting a contributing role of age of acquisition.

On the other hand, others have found that the extent of second-language activation differs considerably from that of the native language (e.g., Chee, Hon, Lee, & Soon, 2001; Dehaene et al., 1997; Ding et al., 2003; Kim, Relkin, Lee, & Hirsch, 1997; Marian, Spivey, & Hirsch, 2003; Newman, Bavelier, Corina, Jezzard, & Neville, 2002; Perani et al., 1998, 2003; Wartenburger et al., 2003). Furthermore, the extent of differential activation may be sensitive to proficiency (e.g., Briellmann et al., 2004; Chee et al., 2001; De Bleser et al., 2003; Meschyan & Hernandez, 2006; Perani et al., 1998; Wartenburger et al., 2003) and age of acquisition (e.g., Mahendra et al., 2003; Perani et al., 1996). For example, Kim et al. (1997) reported that the left inferior frontal gyrus was sensitive to age of acquisition, with L1 and L2 being spatially separated in late bilinguals but largely overlapping in early bilinguals. These two factors may also interact. Perani et al. (2003) assessed the roles of age of acquisition and proficiency using a verbal fluency task and reported that earlier age of acquisition and higher proficiency led to less extensive activation in the left inferior frontal gyrus. The relative influences of proficiency and age of acquisition may also be modulated by the specific linguistic process: for example, Wartenburger et al. (2003) concluded that activation for semantic judgments was more significantly affected by proficiency level, whereas grammaticality processing was more affected by age of acquisition.

Regardless of the amount of neural overlap between the L1 and L2, the L2 generally activates a more extensive region of areas, reflecting the more effortful processing as a consequence of reduced proficiency (e.g., Briellmann et al., 2004; Chee et al., 2001; Ding et al., 2003; Hasegawa et al., 2002; Hernandez & Meschyan, 2006; Marian et al., 2003; Meschyan & Hernandez, 2006; Perani et al., 2003; Rüschemeyer et al., 2006; Vingerhoets et al., 2003; Wartenburger et al., 2003). For example, Briellmann et al. (2004) reported that the extent of activation during a noun generation task correlated with language proficiency. Hernandez and Meschyan (2006) reported more extensive activation for L2 during picture naming, suggesting that naming is more effortful in an L2. The L2 often activates the left inferior frontal gyrus to a greater extent (de Bleser et al., 2003; Kim et al., 1997; Marian et al., 2003; Perani et al., 2003), suggesting that the weaker language places increased cognitive demands on the language-processing system and requires more extensive neural resources. Therefore, this research suggests that although bilinguals activate similar language-processing areas as monolinguals, the amount and sometimes the regions of activation are more extensive in bilinguals, specifically in brain areas related to interference processing (see Section 4.2.3.2).

Above and beyond where and how the L1 and L2 overlap in the bilingual brain, structural neuroimaging studies have demonstrated differences between bilingual and monolingual brains (García-Pentón, Pérez Fernández, Iturria-Medina, Gillon-Dowens, & Carreiras, 2014; Mechelli et al., 2004). García-Pentón et al. (2014) used diffusion-tensor imaging, which measures the flow of water in the brain and is used to map the integrity and connectivity of white matter (WM) tracts, to compare WM connectivity in monolinguals and bilinguals. They reported that two structural subnetworks were more connected by WM tracts in bilinguals compared to monolinguals: a connection between left frontal and parietal/temporal regions and a connection between left occipital and parietal/temporal regions, as well as the right superior frontal gyrus. Importantly, these regions are related to language processing and language monitoring. The authors interpret this as evidence of structural plasticity in bilinguals, suggesting that bilinguals develop specialized networks to deal with two languages. In a now-famous study, Mechelli et al. (2004) used voxel-based morphometry to measure grey matter (GM) density in monolingual and bilingual brains. They reported increased GM density for bilinguals compared to monolinguals in the left inferior parietal cortex; importantly, GM density in this area correlated positively with second-language proficiency and negatively with L2 age of acquisition. Therefore, bilingualism confers both functional and structural changes on the organization of language networks in the brain.

4.2.2 How Are Two Languages Accessed? Selective Versus Nonselective Access to the Bilingual Lexicon

In early bilingualism research, one important issue concerned whether the lexicons for each language were separate, such that semantic representations existed individually in each lexicon, or integrated, such that the same semantic representations were shared between both languages. For example, case studies from bilingual aphasia sometimes show a loss of one language while the other is maintained (see Lorenzen & Murray, 2008 for a review), which would suggest separate representations. If bilingual lexicons were separate, this raised the question of how language switching occurred, i.e., how bilinguals selected the target language or suppressed the nontarget language ("selective" access to separate lexicons). If bilingual lexicons were integrated, the next question was how bilinguals controlled the relative activations of each language and managed to keep them from interfering with each other ("nonselective" access to an integrated lexicon).

Proponents of language-selective models of bilingualism (e.g., Bloem & La Heij, 2003; Bloem, van den Boogaard, & La Heij, 2004) propose that the conceptual specification of a language is represented by a language tag or feature. This allows for language-selective access based on task demands such that only one language ever becomes available for encoding during a particular task. For example, Macnamara and Kushnir (1971) reported that switching between languages during reading led to longer reaction times (RTs). These processing difficulties were thought to arise from the need to turn languages "on" and "off." Some researchers also report that in a dual-language context, the nontarget language has no influence on target language processing (Gerard & Scarborough, 1989; Scarborough, Gerard, & Cortese, 1984). For example, in a lexical decision task in Spanish-English bilinguals, Gerard and Scarborough (1989) modulated lexical frequency between the two languages using homographic noncognates (words between languages that share the same orthographic form but have different meanings: e.g., "fin" in English is a low-frequency word but in Spanish is a high-frequency word meaning "end"). Gerard and Scarborough (1989) found that nontarget-language homographic noncognates did not influence target-language processing, suggesting that the nontarget language was not activated and bilinguals were functioning essentially as monolinguals.

In contrast, other studies provide evidence in favor of nonselective access, demonstrating significant effects of the L2 on L1 processing (see reviews in Brysbaert & Duyck, 2010; Dijkstra & van Heuven, 2002; Kroll, Bobb, & Wodniecka, 2006; Kroll, Dussias, Bogulski, & Valdes Kroff, 2012). These studies commonly assess cross-linguistic influences using cognates or interlingual homographs (e.g., Degani & Tokowicz, 2010; Kerkhofs, Dijkstra, Chwilla, & de Bruijn, 2006; van Hell & Dijkstra, 2002; van Heuven, Schriefers, Dijkstra, & Hagoort, 2008). Cognates are words that have the same spelling and meaning between languages, such as "television" and "televisión" in English and Spanish, respectively. Interlingual homographs are words with the same spelling but different meanings and pronunciations: for example, the Dutch-English interlingual homograph "room" (/ru:m/) is pronounced /ro:m/ in Dutch and means 'cream'. van Heuven et al. (2008) reported that interlingual homographs showed increased activation of areas of the executive control network (see Section 4.2.3.2) during a lexical decision task, even when the task was administered in only one language. This suggests that even in monolingual contexts, both languages are activated and can interfere with each other in the bilingual brain.

The literature has now come to a general consensus that the bilingual lexicon is nonselective in nature, as supported by many additional studies demonstrating cross-linguistic influences in bilinguals (e.g., Brysbaert & Duyck, 2010; Kroll et al., 2006, 2012; Midgley, Holcomb, van Heuven, & Grainger, 2008; Poulisse & Bongaerts, 1994; Soares & Grosjean, 1984; Taft, 2002; Thierry & Wu, 2004; van Heuven, Dijkstra, & Grainger, 1998). Even in completely monolingual contexts, the nontarget language can influence or interfere with the target language during language processing. Therefore, bilinguals cannot completely "turn off" a language: both become activated in parallel and have the potential to interact with each other, to the detriment or advantage of the language-processing system.

4.2.3 How Are Two Languages Controlled? Models of Bilingual Language Control and the Interdependence of Language and Executive Control

4.2.3.1 Models of Bilingual Language Control

Many models of bilingual language processing have attempted to explain how bilinguals control their two languages. Of these, two are most prominent in the current literature. The Bilingual Interactive Activation (BIA+) Model (Dijkstra & van Heuven, 2002) is a model of bilingual word recognition which proposes an integrated bilingual lexicon, in which the words in a target language are selected by means of a higher-level control system. Within the word identification system, orthographic inputs activate associated phonological and semantic representations, as well as associated language nodes, which act as tags specifying a word's language membership. As representations from different languages are activated, the word identification system collects the relative activation and passes this information to a task/decision system, which achieves response selection by weighing the relative activation of language node information and making a decision based on the specific task goal. Thus, this model achieves language selection by accumulating evidence from bottom-up word identification processes. The BIA+ model generalizes across a variety of tasks and modalities and can account for much of the emerging neuroimaging evidence on bilingual comprehension (van Heuven & Dijkstra, 2010), making it one of the primary models of bilingual language processing.

The Inhibitory Control (IC) Model (Green, 1998) is a production model addressing bilingual language control during speech. This model considers languages as task schemas, which compete with each other to control the output from the lexico-semantic system. Inhibitory links within and between the language task schemas ensure that the task goals of a higher-level "supervisory attentional system" (SAS) are met. Word selection in each language is performed via "language tags" at the lemma level (lemmas are conceptual representations which are associated with specific word forms and specify various syntactic properties). During language selection, lemmas in both languages are activated, and all lemmas that do not possess the target language tag are inhibited. Language control in the IC model is cross-linguistically inhibitory (i.e., is exerted by the target task schema over the nontarget task schema) and self-inhibitory (i.e., within the nontarget language task schema). Importantly, the SAS is a reactive system, responding proportionally to the amount of activation in each language: if the nontarget language is activated strongly, the SAS will respond accordingly by implementing strong inhibition. Illustrating this, the IC model predicts that speaking in a less-dominant L2 requires stronger inhibitory control because the L1, the nontarget language, is the stronger language. This model thus proposes that language control over nonselective lexical access is obtained by reactive inhibition at the lemma level (see also Guo, Ma, & Liu, 2013).

4.2.3.2 The Need for Executive Control During Bilingual Language Processing

Neuroimaging studies of language control have found that bilinguals recruit areas of the executive control network during language processing to manage the cross-linguistic influences that arise as a result of nonselective activation. "Cognitive" or "executive" control is an umbrella term referring to a variety of cognitive situations in which a habitual response must be overcome, distracting information must be ignored, or one must switch between varying mental sets. (The terms "executive control," "cognitive control," and "executive processing" are used synonymously here.) These processes require a variety of cognitive functions such as working memory, task maintenance, decision-making, conflict detection/resolution, response selection and/or suppression, and inhibitory control.

Neuroimaging techniques have been immeasurably valuable in understanding the neural mechanisms involved in executive control. Studies using fMRI have identified an extensive network of brain areas involved in executive control, mainly localized to the prefrontal and parietal cortices. This network is reliably activated for a range of executive functions, including working memory, vigilance or sustained attention, inhibition of prepotent behaviors, and the detection and resolution of cognitive conflict (Niendam et al., 2012). The executive control network consists of numerous prefrontal and parietal structures (see Fig. 4.2). The anterior cingulate cortex (ACC) and dorsolateral prefrontal cortex (DLPFC) are thought to be involved in conflict detection and resolution (Botvinick, Braver, Barch, Carter, & Cohen, 2001; Botvinick, Cohen, & Carter, 2004; Melcher & Gruber, 2009; Peterson et al., 1999, 2002; Roelofs, van Turennout, & Coles, 2006; Swick & Turken, 2002). The left inferior frontal gyrus (LIFG), which is often implicated in language processing (e.g., Costafreda et al., 2006; Montant, Schön, Anton, & Ziegler, 2011), is active in both linguistic and nonlinguistic conflict tasks and may execute suppression of irrelevant semantic information (Novick, Kan, Trueswell, & Thompson-Schill, 2009; Novick, Trueswell, & Thompson-Schill, 2005; Ye & Zhou, 2009). The left angular gyrus, also usually implicated in language processes (e.g., Binder et al., 1997; Horwitz, Rumsey, & Donohue, 1998; Penniello et al., 1995; Pugh et al., 2000), has been reported for conflict tasks and may be involved in keeping multiple responses in mind during response selection (Bunge, Hazeltine, Scanlon, Rosen, & Gabrieli, 2002; Fan, Flombaum, McCandliss, Thomas, & Posner, 2003; Schroeder et al., 2002; Ye & Zhou, 2009). The right inferior and superior parietal lobes have been

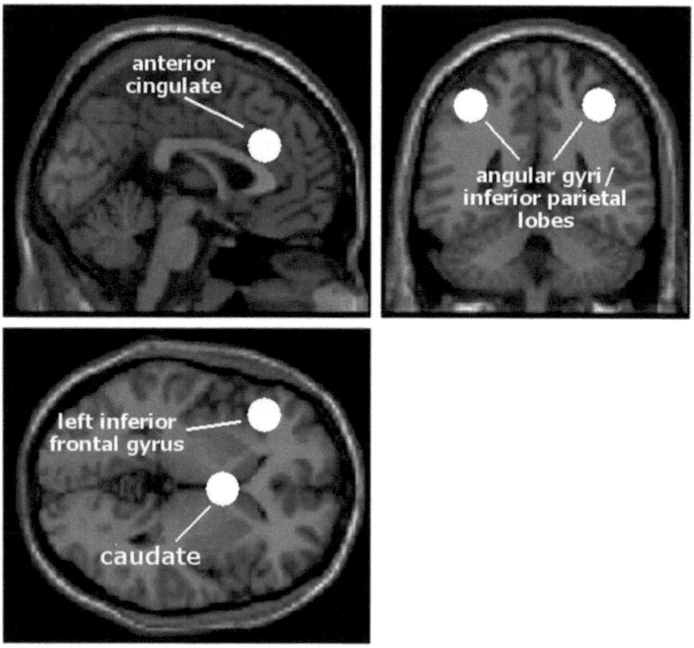

Fig. 4.2 Major areas of the executive control network

associated with visuospatial attention, particularly top-down control of attention toward the task-relevant target (Corbetta, Miezin, Shulman, & Petersen, 1993; Culham & Kanwisher, 2001; Milham, Banich, & Barad, 2003; Rushworth, Ellison, & Walsh, 2001). Finally, subcortical structures like the caudate are also involved in cognitive control (Abutalebi & Green, 2007; Lehtonen et al., 2005; Niendam et al., 2012). This executive control network is activated during general executive functioning such as shifting and updating, initiation and planning of actions, conflict control, and cognitive flexibility (see Nee, Wager, & Jonides, 2007; Niendam et al., 2012 for meta-analyses). Although debates still exist about the contributions of individual structures in this network to executive control and conflicting evidence exists regarding the function of almost all of them, their reliable activation across tasks highlights their involvement in the control network in a general and nonspecific way.

Neuroimaging research demonstrates that bilinguals activate areas of the executive control network during language comprehension and production (e.g., Hernandez & Meschyan, 2006; Parker Jones et al., 2011; Rodriguez-Fornells et al., 2005; van Heuven et al., 2008). For example, in a picture-naming task in Spanish-English bilinguals with fMRI, Hernandez and Meschyan (2006) reported more extensive activation for L2 in areas of the executive control network, suggesting that naming in an L2 requires more cognitive control. As mentioned in Section 4.2.2, van Heuven et al. (2008) found increased activation of the executive control network when Dutch-English bilinguals encountered interlingual homographs, suggesting that they recruited executive control to manage the cross-language conflict. Parker Jones et al. (2011) found that compared to monolinguals, picture naming in both languages for bilinguals elicited activation in left frontal and temporal regions of the executive control network. They interpret this as evidence that word processing is more cognitively demanding for bilinguals than monolinguals. Thus, substantial evidence demonstrates that even during basic language processing, bilinguals recruit executive control mechanisms to manage the cross-linguistic activation resulting from nonselective access.

Many researchers have attempted to understand how bilinguals control their languages by investigating explicit language switching (Christoffels, Firk, & Schiller, 2007; Crinion et al., 2006; Hernandez, 2009; Hernandez et al., 2001, 2000; Kuipers & Thierry, 2010; Magezi, Khateb, Mouthon, Spierer, & Annoni, 2012; Price, Green, & von Studnitz, 1999; Wang, Kuhl, Chen, & Dong, 2009; see Hervais-Adelman, Moser-Mercer, & Golestani, 2011; Luk, Green, Abutalebi, & Grady, 2012; and Rodriguez-Fornells, De Diego Balaguer, & Münte, 2006 for reviews). Language-switching paradigms typically require bilinguals to name pictures in alternating languages, with the target language indicated by an external cue. Switch trials require naming in the alternative language from the previous trial, whereas non-switch trials maintain the same language across consecutive trials. The "switch cost" is the RT difference between switch and non-switch trials. For unbalanced bilinguals, switching into the L1 generates larger switch costs than switching into the L2, whereas balanced bilinguals generally show similar switching costs for each language (Costa, Santesteban, & Ivanova, 2006; Meuter & Allport, 1999; although see also Calabria, Hernández, Branzi, & Costa, 2012). This asymmetry in unbalanced bilinguals is attributed to the need to overcome the stronger inhibition exerted over the L1 (as proposed by the IC model, Green, 1998). Extensive research has demonstrated that bilinguals recruit areas of the executive control network during language switching (Abutalebi & Green, 2008; Abutalebi et al., 2007; Bialystok, Craik, Green, & Gollan, 2009; Crinion et al., 2006; Guo, Liu, Misra, & Kroll, 2011; Hernandez, 2009; Hernandez et al., 2001; Luk et al., 2012; Price et al., 1999; Rodriguez-Fornells et al., 2006; Wang et al., 2009). Activation in these executive control areas is generally enhanced when switching into the weaker language (Abutalebi et al., 2007, 2008; Wang, Xue, Chen, Xue, & Dong, 2007), mirroring the pattern in asymmetric switch costs and suggesting that bilinguals inhibit their L1 to speak the L2. Importantly, the areas recruited by bilinguals to deal with cross-linguistic interference and language switching are also involved more generally in the executive control network found in monolinguals. Therefore, bilinguals use language-nonspecific mechanisms of executive control to manage their languages (see Section 4.3.2 for consequences of this).

4.2.4 How Do Two Languages Interact? Cross-Linguistic Interactions and the Effects of Script

As mentioned, research has documented cross-linguistic influences such that the L1 influences processing in the L2 and vice versa (Brysbaert & Duyck, 2010; Dijkstra & van Heuven, 2002; Kroll et al., 2006, 2012). The majority of this research has used cognates or interlingual homographs, meaning that both languages might be "cued" (e.g., Degani & Tokowicz, 2010; Kerkhofs et al., 2006; van Hell & Dijkstra, 2002; van Heuven et al., 2008). Stronger support for nonselective access and cross-linguistic influences comes from studies demonstrating cross-linguistic effects even in the absence of any nontarget-language cues (e.g., Colomé, 2001; Rodriguez-Fornells et al., 2005; Spalek, Hoshino, Damian, & Thierry, 2011; Thierry & Wu, 2004, 2007; Wu & Thierry, 2011). For example, Rodriguez-Fornells et al. (2005) tested German-Spanish bilinguals on a picture-naming task with an added phonological go/no-go component: participants named pictures in alternating language-naming blocks and were told to respond only when the picture's name in the target language started with a consonant; if it began with a vowel, they had to withhold their response. On critical trials, the picture's name started with a consonant in one of the bilingual's languages but a vowel in the other, leading to conflicting responses between the two languages. Combining behavioral, EEG, and fMRI measures, Rodriguez-Fornells et al. (2005) observed that these critical trials had a higher error rate, showed an increased negativity in the EEG data approximately 200 ms after word presentation, and activated the executive control network in the brain. This indicates that although naming was performed in one language, both languages were activated and interfered with each other, creating clear processing difficulties.

There is even evidence that the grammatical properties of one language can affect how the other is processed (e.g., Dussias & Sagarra, 2007; Hartsuiker, Pickering, & Veltkamp, 2004; Runnqvist, Gollan, Costa, & Ferreira, 2013), demonstrating that cross-linguistic effects are not limited to single-word processing. Hartsuiker et al. (2004) performed a syntactic priming task in Spanish-English bilinguals and reported significant priming effects of the L1 on the L2. Dussias and Sagarra (2007) presented ambiguous sentences to Spanish-English bilinguals and native English speakers. For example, the ambiguous sentence "An armed robber shot the sister of the actor who was on the balcony" could be parsed in two ways. Native Spanish speakers prefer attachment to the first noun (NP1 attachment), interpreting the sentence as the actor being on the balcony. Native English speakers prefer attachment to the second noun (NP2 attachment), interpreting the sister being as on the balcony. Dussias and Sagarra found that native Spanish speakers who had extensive immersion experience in their L2 English environment parsed Spanish sentences with NP1 attachment, i.e., the way native English speakers would. These studies illustrate that not only can the L1 have an influence on the L2 but also that experience in an L2 can also have backward influences on how the L1 is processed.

4.2.4.1 Cross-Linguistic Interactions for Two Languages with Different Writing Systems

Given the literature reviewed above reporting cross-language interactions, a logical question is what happens when languages are from completely different writing systems with very different orthographies. For example, would languages as dissimilar as Chinese and English still activate each other?

The majority of work on cross-linguistic influences has been done with same-script bilinguals. For example, cognate facilitation effects have been documented in Catalan-Spanish (Costa, Caramazza, & Sebastian-Galles, 2000), Spanish-English (Schwartz & Kroll, 2006), Welsh-English (Lallier, Carreiras, Tainturier, Savill, & Thierry, 2013), and Dutch-English (de Groot & Nas, 1991; Dijkstra, Grainger, & van Heuven, 1999; van Hell & Dijkstra, 2002), to name a few. However, a host of studies now report facilitatory cross-linguistic influences for different-script languages as well (Hoshino & Kroll, 2008; Sumiya & Healy, 2004; Zhang, van Heuven, & Conklin, 2011). Across masked priming and translation priming lexical decision and picture-naming tasks, cognate facilitation effects have been found for Greek-French (Voga & Grainger, 2007), Korean-English (Kim & Davis, 2003), Chinese-English (Chen, Zhou, Gao, & Dunlap, 2014), Japanese-English (Allen & Conklin, 2013; Hoshino & Kroll, 2008), and Hebrew-English (Gollan, Forster, & Frost, 1997) bilinguals, indicating that the nontarget language can be active despite having a completely different writing system. Interestingly, Table 1 in Schoonbaert, Duyck, Brysbaert, and Hartsuiker (2009), which quantitates translation priming effects in same-script and different-script bilinguals, suggests that L2-L1 priming may not be as strong in different-script bilinguals as in same-script (see Section 4.2.4.2).

There is further evidence, beyond the use of cognates, that cross-linguistic influences are present even in languages that have different orthographic properties. Thierry and Wu (2007) reported that when Chinese-English bilinguals read English (L2) words, they automatically activate the Chinese translation. They performed an ERP study in which participants had to decide whether two English words were semantically related. Unbeknownst to the bilinguals, some of the English word pairs, when translated into Chinese, repeated a character; these repeated stimuli showed reduced P2 and N400 components compared to English words that did not repeat a character when translated into Chinese. Because the experiment was performed in a completely English context, the English words were not orthographically related, and the languages did not share a script; this suggests that the bilinguals had automatically translated the L2 into the L1. In a similar paradigm, Zhang et al. (2011) found that the priming effect only occurred when the first character, but not the second, was repeated, providing evidence for rapid morphological decomposition in the automatically activated L1. Therefore, despite the fact that a bilingual's languages might have completely different orthographic properties, there is still evidence of cross-linguistic influences from the L1 to the L2 and vice versa.

Cross-linguistic influences also affect how language is represented and processed in the brain. Even in languages from the same writing system, differences in orthographic depth can affect language processing and carry over between languages. For example, Lallier et al. (2013) have shown that the orthographic transparency of the L1 influences ERP indices of L2 reading in Welsh-English bilinguals. Testing a group of English monolinguals and simultaneous Welsh-English bilinguals on early indices of letter-string encoding, they measured visual skills by asking participants to detect whether a target letter was presented in a letter-string probe. They found ERP evidence that the Welsh-English bilinguals more accurately encoded the leftmost letters of a letter string, which is in keeping with the idea of a reduced orthographic grain size. This indicates that visual attention skills underlying orthographic processing are tuned differently for bilinguals with experience in reading both an opaque (English) and transparent (Welsh) language compared to reading acquired in an opaque language only (English monolinguals). Thus, merely the experience of reading in different types of orthographies can tune the cognitive processing system differently and affect how languages are processed.

Orthographic depth also affects processing in languages of different writing systems. Recent work from Koyama, Stein, Stoodley, and Hansen (2013) tested two groups of bilinguals, L1 English/L2 Japanese and L1 Japanese/L2 English, on a word-reading task in English and Japanese kana using fMRI. They found that during English reading, the Japanese-English bilinguals showed increased activation in executive control areas relative to the English-Japanese bilinguals. This was interpreted as an increased cognitive load for L2 English reading, especially because L1 readers of Japanese are used to a very regular orthography in kana and recruit more processes for the greater phonological demands of the more irregular English orthography. In contrast, during kana reading, the English-Japanese bilinguals showed greater recruitment of visual areas to deal with the less-familiar L2 kana symbols. This study further illustrates how the linguistic characteristics of one's native language can influence how the L2 is acquired and processed.

More generally, differences in the linguistic properties of the L1 can influence how the L2 is organized and represented in the brain. Researchers have shown that native English speakers activated a left-lateralized language network when reading English, their native alphabetic language, but activated a more bilateral language network when reading Chinese, their late-acquired L2 (Nelson et al., 2009; Perfetti et al., 2007). This was described as an accommodation strategy in L2 reading, in which participants recruit additional neural structures not usually used for alphabetic languages to process the logographic language. In contrast, Nelson et al. (2009) showed that native Chinese speakers activated a bilateral language network when reading both Chinese, their native logographic language, and English, their late-acquired L2. As English is typically a left-lateralized language, this was interpreted as an assimilation strategy: native Chinese speakers were using language areas already in place for their logographic writing system and were "read[ing] English as if it were Chinese" (Perfetti et al., 2007, p. 136). Few studies have been done on assimilation and accommodation in L2 learning; it may be that the relative ratio of assimilation and accommodation in L2 reading is modulated by the linguistic-specific properties of the orthographic-phonological mappings. For example, Cao, Tao, Liu, Perfetti, and Booth (2013) proposed that the neural substrates of Chinese, being a more complex language system with more arbitrary orthographic-phonological mappings, can handle a relatively simpler language like English. In contrast, English has semi-regular orthographic-phonological mappings, so a more complex language like Chinese recruits additional resources. Other researches on assimilation and accommodation in Hindi-English bilinguals (Das, Padakannaya, Pugh, & Singh, 2011) suggest that even subtle differences in orthographic depth can affect how the L2 is represented and organized in the brain. Additional factors such as L2 proficiency and age of acquisition may also come into play. For example, research has shown that higher proficiency in an alphabetic L2 was associated with more assimilation and less accommodation (Cao et al., 2013). Furthermore, late bilinguals demonstrated more accommodation compared to early bilinguals, who showed native-like lateralization in both languages (Das et al., 2011).

4.2.4.2 How the Magnitude of Script Similarity Affects the Magnitude of Cross-Linguistic Influences

As demonstrated by the literature reviewed previously, cross-linguistic influences are found for both same- and different-script languages. Although the general properties of a language can influence how a second language is represented, the magnitude of cross-linguistic activation may also be modulated by language similarity. For example, in an early review, Koda (1996) reported that more orthographic similarity between L1 and L2 accelerated L2 lexical processing efficiency. The BIA+ model of bilingual word recognition (which is based on alphabetic word recognition and codes letter positions; see Section 4.2.3.1) provides a framework for understanding this and proposes that the amount of cross-linguistic activation depends on the degree of orthographic overlap:

> The larger the overlap between the input string and a representation in the mental lexicon, the more the internal representation is activated…if the two languages differ with respect to their input codes (e.g. letter sets), the activated set of [orthographic] neighbors may become much smaller (Dijkstra & van Heuven, 2002, pp. 182–183).

Therefore, languages with more orthographic overlap (i.e., overlapping letters) should create more cross-linguistic influences on each other. Furthermore, same-script languages contain many orthographic neighbors (words that differ by only one letter), homographs (words that are spelled the same between languages), and cognates, which may exaggerate cross-linguistic influences. As the BIA+ model is based on an alphabetic writing system, it predicts that less letter overlap should lead to the activation of fewer orthographic neighbors. In different-script bilinguals, therefore, this model would predict no cross-linguistic activation of orthography as there is no orthographic correspondence between, for example, alphabetic letters and logographic characters.

However, as reviewed above, cross-linguistic influences do still occur even for bilinguals whose languages do not share a script. To explain this, the BIA+ model also predicts that overlap of phonology can cause cross-linguistic activation. When reading a word, all words that share the input word's orthography or phonology are activated. The activation of these related words feeds backward to the target, increasing its activation. Therefore, shared phonology can also create cross-linguistic influences, even between languages that do not have any overlap in orthography (Allen & Conklin, 2013). Importantly, in languages that do not share an orthography, such as Japanese and English, cognates must be defined by their amount of phonological overlap. For example, the Japanese word ラジオ ("rajio") and the English word "radio" are cognates because they sound similar and share the same meaning, even though the orthography is completely different. The BIA+ model thus predicts that the amount of orthographic and phonological overlap will determine the amount of cross-linguistic activation. This was confirmed by Allen and Conklin (2013), who reported a larger facilitation effect for cognates with a higher degree of phonological similarity. Therefore, the cognate facilitation effect can be modulated by the amount of cross-linguistic overlap.

If greater cross-linguistic activation can create greater cognate facilitation, it should also create greater cross-linguistic interference. Few studies have investigated this proposal, but the extant research supports this hypothesis. In one study, van Heuven, Conklin, Coderre, Guo, and Dijkstra (2011) tested three groups of trilinguals: German-English-Dutch (all alphabetic writing systems), Chinese-English-Malay (Chinese logographic, English and Malay alphabetic), and Chinese-English-Uighur (Chinese and Uighur logographic, English alphabetic). The authors administered a Stroop task, in which a color word is printed in colored ink and participants must ignore the word and respond to the color of the ink (Stroop, 1935). In a bilingual Stroop task, bilinguals must respond to the color of the ink in the nontarget language. For example, seeing the German word "rot" (red) printed in blue ink, a participant may be required to respond "blue" in English. Using such a vocal-response bilingual Stroop task, van Heuven et al. (2011) observed larger between-language interference for trilingual groups with greater orthographic similarity between languages. This suggests that cross-linguistic activation was greater for same-script bilinguals, creating more interference when trying to ignore the influence of the nontarget language and resulting in increased Stroop effects. Few studies have systematically varied script similarity in order to evaluate its effect on language processing, but this will be an interesting avenue for future research.

4.3 Cognitive Effects of Bilingualism

Given the evidence reviewed in the previous two sections, it is apparent that managing multiple languages creates significant processing demands on the bilingual brain. What are the cognitive effects of these increased processing demands for bilinguals? Bilingualism literature has documented both disadvantages and advantages to managing multiple languages in one brain, most notably a general slowing of lexical processing speed, on the downside, and an improvement in cognitive control abilities, on the upside.

4.3.1 Disadvantages: Delayed Lexical Processing

It is commonly reported that throughout the life span, bilinguals are slower in lexical processing tasks and command smaller vocabularies compared to monolinguals (Bialystok, Craik, & Luk, 2008; see Bialystok et al., 2009 and Bialystok, 2009 for reviews). For instance, Bialystok et al. (2008) tested monolinguals and bilinguals on a range of working memory, lexical retrieval, and cognitive control tasks. Bilinguals performed worse on all tasks of verbal ability: they were slower to name pictures, produced fewer exemplars in category fluency, and scored lower on vocabulary measures. In addition to these findings of generally slower lexical processing in bilinguals, behavioral and electrophysiological research also demonstrates delays in both the L1 and L2.

4.3.1.1 Delays in the L2

The majority of bilingualism research has focused on processing difficulties in the L2, as this is usually the weaker, less-dominant language. Reduced proficiency can affect the performance of the language system in a number of ways, particularly in terms of lexical processing speed. One hypothesis incorporated in the BIA+ model is the temporal delay assumption (Dijkstra & van Heuven, 2002; see also van Heuven & Dijkstra, 2010), which proposes that L2 semantic representations have a lower "resting-level" activation than L1 representations. Resting level is determined by subjective frequency, i.e., how often the word is encountered. Reduced experience with the less-proficient language creates a lower resting-level activation of L2 words, requiring more processing time and producing delayed semantic activation in the L2 compared to the L1.

Evidence for this conjecture comes from a range of tasks demonstrating that bilinguals experience difficulties and delays in lexical processing in their L2 (see Moreno, Rodriguez-Fornells, & Laine, 2008; Runnqvist, Strijkers, Sadat, & Costa, 2011; and van Heuven & Dijkstra, 2010 for reviews). For example, in category fluency, Gollan, Montoya, and Werner (2002) reported that bilinguals produced fewer words in their L2 than monolinguals. Bilinguals also experience more tip-of-the-tongue states than monolinguals when naming pictures in their L2, demonstrating word-finding difficulties (Gollan, Montoya, & Bonanni, 2005; Gollan & Silverberg, 2001; Pyers, Gollan, & Emmorey, 2009). In production tasks such as picture naming, bilinguals name pictures in their L2 or less-dominant language more slowly than monolinguals (Gollan, Montoya, Cera, & Sandoval, 2008; Ivanova & Costa, 2008; Kohnert, Hernandez, & Bates, 1998). L2 delays have also been documented in comprehension tasks such as lexical decision, in which bilinguals perform more slowly in their L2 compared to monolinguals (Portin & Laine, 2001; Ransdell & Fischler, 1987).

More compelling support for the L2-processing delay comes from studies using EEG, which provides a more accurate temporal quantification of the delay. Most studies investigating bilingual language processing with EEG have focused on the later ERP components reflecting higher-level linguistic processes (see Moreno et al. 2008 for a review). For example, the N400 is significantly delayed in the bilingual L2 compared to the L1 (Ardal, Donald, Meuter, Muldrew, & Luce, 1990; Hahne, 2001; Moreno & Kutas, 2005; Moreno et al., 2008), indicating that semantic integration strategies operate differently in a native versus a nonnative language. Bilinguals also show differences in syntax processing in the L2, such that many grammar-related ERP components are reduced or even absent in L2 learners (e.g., Bowden, Steinhauer, Sanz, & Ullman, 2013; Hahne, 2001; Hahne & Friederici, 2001; Proverbio, Čok, & Zani, 2002; van Hell & Tokowicz, 2010), demonstrating difficulties with syntax processing in the L2.

Some EEG studies have also reported L2 delays at very early stages of linguistic processing. Differences between native and nonnative languages are observable as early as 150 ms after word presentation, indicating delays in low-level lexical processes such as visual letter decoding or orthographic word recognition (Liu & Perfetti, 2003; Proverbio et al., 2009). For example, Liu and Perfetti (2003) observed later peaks of the N150 (thought to index graphical form) and the N250 (thought to index phonological processing and articulatory preparation) in the L2 than the L1. Proverbio et al. (2009), testing Italian-English-German trilinguals on a letter-detection task, observed different time windows of lexical effects for each language. Orthographic recognition effects occurred in the L1 between 160 and 180 ms (N1 component), in the L2 between 260 and 320 ms (N2 component), and in the L3 between 320 and 380 ms (N3 window), demonstrating nonnative delays at very early stages of linguistic processing.

Processing difficulties in the L2 have thus been documented across a range of language-related ERP components. The L2 processing delay may also be modulated by individual differences such as L2 proficiency and age of acquisition. For example, some have reported that the N400 latency is delayed for late compared to early learners (e.g., Weber-Fox & Neville, 1996) or in nonnatives compared to natives but is not affected by proficiency (e.g., Ardal et al., 1990; Newman, Tremblay, Nichols, Neville, & Ullman, 2012), suggesting a stronger influence of age of acquisition on N400 delays. In contrast, others report later latencies or enhanced amplitudes of semantic and syntactic processing components in low-proficiency bilinguals (e.g., Moreno & Kutas, 2005; Weber-Fox & Neville, 2001), suggesting a primary role of proficiency. Finally, some evidence suggests that proficiency and age of acquisition interact in bilingual language processing (Proverbio et al., 2009).

Therefore, the L2 delay in lexical processing is robust and well documented, both in behavioral and EEG studies and across production and comprehension modalities. However, bilinguals do not always show disadvantages in picture naming: as mentioned earlier, cognate facilitation effects have been documented (e.g., Hoshino and Kroll, 2008), and Gollan, Montoya, and Bonanni (2005) reported no bilingual delays when naming proper nouns. Specifically, the fact that bilinguals do not experience delays when naming cognates—which theoretically should be used across both languages with a frequency equal to that of monolinguals—highlights that the bilingual disadvantage arises from reduced frequency of language use.

4.3.1.2 Delays in the L1

Given that the L2 is often the less-dominant, later-learned language, processing difficulties might be expected. However, it has also been proposed that the bilingual L1 is delayed compared to monolinguals (Gollan, Montoya, Fennema-Notestine, & Morris, 2005). This theory, known as the weaker links hypothesis (Gollan et al., 2005) or the reduced frequency account (Pyers et al., 2009), suggests that compared to monolinguals, bilinguals use each of their languages less often, including their first language. This reduced frequency of use leads to weaker ties between words and concepts and consequently delayed lexical access. Therefore, splitting communication between two languages predicts delays not only in the L2 but also in the L1 compared to monolinguals (Gollan & Acenas, 2004; Gollan et al., 2005, 2008). In other words, bilingual lexical access is slower than that of monolinguals even when both groups are performing in their native language.

Mirroring the L2 disadvantage patterns, evidence shows that bilinguals are slower at picture naming in their L1 or more dominant language compared to monolinguals (Gollan et al., 2005; Ivanova & Costa, 2008). For example, Ivanova and Costa (2008) reported faster picture-naming responses for monolinguals than bilinguals, even when bilinguals were naming in their first and most dominant language. L1 delays have also been reported in comprehension, such that bilinguals demonstrate slower RTs in lexical decision and list recognition tasks (e.g., Lehtonen & Laine, 2003; Ransdell & Fischler, 1987). Only two previous studies have used EEG to investigate processing delays in bilinguals' L1 compared to monolinguals (Ardal et al., 1990; Proverbio et al., 2002). Ardal et al. (1990) reported a small but significant delay in the N400 response to semantically anomalous sentences for bilinguals when performing in their L1 compared to monolinguals. Proverbio et al. (2002) found that bilinguals were slower to make grammaticality judgments on sentences in their native language than monolinguals were, indicating bilingual difficulties with higher-level syntactic processing in the L1. Therefore, EEG evidence also exists for L1 lexical processing delays; however, these previous studies tested higher-level semantic and syntactic processing rather than earlier lexical access, so it remains unclear how early in language processing L1 delays might occur.

4.3.1.3 Effects of Script on Lexical Processing Speed

As discussed in Section 4.1, script-specific lexical factors can affect processes of phonological access (e.g., Perfetti et al., 2005; Saalbach & Stern, 2004), orthographic recognition (e.g., Bar-Kochva, 2011; Meschyan & Hernandez, 2006), language production speed (Bates et al., 2003), and the neural representations of languages in the brain (e.g., Bick et al., 2011; Bolger et al., 2005; Tan et al., 2005). It is therefore also possible that the magnitude of the L1 and/or L2 delay in lexical processing is modulated by language-specific differences. Previous studies have reported lexical processing delays for bilinguals from various language backgrounds, including Chinese (Liu & Perfetti, 2003; Weber-Fox & Neville, 1996), Spanish (Moreno & Kutas, 2005; Newman et al., 2012), French (Ardal et al., 1990), German (Proverbio et al., 2009; Spalek et al., 2011), Italian, and Slovenian (Proverbio et al., 2009). However, there has not yet been an explicit test of this factor and how it might affect the magnitude of the delay.

4.3.2 Advantages: Improved Cognitive Control

As discussed in Section 4.2.2, nonselective access to an integrated bilingual lexicon proposes that both languages are activated in parallel during language processing. Consequently, bilinguals must constantly exert control over their languages, either by inhibiting the nontarget language (as proposed by the IC model; Green, 1998) or engaging advanced selection mechanisms (as proposed by the BIA+ model; Dijkstra & van Heuven, 2002; see Section 4.2.3.1). Accordingly, neuroimaging evidence has shown that bilinguals recruit cognitive control mechanisms to help with language control (Section 4.2.3.2). The need to constantly control multiple languages places extraordinary demands on cognitive resources, especially the executive control system.

The interdependence of cognitive control and language processing in bilingualism is believed to enhance cognitive processing abilities beyond those of monolinguals who do not need to resolve language conflict on a daily basis (see Green & Abutalebi, 2013 for an extended discussion). This is supported by extensive empirical evidence demonstrating that bilinguals outperform their monolingual counterparts across a range of executive control domains (see Bialystok, 2009, 2011; Bialystok et al., 2009; Hilchey & Klein, 2011; Kroll & Bialystok, 2013; and Tao, Marzecová, Taft, Asanowicz, & Wodniecka, 2011 for reviews). The hypothesis that bilinguals experience superior cognitive abilities due to the entwined functions of executive control and language processing is referred to here as the *bilingual cognitive advantage hypothesis*.

Executive control is typically measured using conflict tasks such as the Stroop, Simon, or flanker tasks. All of these tasks consist of incongruent and congruent trials. In incongruent trials, there is conflict between two stimulus dimensions, which must be overcome in order to make a correct response (e.g., the word "red" printed in blue ink, in the Stroop task, requiring a response of "blue"). In congruent trials, the stimulus dimensions are compatible so there is no conflict (e.g., the word "red" printed in red ink), requiring no recruitment of the executive control system. Control trials are also often included, in which the distracting stimulus dimension is neither conflicting nor compatible, containing a neutral stimulus (e.g., a string of symbols "%%%" printed in red ink). The interference effect, the difference between incongruent and control trials, is interpreted as a measure of the degree of cognitive control required.

The bilingual cognitive advantage has been documented across a spectrum of executive control tasks. For instance, bilinguals show smaller conflict effects than monolinguals on the Simon and Stroop tasks, an advantage that is maintained across the life span (Bialystok, 2006; Bialystok & Depape, 2009; Bialystok, Craik, Klein, & Viswanathan, 2004; Bialystok et al., 2008; Martin-Rhee & Bialystok, 2008). The attentional network task (ANT; Fan, McCandliss, Sommer, Raz, & Posner, 2002) is an executive control paradigm assessing three dimensions of executive processing: control (via a flanker task), alerting (response to cueing), and orienting (response to valid cueing). Bilinguals demonstrate not only smaller flanker effects

(Costa, Hernández, Costa-Faidella, & Sebastián-Gallés, 2009; Costa, Hernández, & Sebastián-Gallés, 2008) but also more efficient use of alerting cues (Costa et al., 2008; although see Hernández, Costa, Fuentes, Vivas, & Sebastián-Gallés, 2010) and better attentional control (Carlson & Meltzoff, 2008; Costa et al., 2009). Bilinguals show smaller switch costs than monolinguals on task-switching paradigms than monolinguals due to their experience with language switching (e.g., Garbin et al., 2010; Prior & Macwhinney, 2010), both in cohorts of children (Bialystok & Martin, 2004) and across adulthood (Bialystok, Craik, & Ruocco, 2006; Marzecová et al., 2013). Bialystok and Shapero (2005) have also reported that bilingual children performed better on a reversible figures task, in which one figure may be seen in multiple different ways, demonstrating greater cognitive flexibility with bilingualism. As well as inhibitory control and task switching, bilingual advantages have been documented in theory of mind (Bialystok & Senman, 2004; Goetz, 2003); in inhibiting prepotent responses, as in anti-saccade and go/no-go tasks (Bialystok & Viswanathan, 2009; although see Luk, Anderson, Craik, Grady, & Bialystok, 2010); and even in learning new words (Bartolotti, Marian, Schroeder, & Shook, 2011). Therefore, bilinguals demonstrate superior executive processing abilities across a wide range of tasks, not just in inhibitory control but in cognitive flexibility and monitoring as well.

4.3.2.1 Neuroimaging Evidence for the Bilingual Advantage

Although the bilingual advantage is behaviorally well established, there is a paucity of research examining this phenomenon using neuroimaging methods (see Bialystok et al., 2009; Hilchey & Klein, 2011; Hervais-Adelman et al., 2011; Moreno et al., 2008; and Rodriguez-Fornells et al., 2006 for reviews).

Only a handful of EEG studies have investigated the electrophysiological indices of the bilingual cognitive advantage. Kousaie and Phillips (2012) used Stroop, flanker, and Simon tasks to evaluate the N2, an ERP component that occurs at approximately 200–300 ms at frontocentral locations and indexes cognitive conflict (e.g., Melara, Wang, Vu, & Proctor, 2008; Tillman & Wiens, 2011). They also evaluated the P3 component, thought to reflect resource allocation and schema updating, and the error-related negativity (ERN), which is elicited by errors and reflects error detection or post-response conflict. Although there were no differences in the behavioral interference effects, bilinguals showed a smaller N2 and ERN in the Stroop task, a larger ERN in the flanker task, and a smaller P3 in the Simon task. This indicates group differences in conflict monitoring, error-related processing, and resource allocation; however, these effects were not consistent across all paradigms, which could suggest that the particular task modulated how bilinguals and monolinguals responded to conflict.

Coderre & van Heuven (2014a) recorded ERPs while Chinese-English bilinguals (L1 Chinese, L2 English) and English monolinguals performed a variation of a Stroop task that varied the stimulus onset asynchrony (SOA) of color and word onset. The irrelevant distractor word could either appear just before the target color (a 'negative SOA'); just after (a 'positive SOA'; or at the same time (a '0 ms SOA'). Coderre & van Heuven (2014a) found some significant, but inconsistent, differences between monolinguals and bilinguals in a conflict-related ERP component (the Ninc): bilinguals showed differences compared to monolinguals when performing this Stroop task in their L1 but not their L2. This provides limited evidence for the presence of a conflict-specific bilingual advantage. However, Coderre & van Heuven (2014a) also observed group differences between bilinguals (in both languages) and monolinguals when comparing control trials, which contained no conflicting or facilitating information. Furthermore, these group differences observed in the negative SOA before the color had been presented, i.e. before the onset of conflict. This suggests a more general advantage in bilingual executive control that is not limited to the presence of conflict but affects monitoring abilities more generally.

As reviewed in Section 4.2.3.2, bilinguals recruit executive control areas of the brain during language processing. Importantly, these areas recruited by bilinguals for language control correspond to those recruited by monolinguals for nonlinguistic executive control. This indicates that bilinguals use language-nonspecific executive control areas for language control and implies that the functional organization and overlap of the executive control and language networks may be altered by bilingualism. For example, Hernandez (2009) has suggested that the language networks of early bilinguals may develop differently than those of late bilinguals or monolinguals because they are integrated from early in development with areas of the executive control network. Therefore, bilinguals experience a fundamental restructuring of the neural networks dedicated to language and executive processing due to their codependency and co-activation.

This proposal is supported by evidence from fMRI and MEG studies demonstrating differences between monolinguals and bilinguals in the extent and location of brain activation during nonlinguistic cognitive control tasks (Abutalebi et al., 2012; Bialystok et al., 2005; Garbin et al., 2010; Luk et al., 2010). For example, Bialystok et al. (2005) reported that better performance on a Simon task correlated with increased cingulate and superior/inferior frontal activation for bilinguals but with left middle frontal regions in monolinguals, suggesting different underlying executive control mechanisms. Abutalebi et al. (2012) reported that although both monolinguals and bilinguals recruited the anterior cingulate for a task-switching and flanker task, bilinguals used this structure more efficiently than monolinguals: better behavioral performance correlated with reduced BOLD signal in this area. Luk et al. (2010) observed that during interference suppression in a flanker task, monolinguals activated the left temporal pole and left superior parietal lobe, whereas bilinguals activated a more extensive network including bilateral frontal, temporal, and subcortical regions. Therefore, bilinguals use different functional networks than monolinguals for executive control tasks, suggesting an interdependence of executive control and language processing that confers functional differences in the organizations and recruitments of these networks.

4.3.2.2 The Elusiveness and Sensitivity of the Bilingual Advantage

Although prevalent in the literature, the bilingual cognitive advantage is sensitive to a number of individual and experimental factors. For example, although advantages have been documented across the life span (Bialystok, 1999, 2010; Bialystok, Craik, & Ryan, 2006; Bialystok et al., 2004, 2008; Costa et al., 2008), the magnitude of the advantage is affected by development, with larger bilingual advantages occurring in children and older adults who have underdeveloped or declining cognitive control (Craik & Bialystok, 2006). In young adults who are at the peak of their cognitive abilities, the bilingual advantage may only be observed in cognitively demanding situations (Costa et al., 2009; Martin-Rhee & Bialystok, 2008). The bilingual cognitive advantage is also sensitive to the type of executive control being tested. For instance, bilinguals typically show advantages on conflict tasks requiring management of conflicting attentional demands or interference suppression (e.g., ignoring a conflicting stimulus), but not on impulse control or response inhibition (e.g., withholding a button press; Bialystok & Viswanathan, 2009; Carlson & Meltzoff, 2008; Luk et al., 2010; Martin-Rhee & Bialystok, 2008).

In some cases, a bilingual advantage is not found unless controlling for other variables. For example, Bialystok and Feng (2009), using a linguistic executive control task (proactive interference), observed a bilingual advantage only when controlling for vocabulary knowledge. Similarly, using a battery of executive function tasks in children, Carlson and Meltzoff (2008) found a bilingual advantage only when controlling for verbal ability, age, and socioeconomical status (SES). Morton and Harper (2007) also identified SES as a critical factor, reporting identical performance for monolingual and bilingual children but larger cognitive advantages for children from high-SES families. Furthermore, the bilingual advantage is sensitive to individual differences such as proficiency level (e.g., superior cognitive control with higher L2 proficiency: Bialystok, Craik, & Ruocco, 2006) and the similarity of the bilingual's two languages (e.g., larger bilingual advantages for speakers of two orthographically similar languages: Linck, Hoshino, & Kroll, 2008; see Section 4.3.2.3). The subjective frequency of language switching may also play a role: for example, Soveri, Rodriguez-Fornells, and Laine (2011) found that bilinguals who reported frequent language switching in daily conversations performed better on a task-switching paradigm than bilinguals who rarely switched. In contrast, Festman, Rodriguez-Fornells, and Münte (2010) reported that frequent language switchers performed worse on tasks of inhibition, self-monitoring, problem-solving, and generative fluency, suggesting that frequent switching may be indicative of weaker language control. Individual differences can therefore significantly influence the magnitude of the bilingual cognitive advantage and are important factors to consider.

Therefore, although the bilingual cognitive advantage is well documented, in actuality, it is an elusive phenomenon and is sensitive to a number of factors. The topic has become a contentious issue in the literature, and some have questioned whether the phenomenon truly exists (Duñabeitia et al., 2014; Hilchey & Klein, 2011; Paap & Greenberg, 2013). In a review, Hilchey and Klein (2011) provided a critical and thorough quantification of the bilingual advantage, concluding that the bilingual "interference advantage" on conflict tasks (i.e., smaller conflict effects when comparing incongruent and congruent trials) is a weak effect that is often not found at all. Far more common is the finding of a "global reaction time" advantage, such that bilinguals are faster than monolinguals on all trials, both incongruent and congruent (e.g., Bialystok et al., 2004, 2005; Costa et al., 2009; Martin-Rhee & Bialystok, 2008; although see Bialystok et al., 2008). To distinguish these different effects, Hilchey and Klein (2011) outlined two hypotheses regarding bilingual executive processing: the "bilingual inhibitory control advantage" or BICA hypothesis and the "bilingual executive processing advantage" or BEPA hypothesis.

The BICA hypothesis refers to the phenomenon of smaller interference effects for bilinguals compared to monolinguals, which is thought to occur arising from the use of inhibitory control mechanisms to control cross-linguistic interference (see the IC model, Section 4.2.3.1; Green, 1998). This predicts more efficient inhibitory processes for bilinguals in the presence of conflict (i.e., incongruent trials), resulting in reduced interference effects. However, because the BICA hypothesis places the locus of the advantage on conflict control, it predicts no difference between groups in the absence of conflict and therefore cannot explain why bilinguals are often faster on all trial types compared to monolinguals. In contrast, the BEPA hypothesis proposes an advantage for bilinguals in domain-general executive processing. Importantly, this is not restricted to conflict processing but extends to a more general advantage in cognitive monitoring (Costa et al., 2009; Martin-Rhee & Bialystok, 2008). For example, bilinguals may have a more efficient mechanism for monitoring the environment for conflict (Costa et al., 2009) or more efficient top-down guidance of attention (Hernández, Costa, & Humphreys, 2012). Such an enhancement of domain-general executive processing, not limited to the presence of conflict, would lead to faster processing on all trial types, explaining the global RT advantage.

Bilingual cognitive control is an emergent field, with numerous unanswered questions and unidentified influences. The precise cognitive and neural loci of the bilingual advantage are still not fully understood, and much research in the past few years has been dedicated to exploring this field (see Kroll & Bialystok, 2013 for a recent review). For example, Morales et al. (2013) have recently suggested that focusing just on inhibition or monitoring in exploring the bilingual advantage may be too specific; instead, they provide evidence of a dynamic combination and interaction of proactive (monitoring) and reactive (inhibition) control that seems to be enhanced in bilinguals, suggesting a more global effect of bilingualism on cognitive function. Other questions also remain largely unexplored: for example, what level of proficiency is needed to confer an advantage? Does the number of languages one speaks affect the magnitude of the advantage? Does the advantage decay over time if bilingualism is not maintained? The opportunities for future work are ripe and many questions are already beginning to be investigated. Neuroimaging will in particular be a valuable technique in the future for understanding the nature of the bilingual advantage.

4.3.2.3 The Effects of Script Similarity on Cognitive Control Abilities

As addressed in Section 4.2.4.2, nonselective access in two orthographically similar languages predicts that bilinguals who use two very similar languages (e.g., German and English) experience more cross-linguistic activation on a daily basis than bilinguals who use two very different languages (e.g., Chinese and English; Dijkstra & van Heuven, 2002). If same-script bilinguals experience more cross-linguistic activation on a daily basis than different-script bilinguals, they should presumably need to recruit their executive control skills to a greater extent and should therefore show larger bilingual advantages. Script similarity is therefore an important factor, which could potentially affect the magnitude of the bilingual cognitive advantage. However, this variable is rarely considered in bilingualism research.

A handful of studies have explicitly investigated the factor of script in bilingual cognitive control (Bialystok et al., 2005; Linck et al., 2008). Bialystok et al. (2005) tested two groups of bilinguals on the Simon task, one French-English and one Cantonese-English (although they did not explain why they tested two different groups nor why those particular languages were chosen). Behaviorally, the monolinguals and French-English bilinguals did not differ, but the Cantonese-English bilinguals demonstrated a global RT advantage, indicating that different-script bilinguals outperformed same-script bilinguals. However, Bialystok et al. (2005) did not interpret this difference between bilingual groups beyond attributing it to sampling variability due to the small number of participants. Linck et al. (2008) hypothesized that different-script bilinguals are able to use script as a cue to restrict lexical selection (Guo et al., 2005; Hoshino & Kroll, 2008), whereas same-script bilinguals cannot use this strategy and must rely on executive control to manage linguistic competition. They therefore predicted greater inhibitory control abilities for same-script bilinguals. Testing Japanese-English and Spanish-English bilinguals on a Simon task, Linck et al. (2008) found no overall differences in Simon effects between the groups. However, when analyzing only the bilinguals tested in an L2 context, different-script bilinguals showed greater inhibitory control abilities. As reported in Bialystok et al. (2005), this again suggests that different-script bilinguals experience superior executive control. Linck et al. (2005, 2008) proposed that this effect was due to language-switching frequency: as Japanese-English bilinguals switch languages within discourse (called "code-switching") less often, they demonstrated greater language control, whereas Spanish-English bilinguals, who code-switch more frequently, have less experience in language inhibition. Coderre & van Heuven (2014b) investigated this issue by testing three groups of bilinguals whose languages had varying overlap of script: German-English (high overlap), Polish-English (moderate overlap), and Arabic-English (low overlap). Along with a group of English monolinguals, these bilinguals performed a Stroop task with SOA manipulation (see Section 4.3.2.1). In contrast to Bialystok et al. (2005) and Linck et al. (2008), Coderre & van Heuven (2014b) found evidence for more effective domain-general executive control in similar-script bilinguals, which was attributed to greater cross-linguistic activation as a result of greater script overlap, and therefore an increase in the daily demands on cognitive control. The influences of script similarity have not been extensively tested in the bilingual cognitive control literature, and discrepancies still exist; however, this is an interesting avenue for future research and a potentially important factor to consider in future studies.

4.4 Implications for Intercultural Relations

This chapter has provided a broad overview of the fascinating situation of bilingualism, discussing the advances that have been made in the past few decades regarding how and where two languages are represented in the brain, how they are controlled and interact, and how linguistic properties such as script similarity can influence the neural and cognitive representations of languages. Also discussed were the cognitive effects of bilingualism, both detrimental, such as delayed language processing, and beneficial, such as enhanced cognitive control.

The past few decades of research into the cognitive effects of bilingualism have illustrated that a culture's language(s) can shape the way its people speak, think, and learn. A better understanding of these effects has the potential to improve intercultural communication and understanding. For example, teachers of international students who are learning in their L2 might ease their students' processing load by speaking slightly more slowly during lectures, to allow for delayed language processing in the L2 to catch up. People listening to a foreign coworker's grammatical mistakes in English would understand that it is not necessarily that their English is poor but that the grammatical rules of their L1 may be intruding. Understanding how much effort goes into processing a second language and controlling multiple languages on a daily basis would increase empathy between members of different cultures. Bilingualism research also has the potential to influence how foreign languages are taught in the classroom. For example, given the specific linguistic properties of the language being taught, instructors might focus a foreign-language curriculum more toward an assimilation or accommodation strategy (see Section 4.2.4.1).

One example of how bilingualism research is impacting education is translanguaging. Translanguaging consists of using a bilingual's two languages in the same context to enhance learning and retention. This concept has been mainly discussed in the context of bilingual education to describe the integration of languages in the bilingual classroom (e.g., Creese & Blackledge, 2010). However, recent experimental investigations have been made to understand its neuropsychological underpinnings and to test its effects on learning more systematically (Beres, Jones, Boutonnet, Davis, & Thierry, 2013). This research is based on the idea that before language can be used successfully, it must be understood; applications of translanguaging to learning thus focus on the dual use of two languages to solidify new concepts. Beres et al. (2013) taught English-Welsh bilinguals to learn the names and definitions of novel objects. In the monolingual context, they were asked to say the name of the new object in the same language as the definition provided; in the translanguaging context, they were asked to produce the name in their other language. Behaviorally, learning occurred slightly faster in the translanguaging context, although this was not statistically significant. However, the authors also showed ERP evidence of a benefit of translanguaging, such that the N400 effect was reduced upon presentation of a word that was learned in the translanguaging context versus the monolingual context, suggesting a facilitation of semantic integration. These ERP effects also persisted upon follow-up testing 2–4 weeks later. Translanguaging is an intriguing new line of research, especially as code-switching (switching between two languages in conversation) can be a contentious and censured act in the bilingual classroom (see the review in Creese & Blackledge, 2010). Yet this research suggests that code-switching and the integrated use of both languages actually have beneficial effects on learning. Code-switching and translanguaging may also have additional social and pedagogic effects on classroom relationships, communication, and performance.

Moving beyond the basic construction and implementation of language, and even beyond the levels of code-switching and translanguaging, Creese and Blackledge (2010) write that "within every utterance there are traces of the social, political, and historical forces that have shaped it" (p. 106). The concept of *heteroglossia* takes into account these social and historic contexts of language use (Bailey, 2007) and brings an additional aspect of intercultural relations to the study of language and multiculturalism. The study of bilingualism thus extends from the level of the underlying neural substrates to the overarching influence of super-linguistic factors, providing a framework for understanding how humans, from every different culture of the world, communicate with each other.

Conclusions

In one of the first experimental forays into the cognitive effects of bilingualism, Peal and Lambert (1962) wrote of their bilingual population: "The picture that emerges of the French-English bilingual in Montreal is that of a youngster whose wider experiences in two cultures have given him advantages which a monolingual does not enjoy" (p. 20). Notice the choice of words here: it is not experiences in two *languages,* but experiences in two *cultures.* Bilingualism is a multifaceted and dynamic experience, which is not limited to linguistic influences. The impact of being fluent in two cultures, as well as two languages, may confer additional cognitive changes that have yet to be explored. This fascinating field is still in its infancy, with a vast number of questions still to be addressed. With the advent of new and advancing neuroimaging techniques, our understanding of the cognitive effects of bilingualism will inevitably expand in the future. What is clear from the extant research is that the art of managing multiple languages and bridging multiple cultures leads to a range of cognitive, neural, and structural effects on the brain and affects everyday life in intricate and widespread ways.

References

Abutalebi, J. (2008). Neural aspects of second language representation and language control. *Acta Psychologica, 128*(3), 466–478.

Abutalebi, J., Annoni, J.-M., Zimine, I., Pegna, A. J., Seghier, M. L., Lee-Jahnke, H., et al. (2008). Language control and lexical competition in bilinguals: An event-related fMRI study. *Cerebral Cortex, 18*(7), 1496–1505.

Abutalebi, J., Brambati, S. M., Annoni, J.-M., Moro, A., Cappa, S. F., & Perani, D. (2007). The neural cost of the auditory perception of language switches: An event-related functional magnetic resonance imaging study in bilinguals. *The Journal of Neuroscience, 27*(50), 13762–13769.

Abutalebi, J., Della Rosa, P. A., Green, D. W., Hernandez, M., Scifo, P., Keim, R., et al. (2012). Bilingualism tunes the anterior cingulate cortex for conflict monitoring. *Cerebral Cortex, 22*(9), 2076–2086.

Abutalebi, J., & Green, D. W. (2007). Bilingual language production: The neurocognition of language representation and control. *Journal of Neurolinguistics, 20*, 242–275.

Abutalebi, J., & Green, D. W. (2008). Control mechanisms in bilingual language production: Neural evidence from language switching studies. *Language and Cognitive Processes, 23*(4), 557–582.

Allen, D. B., & Conklin, K. (2013). Cross-linguistic similarity and task demands in Japanese-English bilingual processing. *PLoS One, 8*(8), e72631.

Appelbaum, L. G., Liotti, M., Perez, R., Fox, S. P., & Woldorff, M. G. (2009). The temporal dynamics of implicit processing of non-letter, letter, and word-forms in the human visual cortex. *Frontiers in Human Neuroscience, 3*(56), 1–11.

Ardal, S., Donald, M. W., Meuter, R., Muldrew, S., & Luce, M. (1990). Brain responses to semantic incongruity in bilinguals. *Brain and Language, 39*(2), 187–205.

Bailey, B. (2007). Heteroglossia and boundaries. In M. Heller (Ed.), *Bilingualism: A social approach* (pp. 257–276). Basingstoke, UK: Palgrave.

Bar-Kochva, I. (2011). Does processing a shallow and a deep orthography produce different brain activity patterns? An ERP study conducted in Hebrew. *Developmental Neuropsychology, 36*(7), 933–938.

Bartolotti, J., Marian, V., Schroeder, S. R., & Shook, A. (2011). Bilingualism and inhibitory control influence statistical learning of novel word forms. *Frontiers in Psychology, 2*(324), 1–10.

Bates, E., D'Amico, S., Jacobsen, T., Székely, A., Andonova, E., Devescovi, A., et al. (2003). Timed picture naming in seven languages. *Psychonomic Bulletin & Review, 10*(2), 344–380.

Bentin, S., Mouchetant-Rostaing, Y., Giard, M. H., Echallier, J. F., & Pernier, J. (1999). ERP manifestations of processing printed words at different psycholinguistic levels: Time course and scalp distribution. *Journal of Cognitive Neuroscience, 11*(3), 235–260.

Beres, A., Jones, M., Boutonnet, B., Davis, N., & Thierry, G. (2013). Translanguaging: Boosting the acquisition of new knowledge using bilingualism. Talk presented at the Society for the Neurobiology of Language, San Diego, CA.

Bialystok, E. (1999). Cognitive complexity and attentional control in the bilingual mind. *Child Development, 70*(3), 636–644.

Bialystok, E. (2006). Effect of bilingualism and computer video game experience on the Simon task. *Canadian Journal of Experimental Psychology, 60*(1), 68–79.

Bialystok, E. (2009). Bilingualism: The good, the bad, and the indifferent. *Bilingualism: Language and Cognition, 12*(01), 3–11.

Bialystok, E. (2010). Global-local and trail-making tasks by monolingual and bilingual children: Beyond inhibition. *Developmental Psychology, 46*(1), 93–105.

Bialystok, E. (2011). Reshaping the mind: The benefits of bilingualism. *Canadian Journal of Experimental Psychology, 65*(4), 229–235.

Bialystok, E., Craik, F. I. M., Grady, C., Chau, W., Ishii, R., Gunji, A., et al. (2005). Effect of bilingualism on cognitive control in the Simon task: Evidence from MEG. *NeuroImage, 24*(1), 40–49.

Bialystok, E., Craik, F. I. M., Green, D. W., & Gollan, T. H. (2009). Bilingual minds. *Psychological Science in the Public Interest, 10*(3), 89–129.

Bialystok, E., Craik, F. I. M., Klein, R., & Viswanathan, M. (2004). Bilingualism, aging, and cognitive control: Evidence from the Simon task. *Psychology and Aging, 19*(2), 290–303.

Bialystok, E., Craik, F. I. M., & Luk, G. (2008). Cognitive control and lexical access in younger and older bilinguals. *Journal of Experimental Psychology: Learning, Memory, and Cognition, 34*(4), 859–873.

Bialystok, E., Craik, F. I. M., & Ruocco, A. C. (2006). Dual-modality monitoring in a classification task: The effects of bilingualism and ageing. *Quarterly Journal of Experimental Psychology, 59*(11), 1968–1983.

Bialystok, E., Craik, F. I. M., & Ryan, J. (2006). Executive control in a modified antisaccade task: Effects of aging and bilingualism. *Journal of Experimental Psychology: Learning, Memory, and Cognition, 32*(6), 1341–1354.

Bialystok, E., & Depape, A.-M. (2009). Musical expertise, bilingualism, and executive functioning. *Journal of Experimental Psychology: Human Perception and Performance, 35*(2), 565–574.

Bialystok, E., & Feng, X. (2009). Language proficiency and executive control in proactive interference: Evidence from monolingual and bilingual children and adults. *Brain and Language, 109*(2–3), 93–100.

Bialystok, E., & Martin, M. M. (2004). Attention and inhibition in bilingual children: Evidence from the dimensional change card sort task. *Developmental Science, 7*(3), 325–339.

Bialystok, E., & Senman, L. (2004). Executive processes in appearance-reality tasks: The role of inhibition of attention and symbolic representation. *Child Development, 75*(2), 562–579.

Bialystok, E., & Shapero, D. (2005). Ambiguous benefits: The effect of bilingualism on reversing ambiguous figures. *Developmental Science, 8*(6), 595–604.

Bialystok, E., & Viswanathan, M. (2009). Components of executive control with advantages for bilingual children in two cultures. *Cognition, 112*(3), 494–500.

Bick, A. S., Goelman, G., & Frost, R. (2011). Hebrew brain vs. English brain: Language modulates the way it is processed. *Journal of Cognitive Neuroscience, 23*(9), 2280–2290.

Binder, J. R., Frost, J. A., Hammeke, T. A., Cox, R. W., Rao, S. M., & Prieto, T. (1997). Human brain language areas identified by functional magnetic resonance imaging. *The Journal of Neuroscience, 17*(1), 353–362.

Bloem, I., & La Heij, W. (2003). Semantic facilitation and semantic interference in word translation: Implications for models of lexical access in language production. *Journal of Memory and Language, 48*(3), 468–488.

Bloem, I., van den Boogaard, S., & La Heij, W. (2004). Semantic facilitation and semantic interference in language production: Further evidence for the conceptual selection model of lexical access. *Journal of Memory and Language, 51*(2), 307–323.

Bolger, D. J., Perfetti, C. A., & Schneider, W. (2005). Cross-cultural effect on the brain revisited: Universal structures plus writing system variation. *Human Brain Mapping, 25*(1), 92–104.

Botvinick, M. M., Braver, T. S., Barch, D. M., Carter, C. S., & Cohen, J. D. (2001). Conflict monitoring and cognitive control. *Psychological Review, 108*(3), 624–652.

Botvinick, M. M., Cohen, J. D., & Carter, C. S. (2004). Conflict monitoring and anterior cingulate cortex: An update. *Trends in Cognitive Sciences, 8*(12), 539–546.

Bowden, H. W., Steinhauer, K., Sanz, C., & Ullman, M. T. (2013). Native-like brain processing of syntax can be attained by university foreign language learners. *Neuropsychologia, 51*(13), 2492–2511.

Briellmann, R. S., Saling, M. M., Connell, A. B., Waites, A. B., Abbott, D. F., & Jackson, G. D. (2004). A high-field functional MRI study of quadri-lingual subjects. *Brain and Language, 89*, 531–542.

Brysbaert, M., & Duyck, W. (2010). Is it time to leave behind the Revised Hierarchical Model of bilingual language processing after fifteen years of service? *Bilingualism: Language and Cognition, 13*(03), 359–371.

Bunge, S. A., Hazeltine, E., Scanlon, M. D., Rosen, A. C., & Gabrieli, J. D. E. (2002). Dissociable contributions of prefrontal and parietal cortices to response selection. *NeuroImage, 17*(3), 1562–1571.

Calabria, M., Hernández, M., Branzi, F. M., & Costa, A. (2012). Qualitative differences between bilingual language control and executive control: Evidence from task-switching. *Frontiers in Psychology, 2*(399), 1–10.
Cao, F., Tao, R., Liu, L., Perfetti, C., & Booth, J. (2013). High proficiency in a second language is characterized by greater involvement of the first language network: Evidence from Chinese learners of English. *Journal of Cognitive Neuroscience, 25*(10), 1649–1663.
Carlson, S. M., & Meltzoff, A. N. (2008). Bilingual experience and executive functioning in young children. *Developmental Science, 11*(2), 282–298.
Chee, M. W., Hon, N., Lee, H. L., & Soon, C. S. (2001). Relative language proficiency modulates BOLD signal change when bilinguals perform semantic judgments. *NeuroImage, 13*, 1155–1163.
Chen, B., Zhou, H., Gao, Y., & Dunlap, S. (2014). Cross-language translation priming asymmetry with Chinese-English bilinguals: A test of the sense model. *Journal of Psycholinguistic Research, 43*, 225–240.
Christoffels, I. K., Firk, C., & Schiller, N. O. (2007). Bilingual language control: An event-related brain potential study. *Brain Research, 1147*, 192–208.
Chua, F. K. (1999). Phonological recoding in Chinese logograph recognition. *Journal of Experimental Psychology: Learning, Memory, and Cognition, 25*(4), 876–891.
Coderre, E. L., Filippi, C. G., Newhouse, P. A., & Dumas, J. A. (2008). The Stroop effect in kana and kanji scripts in native Japanese speakers: An fMRI study. *Brain and Language, 107*(2), 124–132.
Coderre, E.L., & van Heuven, W.J.B. (2014a). Electrophysiological explorations of the bilingual advantage: evidence from a stroop task. *PloS One, 9*(7), e103424.
Coderre, E.L., & van Heuven, W.J.B. (2014b). The effect of script similarity on executive control in bilinguals. *Frontiers in Psychology, 5*(1070), 1–16.
Cohen, L., Dehaene, S., Naccache, L., Lehéricy, S., Dehaene-Lambertz, G., Hénaff, M. A., et al. (2000). The visual word form area: Spatial and temporal characterization of an initial stage of reading in normal subjects and posterior split-brain patients. *Brain, 123*, 291–307.
Colomé, À. (2001). Lexical activation in bilinguals' speech production: Language-specific or language-independent? *Journal of Memory and Language, 45*(4), 721–736.
Consonni, M., Cafiero, R., Marin, D., Tettamanti, M., Iadanza, A., Fabbro, F., et al. (2013). Neural convergence for language comprehension and grammatical class production in highly proficient bilinguals is independent of age of acquisition. *Cortex, 49*(5), 1252–1258.
Corbetta, M., Miezin, F. M., Shulman, G. L., & Petersen, S. E. (1993). A PET study of visuospatial attention. *The Journal of Neuroscience, 13*(3), 1202–1226.
Costa, A., Caramazza, A., & Sebastian-Galles, N. (2000). The cognate facilitation effect: Implications for models of lexical access. *Journal of Experimental Psychology: Learning, Memory, and Cognition, 26*(5), 1283–1296.
Costa, A., Hernández, M., Costa-Faidella, J., & Sebastián-Gallés, N. (2009). On the bilingual advantage in conflict processing: Now you see it, now you don't. *Cognition, 113*(2), 135–149.
Costa, A., Hernández, M., & Sebastián-Gallés, N. (2008). Bilingualism aids conflict resolution: Evidence from the ANT task. *Cognition, 106*(1), 59–86.
Costa, A., Santesteban, M., & Ivanova, I. (2006). How do highly proficient bilinguals control their lexicalization process? Inhibitory and language-specific selection mechanisms are both functional. *Journal of Experimental Psychology: Learning, Memory, and Cognition, 32*(5), 1057–1074.
Costafreda, S. G., Fu, C. H. Y., Lee, L., Everitt, B., Brammer, M. J., & David, A. S. (2006). A systematic review and quantitative appraisal of fMRI studies of verbal fluency: Role of the left inferior frontal gyrus. *Human Brain Mapping, 27*, 799–810.
Craik, F. I. M., & Bialystok, E. (2006). Cognition through the lifespan: Mechanisms of change. *Trends in Cognitive Sciences, 10*(3), 131–138.
Creese, A., & Blackledge, A. (2010). Translanguaging in the bilingual classroom: A pedagogy for learning and teaching? *The Modern Language Journal, 94*, 103–115.
Crinion, J., Turner, R., Grogan, A., Hanakawa, T., Noppeney, U., Devlin, J. T., et al. (2006). Language control in the bilingual brain. *Science, 312*(5779), 1537–1540.

Culham, J. C., & Kanwisher, N. G. (2001). Neuroimaging of cognitive functions in human parietal cortex. *Current Opinion in Neurobiology, 11*(2), 157–163.

Das, T., Padakannaya, P., Pugh, K. R., & Singh, N. C. (2011). Neuroimaging reveals dual routes to reading in simultaneous proficient readers of two orthographies. *NeuroImage, 54*(2), 1476–1487.

de Bleser, R., Dupont, P., Postler, J., Bormans, G., Speelman, D., Mortelmans, L., et al. (2003). The organisation of the bilingual lexicon: A PET study. *Journal of Neurolinguistics, 16*(4–5), 439–456.

de Groot, A. M. B., & Nas, G. (1991). Lexical representation of cognates and noncognates in compound bilinguals. *Journal of Memory and Language, 30*, 90–123.

Degani, T., & Tokowicz, N. (2010). Semantic ambiguity within and across languages: An integrative review. *Quarterly Journal of Experimental Psychology, 63*(7), 1266–1303.

Dehaene, S., Dupoux, E., Mehler, J., Cohen, L., Paulesu, E., Perani, D., et al. (1997). Anatomical variability in the cortical representation of first and second language. *Neuroreport, 8*(17), 3809–3815.

Dijkstra, T., Grainger, J., & van Heuven, W. J. B. (1999). Recognition of cognates and interlingual homographs: The neglected role of phonology. *Journal of Memory and Language, 41*(4), 496–518.

Dijkstra, T., & van Heuven, W. J. B. (2002). The architecture of the bilingual word recognition system: From identification to decision. *Bilingualism: Language and Cognition, 5*(03), 175–197.

Ding, G., Perry, C., Peng, D., Ma, L., Li, D., Xu, S., et al. (2003). Neural mechanisms underlying semantic and orthographic processing in Chinese-English bilinguals. *Neuroreport, 14*(12), 1557–1562.

Duñabeitia, J. A., Hernández, J. A., Antón, E., Macizo, P., Estévez, A., Fuentes, L. J., et al. (2014). The inhibitory advantage in bilingual children revisited: myth or reality? *Exerimental Psychology, 61*(3), 234–251.

Dussias, P. E., & Sagarra, N. (2007). The effect of exposure on syntactic parsing in Spanish–English bilinguals. *Bilingualism: Language and Cognition, 10*(01), 101–116.

Fan, J., Flombaum, J. I., McCandliss, B. D., Thomas, K. M., & Posner, M. I. (2003). Cognitive and brain consequences of conflict. *NeuroImage, 18*, 42–57.

Fan, J., McCandliss, B. D., Sommer, T., Raz, A., & Posner, M. I. (2002). Testing the efficiency and independence of attentional networks. *Journal of Cognitive Neuroscience, 14*(3), 340–347.

Ferstl, E., Neumann, J., Bogler, C., & Von Cramon, D. Y. (2008). The extended language network: A meta-analysis of neuroimaging studies on text comprehension. *Human Brain Mapping, 29*(5), 581–593.

Festman, J., Rodriguez-Fornells, A., & Münte, T. F. (2010). Individual differences in control of language interference in late bilinguals are mainly related to general executive abilities. *Behavioral and Brain Functions, 6*(5), 1–12.

Friederici, A. D., & Meyer, M. (2004). The brain knows the difference: Two types of grammatical violations. *Brain Research, 1000*(1–2), 72–77.

Frost, R. (1998). Toward a strong phonological theory of visual word recognition: True issues and false trails. *Psychological Bulletin, 123*(1), 71–99.

Gandour, J., Wong, D., Hsieh, L., Weinzapfel, B., Van Lancker, D., & Hutchins, G. D. (2000). A crosslinguistic PET study of tone perception. *Journal of Cognitive Neuroscience, 12*(1), 207–222.

Garbin, G., Sanjuan, A., Forn, C., Bustamante, J. C., Rodriguez-Pujadas, A., Belloch, V., et al. (2010). Bridging language and attention: Brain basis of the impact of bilingualism on cognitive control. *NeuroImage, 53*(4), 1272–1278.

García-Pentón, L., Pérez Fernández, A., Iturria-Medina, Y., Gillon-Dowens, M., & Carreiras, M. (2014). Anatomical connectivity changes in the bilingual brain. *NeuroImage, 84*, 495–504.

Gerard, L., & Scarborough, D. L. (1989). Language-specific lexical access of homographs by bilinguals. *Journal of Experimental Psychology: Learning, Memory, and Cognition, 15*(2), 305–315.

Gitelman, D. R., Nobre, A. C., Sonty, S., Parrish, T. B., & Mesulam, M.-M. (2005). Language network specializations: An analysis with parallel task designs and functional magnetic resonance imaging. *NeuroImage, 26*(4), 975–985.

Goetz, P. J. (2003). The effects of bilingualism on theory of mind development. *Bilingualism: Language and Cognition, 6*(1), 1–15.

Gollan, T. H., & Acenas, L. A. R. (2004). What is a TOT? Cognate and translation effects on tip-of-the-tongue states in Spanish-English and Tagalog-English bilinguals. *Journal of Experimental Psychology: Learning, Memory, and Cognition, 30*(1), 246–269.

Gollan, T. H., Forster, K. I., & Frost, R. (1997). Translation priming with different scripts: Masked priming with cognates and noncognates in Hebrew-English bilinguals. *Journal of Experimental Psychology: Learning, Memory, and Cognition, 23*(5), 1122–1239.

Gollan, T. H., Montoya, R. I., Cera, C., & Sandoval, T. C. (2008). More use almost always means a smaller frequency effect: Aging, bilingualism, and the weaker links hypothesis. *Journal of Memory and Language, 58*(3), 787–814.

Gollan, T. H., Montoya, R. I., Fennema-Notestine, C., & Morris, S. K. (2005). Bilingualism affects picture naming but not picture classification. *Memory & Cognition, 33*(7), 1220–1234.

Gollan, T. H., Montoya, R. I., & Bonanni, M. P. (2005). Proper names get stuck on bilingual and monolingual speakers' tip of the tongue equally often. *Neuropsychology, 19*(3), 278–287.

Gollan, T. H., Montoya, R. I., & Werner, G. A. (2002). Semantic and letter fluency in Spanish-English bilinguals. *Neuropsychology, 16*(4), 562–576.

Gollan, T. H., & Silverberg, N. B. (2001). Tip-of-the-tongue states in Hebrew-English bilinguals. *Bilingualism: Language and Cognition, 4*(1), 63–83.

Green, D. W. (1998). Mental control of the bilingual lexico-semantic system. *Bilingualism: Language and Cognition, 1*, 67–81.

Green, D. W., & Abutalebi, J. (2013). Language control in bilinguals: The adaptive control hypothesis. *Journal of Cognitive Psychology, 25*(5), 515–530.

Grossi, G., Savill, N., Thomas, E., & Thierry, G. (2010). Posterior N1 asymmetry to English and Welsh words in Early and Late English-Welsh bilinguals. *Biological Psychology, 85*(1), 124–133.

Guo, T., Liu, H., Misra, M., & Kroll, J. F. (2011). Local and global inhibition in bilingual word production: fMRI evidence from Chinese-English bilinguals. *NeuroImage, 56*(4), 2300–2309.

Guo, T., Ma, F., & Liu, F. (2013). An ERP study of inhibition of non-target languages in trilingual word production. *Brain and Language, 127*, 12–20.

Guo, T., Peng, D., & Liu, Y. (2005). The role of phonological activation in the visual semantic retrieval of Chinese characters. *Cognition, 98*(2), B21–B34.

Hahne, A. (2001). What's different in second-language processing? Evidence from event-related brain potentials. *Journal of Psycholinguistic Research, 30*(3), 251–266.

Hahne, A., & Friederici, A. D. (2001). Processing a second language: Late learners' comprehension mechanisms as revealed by event-related brain potentials. *Bilingualism: Language and Cognition, 4*(2), 123–141.

Hansen, L. (2001). Language attrition: The fate of the start. *Annual Review of Applied Linguistics, 21*, 60–73.

Hartsuiker, R. J., Pickering, M. J., & Veltkamp, E. (2004). Is syntax separate or shared between languages? Cross-linguistic syntactic priming in Spanish-English bilinguals. *Psychological Science, 15*(6), 409–414.

Hasegawa, M., Carpenter, P. A., & Just, M. A. (2002). An fMRI study of bilingual sentence comprehension and workload. *NeuroImage, 15*(3), 647–660.

Hernandez, A. E. (2009). Language switching in the bilingual brain: What's next? *Brain and Language, 109*, 133–140.

Hernández, M., Costa, A., Fuentes, L. J., Vivas, A. B., & Sebastián-Gallés, N. (2010). The impact of bilingualism on the executive control and orienting networks of attention. *Bilingualism: Language and Cognition, 13*(03), 315–325.

Hernández, M., Costa, A., & Humphreys, G. W. (2012). Escaping capture: Bilingualism modulates distraction from working memory. *Cognition, 122*, 37–50.

Hernandez, A. E., Dapretto, M., Mazziotta, J., & Bookheimer, S. (2001). Language switching and language representation in Spanish-English bilinguals: An fMRI study. *NeuroImage, 14*(2), 510–520.

Hernandez, A. E., Martinez, A., & Kohnert, K. (2000). In search of the language switch: An fMRI study of picture naming in Spanish-English bilinguals. *Brain and Language, 73*(3), 421–431.

Hernandez, A. E., & Meschyan, G. (2006). Executive function is necessary to enhance lexical processing in a less proficient L2: Evidence from fMRI during picture naming. *Bilingualism: Language and Cognition, 9*(2), 177–188.

Hervais-Adelman, A. G., Moser-Mercer, B., & Golestani, N. (2011). Executive control of language in the bilingual brain: Integrating the evidence from neuroimaging to neuropsychology. *Frontiers in Psychology, 2*(234), 1–8.

Hilchey, M. D., & Klein, R. M. (2011). Are there bilingual advantages on nonlinguistic interference tasks? Implications for the plasticity of executive control processes. *Psychonomic Bulletin & Review, 18*(4), 625–658.

Horwitz, B., Rumsey, J. M., & Donohue, B. C. (1998). Functional connectivity of the angular gyrus in normal reading and dyslexia. *Proceedings of the National Academy of Sciences, 95*(15), 8939–8944.

Hoshino, N., & Kroll, J. F. (2008). Cognate effects in picture naming: Does cross-language activation survive a change of script? *Cognition, 106*(1), 501–511.

Illes, J., Francis, W. S., Desmond, J. E., Gabrieli, J. D., Glover, G. H., Poldrack, R., et al. (1999). Convergent cortical representation of semantic processing in bilinguals. *Brain and Language, 70*(3), 347–363.

Indefrey, P. (2006). A meta-analysis of hemodynamic studies on first and second language processing: Which suggested differences can we trust and what do they mean? *Language Learning, 56*, 279–304.

Ivanova, I., & Costa, A. (2008). Does bilingualism hamper lexical access in speech production? *Acta Psychologica, 127*(2), 277–288.

Kawabata Duncan, K. J., Twomey, T., Parker Jones, O., Seghier, M. L., Haji, T., Sakai, K., et al. (2014). Inter- and intrahemispheric connectivity differences when reading Japanese Kanji and Hiragana. *Cerebral Cortex, 24*(6), 1601–1608.

Kerkhofs, R., Dijkstra, T., Chwilla, D. J., & de Bruijn, E. R. A. (2006). Testing a model for bilingual semantic priming with interlingual homographs: RT and N400 effects. *Brain Research, 1068*(1), 170–183.

Kim, J., & Davis, C. (2003). Task effects in masked cross-script translation and phonological priming. *Journal of Memory and Language, 49*(4), 484–499.

Kim, K. H., Relkin, N. R., Lee, K. M., & Hirsch, J. (1997). Distinct cortical areas associated with native and second languages. *Nature, 388*, 171–174.

Klein, D., Milner, B., Zatorre, R. J., Zhao, V., & Nikelski, J. (1999). Cerebral organization in bilinguals: A PET study of Chinese-English verb generation. *NeuroReport, 10*, 2841–2846.

Koda, K. (1996). L2 word recognition research: A critical review. *The Modern Language Journal, 80*, 450–460.

Kohnert, K. J., Hernandez, A. E., & Bates, E. (1998). Bilingual performance on the boston naming test: Preliminary norms in Spanish and English. *Brain and Language, 65*(3), 422–440.

Kousaie, S., & Phillips, N. A. (2012). Conflict monitoring and resolution: Are two languages better than one? Evidence from reaction time and event-related brain potentials. *Brain Research, 1446*, 71–90.

Kovelman, I., Baker, S. A., & Petitto, L.-A. (2008). Bilingual and monolingual brains compared: a functional magnetic resonance imaging investigation of syntactic processing and a possible "neural signature" of bilingualism. *Journal of Cognitive Neuroscience, 20*(1), 153–169.

Kovelman, I., Shalinsky, M. H., Berens, M. S., & Petitto, L. A. (2008). Shining new light on the brain's "Bilingual Signature": A functional near infrared spectroscopy investigation of semantic processing. *NeuroImage, 39*(3), 1457–1471.

Koyama, M. S., Stein, J. F., Stoodley, C. J., & Hansen, P. C. (2013). Cerebral mechanisms for different second language writing systems. *Neuropsychologia, 51*(11), 2261–2270.

Kroll, J. F., & Bialystok, E. (2013). Understanding the consequences of bilingualism for language processing and cognition. *Journal of Cognitive Psychology, 25*(5), 497–514.
Kroll, J. F., Bobb, S. C., & Wodniecka, Z. (2006). Language selectivity is the exception, not the rule: Arguments against a fixed locus of language selection in bilingual speech. *Bilingualism: Language and Cognition, 9*(02), 119–135.
Kroll, J. F., Dussias, P. E., Bogulski, C. A., & Valdes Kroff, J. R. (2012). Juggling two languages in one mind: What bilinguals tell us about language processing and its consequences for cognition. In B. H. Ross (Ed.), *The psychology of learning and motivation* (Vol. 56, pp. 229–262). San Diego, CA, USA: Elsevier.
Kuipers, J.-R., & Thierry, G. (2010). Event-related brain potentials reveal the time-course of language change detection in early bilinguals. *NeuroImage, 50*(4), 1633–1638.
Kutas, M., & Hillyard, S. (1980). Reading senseless sentences: Brain potentials reflect semantic incongruity. *Science, 207*(4427), 203–205.
Lallier, M., Carreiras, M., Tainturier, M.-J., Savill, N., & Thierry, G. (2013). Orthographic transparency modulates the grain size of orthographic processing: Behavioral and ERP evidence from bilingualism. *Brain Research, 10*, 47–60.
Lau, E. F., Phillips, C., & Poeppel, D. (2008). A cortical network for semantics: (de)constructing the N400. *Nature Reviews Neuroscience, 9*(12), 920–933.
Lehtonen, M., & Laine, M. (2003). How word frequency affects morphological processing in monolinguals and bilinguals. *Bilingualism: Language and Cognition, 6*(3), 213–225.
Lehtonen, M., Laine, M., Niemi, J., Thomsen, T., Vorobyev, V. A., & Hugdahl, K. (2005). Brain correlates of sentence translation in Finnish-Norwegian bilinguals. *Neuroreport, 16*(6), 607–610.
Lin, S. E., Chen, H. C., Zhao, J., Li, S., He, S., & Weng, X. C. (2011). Left-lateralized N170 response to unpronounceable pseudo but not false Chinese characters-the key role of orthography. *Neuroscience, 190*, 200–206.
Linck, J. A., Hoshino, N., & Kroll, J. F. (2008). Cross-language lexical processes and inhibitory control. *The Mental Lexicon, 3*(3), 349–374.
Linck, J. A., Kroll, J. F., & Sunderman, G. (2009). Losing access to the native language while immersed in a second language: Evidence for the role of inhibition in second-language learning. *Psychological Science, 20*(12), 1507–1515.
Liu, Y., & Perfetti, C. A. (2003). The time course of brain activity in reading English and Chinese: An ERP study of Chinese bilinguals. *Human Brain Mapping, 18*(3), 167–175.
Liu, Y., Perfetti, C. A., & Hart, L. (2003). ERP evidence for the time course of graphic, phonological, and semantic information in Chinese meaning and pronunciation decisions. *Journal of Experimental Psychology: Learning, Memory, and Cognition, 29*(6), 1231–1247.
Lorenzen, B., & Murray, L. L. (2008). Bilingual aphasia: A theoretical and clinical review. *American Journal of Speech-Language Pathology, 17*(3), 299–317.
Luck, S., Heinze, H., Mangun, G., & Hillyard, S. (1990). Visual event-related potentials index focused attention within bilateral stimulus arrays. II. Functional dissociation of P1 and N1 components. *Electroencephalography and Clinical Neurophysiology, 75*, 528–542.
Luk, G., Anderson, J. A. E., Craik, F. I. M., Grady, C., & Bialystok, E. (2010). Distinct neural correlates for two types of inhibition in bilinguals: Response inhibition versus interference suppression. *Brain and Cognition, 74*(3), 347–357.
Luk, G., & Bialystok, E. (2013). Bilingualism is not a categorical variable: Interaction between language proficiency and usage. *Journal of Cognitive Psychology, 25*(5), 605–621.
Luk, G., Green, D. W., Abutalebi, J., & Grady, C. (2012). Cognitive control for language switching in bilinguals: A quantitative meta-analysis of functional neuroimaging studies. *Language and Cognitive Processes, 27*(10), 1479–1488.
Macnamara, J., & Kushnir, S. L. (1971). Linguistic independence of bilinguals: The input switch. *Journal of Verbal Learning and Verbal Behavior, 87*, 480–487.
Magezi, D. A., Khateb, A., Mouthon, M., Spierer, L., & Annoni, J.-M. (2012). Cognitive control of language production in bilinguals involves a partly independent process within the domain-general cognitive control network: Evidence from task-switching and electrical brain activity. *Brain and Language, 122*(1), 55–63.

Mahendra, N., Plante, E., Magloire, J., Milman, L., & Trouard, T. P. (2003). fMRI variability and the localization of languages in the bilingual brain. *NeuroReport, 14*(9), 1225–1228.
Mangun, G. R., Buonocore, M. H., Girelli, M., & Jha, A. P. (1998). ERP and fMRI measures of visual spatial selective attention. *Human Brain Mapping, 6*(5–6), 383–389.
Marian, V., Spivey, M., & Hirsch, J. (2003). Shared and separate systems in bilingual language processing: Converging evidence from eyetracking and brain imaging. *Brain and Language, 86*(1), 70–82.
Martin-Rhee, M. M., & Bialystok, E. (2008). The development of two types of inhibitory control in monolingual and bilingual children. *Bilingualism: Language and Cognition, 11*(1), 81–93.
Marzecová, A., Bukowski, M., Correa, Á., Boros, M., Lupiáñez, J., & Wodniecka, Z. (2013). Tracing the bilingual advantage in cognitive control: The role of flexibility in temporal preparation and category switching. *Journal of Cognitive Psychology, 25*(5), 586–604.
Maurer, U., Brem, S., Bucher, K., & Brandeis, D. (2005). Emerging neurophysiological specialization for letter strings. *Journal of Cognitive Neuroscience, 17*(10), 1532–1552.
McCandliss, B. D., Cohen, L., & Dehaene, S. (2003). The visual word form area: expertise for reading in the fusiform gyrus. *Trends in Cognitive Sciences, 7*(7), 293–299.
Mechelli, A., Crinion, J., Noppeney, U., O'Doherty, J., Ashburner, J., Frackowiak, R., et al. (2004). Structural plasticity in the bilingual brain. *Nature, 431*, 757.
Melara, R. D., Wang, H., Vu, K.-P. L., & Proctor, R. W. (2008). Attentional origins of the Simon effect: Behavioral and electrophysiological evidence. *Brain Research, 1215*, 147–159.
Melcher, T., & Gruber, O. (2009). Decomposing interference during Stroop performance into different conflict factors: An event-related fMRI study. *Cortex, 45*(2), 189–200.
Meschyan, G., & Hernandez, A. E. (2006). Impact of language proficiency and orthographic transparency on bilingual word reading: An fMRI investigation. *NeuroImage, 29*(4), 1135–1140.
Meuter, R. F. I., & Allport, A. (1999). Bilingual language switching in naming: Asymmetrical costs of language selection. *Journal of Memory and Language, 40*, 25–40.
Midgley, K., Holcomb, P., van Heuven, W. J. B., & Grainger, J. (2008). An electrophysiological investigation of cross-language effects of orthographic neighborhood. *Brain Research, 1246*, 123–135.
Milham, M. P., Banich, M. T., & Barad, V. (2003). Competition for priority in processing increases prefrontal cortex's involvement in top-down control: An event-related fMRI study of the Stroop task. *Cognitive Brain Research, 17*, 212–222.
Montant, M., Schön, D., Anton, J.-L., & Ziegler, J. C. (2011). Orthographic contamination of Broca's area. *Frontiers in Psychology, 2*(378), 1–10.
Morales, J., Gómez-Ariza, C. J., & Bajo, T.M. (2013). Dual mechanisms of cognitive control in bilinguals and monolinguals. *Journal of Cognitive Psychology, 25*(5), 531–546.
Moreno, E. M., & Kutas, M. (2005). Processing semantic anomalies in two languages: An electrophysiological exploration in both languages of Spanish-English bilinguals. *Cognitive Brain Research, 22*(2), 205–220.
Moreno, E. M., Rodriguez-Fornells, A., & Laine, M. (2008). Event-related potentials (ERPs) in the study of bilingual language processing. *Journal of Neurolinguistics, 21*(6), 477–508.
Morton, J. B., & Harper, S. N. (2007). What did Simon say? Revisiting the bilingual advantage. *Developmental Science, 10*(6), 719–726.
Nee, D. E., Wager, T. D., & Jonides, J. (2007). Interference resolution: Insights from a meta-analysis of neuroimaging tasks. *Cognitive, Affective & Behavioral Neuroscience, 7*(1), 1–17.
Nelson, J. R., Liu, Y., Fiez, J., & Perfetti, C. A. (2009). Assimilation and accommodation patterns in ventral occipitotemporal cortex in learning a second writing system. *Human Brain Mapping, 30*(3), 810–820.
Newman, A. J., Bavelier, D., Corina, D., Jezzard, P., & Neville, H. J. (2002). A critical period for right hemisphere recruitment in American Sign Language processing. *Nature Neuroscience, 5*(1), 76–80.
Newman, A. J., Tremblay, A., Nichols, E., Neville, H. J., & Ullman, M. T. (2012). The influence of language proficiency on lexical semantic processing in native and late learners of English. *Journal of Cognitive Neuroscience, 24*(5), 1205–1223.

Niendam, T. A., Laird, A. R., Ray, K. L., Dean, Y. M., Glahn, D. C., & Carter, C. S. (2012). Meta-analytic evidence for a superordinate cognitive control network subserving diverse executive functions. *Cognitive, Affective & Behavioral Neuroscience, 12*(2), 241–268.

Novick, J. M., Kan, I. P., Trueswell, J. C., & Thompson-Schill, S. L. (2009). A case for conflict across multiple domains: Memory and language impairments following damage to ventrolateral prefrontal cortex. *Cognitive Neuropsychology, 26*(6), 527–567.

Novick, J. M., Trueswell, J. C., & Thompson-Schill, S. L. (2005). Cognitive control and parsing: Reexamining the role of Broca's area in sentence comprehension. *Cognitive, Affective, & Behavioral Neuroscience, 5*(3), 263–281.

Paap, K. R., & Greenberg, Z. I. (2013). There is no coherent evidence for a bilingual advantage in executive processing. *Cognitive Psychology, (66)*, 232–258.

Parker Jones, O., Green, D. W., Grogan, A., Pliatsikas, C., Filippopolitis, K., Ali, N., et al. (2011). Where, when and why brain activation differs for bilinguals and monolinguals during picture naming and reading aloud. *Cerebral Cortex, 22*(4), 892–902.

Paulesu, E., McCrory, E., Fazio, F., Menoncello, L., Brunswick, N., Cappa, S. F., et al. (2000). A cultural effect on brain function. *Nature Neuroscience, 3*(1), 91–96.

Peal, E., & Lambert, W. (1962). The relation of bilingualism to intelligence. *Psychological Monographs: General and Applied, 76*(27), 1–23.

Penniello, M.-J., Lambert, J., Eustache, F., Petit-Taboué, M. C., Barré, L., Viader, F., et al. (1995). A PET study of the functional neuroanatomy of writing impairment in Alzheimer's disease: The role of the left supramarginal and left angular gyri. *Brain, 118*(Pt 3), 697–706.

Perani, D., Abutalebi, J., Paulesu, E., Brambati, S., Scifo, P., Cappa, S. F., et al. (2003). The role of age of acquisition and language usage in early, high-proficient bilinguals: An fMRI study during verbal fluency. *Human Brain Mapping, 19*(3), 170–182.

Perani, D., Dehaene, S., Grassi, F., Cohen, L., Cappa, S. F., Dupoux, E., et al. (1996). Brain processing of native and foreign languages. *NeuroReport, 7*, 2349–2444.

Perani, D., Paulesu, E., Galles, N. S., Dupoux, E., Dehaene, S., Bettinardi, V., et al. (1998). The bilingual brain: Proficiency and age of acquisition of the second language. *Brain, 121*, 1841–1852.

Perfetti, C. A., Liu, Y., Fiez, J., Nelson, J., Bolger, D. J., & Tan, L. H. (2007). Reading in two writing systems: Accommodation and assimilation of the brain's reading network. *Bilingualism: Language and Cognition, 10*(02), 131–146.

Perfetti, C. A., Liu, Y., & Tan, L. H. (2005). The lexical constituency model: Some implications of research on Chinese for general theories of reading. *Psychological Review, 112*(1), 43–59.

Peterson, B. S., Kane, M. J., Alexander, G. M., Lacadie, C., Skudlarski, P., Leung, H. C., et al. (2002). An event-related functional MRI study comparing interference effects in the Simon and Stroop tasks. *Cognitive Brain Research, 13*(3), 427–440.

Peterson, B. S., Skudlarski, P., Gatenby, J. C., Zhang, H., Anderson, A. W., & Gore, J. C. (1999). An fMRI study of Stroop word-color interference: evidence for cingulate subregions subserving multiple distributed attentional systems. *Biological Psychiatry, 45*(10), 1237–1258.

Portin, M., & Laine, M. (2001). Processing cost associated with inflectional morphology in bilingual speakers. *Bilingualism: Language and Cognition, 4*(1), 55–62.

Poulisse, N., & Bongaerts, T. (1994). First language use in second language production. *Applied Linguistics, 15*(1), 36–57.

Price, C. J. (2010). The anatomy of language: a review of 100 fMRI studies published in 2009. *Annals of the New York Academy of Sciences, 1191*, 62–88.

Price, C. J., Green, D. W., & von Studnitz, R. (1999). A functional imaging study of translation and language switching. *Brain, 122*, 2221–2235.

Prior, A., & Macwhinney, B. (2010). A bilingual advantage in task switching. *Bilingualism: Language and Cognition, 13*(2), 253–262.

Proverbio, A. M., Adorni, R., & Zani, A. (2009). Inferring native language from early bio-electrical activity. *Biological Psychology, 80*(1), 52–63.

Proverbio, A. M., Čok, B., & Zani, A. (2002). Electrophysiological measures of language processing in bilinguals. *Journal of Cognitive Neuroscience, 14*(7), 994–1017.

Pugh, K., Mencl, W., Shaywitz, B., Shaywitz, S., Fulbright, R., Constable, R., et al. (2000). The angular gyrus in developmental dyslexia: Task-specific differences in functional connectivity within posterior cortex. *Psychological Science, 11*(1), 51–56.

Pyers, J. E., Gollan, T. H., & Emmorey, K. (2009). Bimodal bilinguals reveal the source of tip-of-the-tongue states. *Cognition, 112*(2), 323–329.

Ransdell, S., & Fischler, I. (1987). Memory in a monolingual mode: When are bilinguals at a disadvantage? *Journal of Memory and Language, 26*, 392–405.

Richardson, F. M., Seghier, M. L., Leff, A. P., Thomas, M. S. C., & Price, C. J. (2011). Multiple routes from occipital to temporal cortices during reading. *The Journal of Neuroscience, 31*(22), 8239–8247.

Rodriguez-Fornells, A., De Diego Balaguer, R., & Münte, T. F. (2006). Executive control in bilingual language processing. *Language Learning, 56*, 133–190.

Rodriguez-Fornells, A., van der Lugt, A., Rotte, M., Britti, B., Heinze, H.-J., & Münte, T. F. (2005). Second language interferes with word production in fluent bilinguals: Brain potential and functional imaging evidence. *Journal of Cognitive Neuroscience, 17*(3), 422–433.

Roelofs, A., van Turennout, M., & Coles, M. G. H. (2006). Anterior cingulate cortex activity can be independent of response conflict in Stroop-like tasks. *Proceedings of the National Academy of Sciences, 103*(37), 13884–13889.

Runnqvist, E., Gollan, T. H., Costa, A., & Ferreira, V. S. (2013). A disadvantage in bilingual sentence production modulated by syntactic frequency and similarity across languages. *Cognition, 129*(2), 256–263.

Runnqvist, E., Strijkers, K., Sadat, J., & Costa, A. (2011). On the temporal and functional origin of l2 disadvantages in speech production: A critical review. *Frontiers in Psychology, 2*(379), 1–8.

Rüschemeyer, S.-A., Zysset, S., & Friederici, A. D. (2006). Native and non-native reading of sentences: An fMRI experiment. *NeuroImage, 31*(1), 354–365.

Rushworth, M. F., Ellison, A., & Walsh, V. (2001). Complementary localization and lateralization of orienting and motor attention. *Nature Neuroscience, 4*(6), 656–661.

Ruz, M., & Nobre, A. C. (2008). Attention modulates initial stages of visual word processing. *Journal of Cognitive Neuroscience, 20*(9), 1727–1736.

Saalbach, H., & Stern, E. (2004). Differences between Chinese morphosyllabic and German alphabetic readers in the Stroop interference effect. *Psychonomic Bulletin & Review, 11*(4), 709–715.

Sakurai, Y., Momose, T., Iwata, M., Sudo, Y., Ohtomo, K., & Kanazawa, I. (2000). Different cortical activity in reading of Kanji words, Kana words and Kana nonwords. *Cognitive Brain Research, 9*(1), 111–115.

Scarborough, D. L., Gerard, L., & Cortese, C. (1984). Independence of lexical access in bilingual word recognition. *Journal of Verbal Learning and Verbal Behavior, 23*(1), 84–99.

Schoonbaert, S., Duyck, W., Brysbaert, M., & Hartsuiker, R. J. (2009). Semantic and translation priming from a first language to a second and back: Making sense of the findings. *Memory & Cognition, 37*(5), 569–586.

Schroeder, U., Kuehler, A., Haslinger, B., Erhard, P., Fogel, W., Tronnier, V. M., et al. (2002). Subthalamic nucleus stimulation affects striato-anterior cingulate cortex circuit in a response conflict task: A PET study. *Brain, 125*, 1995–2004.

Schwartz, A. I., & Kroll, J. F. (2006). Bilingual lexical activation in sentence context. *Journal of Memory and Language, 55*(2), 197–212.

Segalowitz, S. J., & Zheng, X. (2009). An ERP study of category priming: Evidence of early lexical semantic access. *Biological Psychology, 80*(1), 122–129.

Sereno, S. C., Rayner, K., & Posner, M. I. (1998). Establishing a time-line of word recognition: Evidence from eye movements and event-related potentials. *Neuroreport, 9*(10), 2195–2200.

Simon, G., Bernard, C., Lalonde, R., & Rebaï, M. (2006). Orthographic transparency and grapheme-phoneme conversion: An ERP study in Arabic and French readers. *Brain Research, 1104*(1), 141–152.

Simon, G., Petit, L., Bernard, C., & Rebaï, M. (2007). N170 ERPs could represent a logographic processing strategy in visual word recognition. *Behavioral and Brain Functions, 3*(21), 1–11.
Siok, W. T., Spinks, J. A., Jin, Z., & Tan, L. H. (2009). Developmental dyslexia is characterized by the co-existence of visuospatial and phonological disorders in Chinese children. *Current Biology, 19*(19), R890–R892.
Soares, C., & Grosjean, F. (1984). Bilinguals in a monolingual and a bilingual speech mode: The effect on lexical access. *Memory & Cognition, 12*(4), 380–386.
Soveri, A., Rodriguez-Fornells, A., & Laine, M. (2011). Is there a relationship between language switching and executive functions in bilingualism? Introducing a within group analysis approach. *Frontiers in Psychology, 2*(183), 1–8.
Spalek, K., Hoshino, N., Damian, M. F., & Thierry, G. (2011). Phonological co-activation of both languages in bilingual speech production. Poster presented at the Cognitive Neuroscience Society, San Francisco, CA.
Spinks, J. A., Liu, Y., Perfetti, C. A., & Tan, L. H. (2000). Reading Chinese characters for meaning: The role of phonological information. *Cognition, 76*(1), B1–B11.
Stowe, L. A., & Sabourin, L. (2005). Imaging the processing of a second language: Effects of maturation and proficiency on the neural processes involved. *International Review of Applied Linguistics in Language Teaching, 43*(4), 329–353.
Stroop, J. (1935). Studies of interference in serial verbal reactions. *Journal of Experimental Psychology: General, 18*, 643–662.
Sumiya, H., & Healy, A. F. (2004). Phonology in the bilingual Stroop effect. *Memory & Cognition, 32*(5), 752–758.
Swick, D., & Turken, A. U. (2002). Dissociation between conflict detection and error monitoring in the human anterior cingulate cortex. *Proceedings of the National Academy of Sciences, 99*(25), 16354–16359.
Taft, M. (2002). Orthographic processing of polysyllabic words by native and nonnative English speakers. *Brain and Language, 81*(1–3), 532–544.
Taft, M., & van Graan, F. (1998). Lack of phonological mediation in a semantic categorization task. *Journal of Memory and Language, 224*(38), 203–224.
Tan, L. H., Laird, A. R., Li, K., & Fox, P. T. (2005). Neuroanatomical correlates of phonological processing of Chinese characters and alphabetic words: A meta-analysis. *Human Brain Mapping, 25*(1), 83–91.
Tan, L. H., Liu, H. L., Perfetti, C. A., Spinks, J. A., Fox, P. T., & Gao, J. H. (2001). The neural system underlying Chinese logograph reading. *NeuroImage, 13*(5), 836–846.
Tan, L. H., & Perfetti, C. A. (1997). Visual Chinese character recognition: Does phonological information mediate access to meaning? *Journal of Memory and Language, 37*(1), 41–57.
Tan, L. H., Spinks, J. A., Eden, G. F., Perfetti, C. A., & Siok, W. T. (2005). Reading depends on writing, in Chinese. *Proceedings of the National Academy of Sciences, 102*(24), 8781–8785.
Tao, L., Marzecová, A., Taft, M., Asanowicz, D., & Wodniecka, Z. (2011). The efficiency of attentional networks in early and late bilinguals: The role of age of acquisition. *Frontiers in Psychology, 2*(123), 1–19.
Thierry, G., & Wu, Y. J. (2004). Electrophysiological evidence for language interference in late bilinguals. *NeuroReport, 15*(10), 1555–1558.
Thierry, G., & Wu, Y. J. (2007). Brain potentials reveal unconscious translation during foreign-language comprehension. *Proceedings of the National Academy of Sciences, 104*(30), 12530–12535.
Tillman, C. M., & Wiens, S. (2011). Behavioral and ERP indices of response conflict in Stroop and flanker tasks. *Psychophysiology, 48*(10), 1405–1411.
Tokowicz, N., Michael, E. B., & Kroll, J. F. (2004). The roles of study-abroad experience and working-memory capacity in the types of errors made during translation. *Bilingualism: Language and Cognition, 7*(3), 255–272.
van Hell, J. G., & Dijkstra, T. (2002). Foreign language knowledge can influence native language performance in exclusively native contexts. *Psychonomic Bulletin & Review, 9*(4), 780–789.

van Hell, J. G., & Tokowicz, N. (2010). Event-related brain potentials and second language learning: Syntactic processing in late L2 learners at different L2 proficiency levels. *Second Language Research, 26*(1), 43–74.

van Heuven, W. J. B., Conklin, K., Coderre, E. L., Guo, T., & Dijkstra, T. (2011). The influence of cross-language similarity on within- and between-language Stroop effects in trilinguals. *Frontiers in Psychology, 2*(374), 1–15.

van Heuven, W. J. B., & Dijkstra, T. (2010). Language comprehension in the bilingual brain: fMRI and ERP support for psycholinguistic models. *Brain Research Reviews, 64*(1), 104–122.

van Heuven, W. J. B., Dijkstra, T., & Grainger, J. (1998). Orthographic neighborhood effects in bilingual word recognition. *Journal of Memory and Language, 39*(3), 458–483.

van Heuven, W. J. B., Schriefers, H., Dijkstra, T., & Hagoort, P. (2008). Language conflict in the bilingual brain. *Cerebral Cortex, 18*(11), 2706–2716.

Vingerhoets, G., Van Borsel, J., Tesink, C., van den Noort, M., Deblaere, K., Seurinck, R., et al. (2003). Multilingualism: An fMRI study. *NeuroImage, 20*(4), 2181–2196.

Voga, M., & Grainger, J. (2007). Cognate status and cross-script translation priming. *Memory & Cognition, 35*(5), 938–952.

Vogel, A. C., Church, J. A., Power, J. D., Miezin, F. M., Petersen, S. E., & Schlaggar, B. L. (2013). Functional network architecture of reading-related regions across development. *Brain and Language, 125*(2), 231–243.

Wang, Y., Kuhl, P. K., Chen, C., & Dong, Q. (2009). Sustained and transient language control in the bilingual brain. *NeuroImage, 47*(1), 414–422.

Wang, Y., Xue, G., Chen, C., Xue, F., & Dong, Q. (2007). Neural bases of asymmetric language switching in second-language learners: An ER-fMRI study. *NeuroImage, 35*(2), 862–870.

Wartenburger, I., Heekeren, H. R., Abutalebi, J., Cappa, S. F., Villringer, A., & Perani, D. (2003). Early setting of grammatical processing in the bilingual brain. *Neuron, 37*(1), 159–170.

Weber-Fox, C., & Neville, H. (1996). Maturational constraints on functional specializations for language processing: ERP and behavioral evidence in bilingual speakers. *Journal of Cognitive Neuroscience, 8*(3), 231–256.

Weber-Fox, C., & Neville, H. J. (2001). Sensitive periods differentiate processing of open-and closed-class words: An ERP study of bilinguals. *Journal of Speech, Language, and Hearing Research, 44*, 1338–1353.

Workman, L., Brookman, F., Mayer, P., Rees, V., & Bellin, W. (2000). Language laterality in English/Welsh bilinguals: Language-acquisitional and language-specific factors in the development of lateralisation. *Laterality, 5*(4), 289–313.

Wu, Y. J., & Thierry, G. (2011). Unconscious translation during incidental foreign language processing. *NeuroImage, 59*(4), 3468–3473.

Xu, Y., Pollatsek, A., & Potter, M. C. (1999). The activation of phonology during silent Chinese word reading. *Journal of Experimental Psychology: Learning, Memory, and Cognition, 25*(4), 838–857.

Ye, Z., & Zhou, X. (2009). Conflict control during sentence comprehension: fMRI evidence. *NeuroImage, 48*(1), 280–290.

Zhang, M., Li, J., Chen, C., Xue, G., Lu, Z., Mei, L., et al. (2014). Resting-state functional connectivity and reading abilities in first and second languages. *NeuroImage, 84*, 546–553.

Zhang, T., van Heuven, W. J. B., & Conklin, K. (2011). Fast automatic translation and morphological decomposition in Chinese-English bilinguals. *Psychological Science, 22*(10), 1237–1242.

Chapter 5
Cross-Cultural Reading the Mind in the Eyes and Its Consequences for International Relations

Robert G. Franklin Jr., Michael T. Stevenson, Nalini Ambady, and Reginald B. Adams Jr.

Abstract Franklin, Stevenson, Ambady, and Adams observe that there is little cross-cultural research in the ability to perceive information (emotion, cognitions, etc.) from the eyes of an individual. The authors argue that the eyes play a major role in social interaction and looking at within- and between-culture use of information from the eyes, what is referred to as mind reading, is important to understanding nonverbal communication.

The authors report on their research on this topic where they studied participants' ability to mind read within and between cultures. While participants were accurate in their mind reading ability for both groups, they did show an advantage for their own culture. This advantage was marked by an increase in activity in the posterior superior temporal sulcus. This is consistent with other research on emotion recognition.

The authors also report on their research looking at how gender can play a role in mind reading accuracy. Additionally, they studied how multiple category differences, gender and culture, may affect mind reading ability. They found that women were more accurate in the mind reading task than men regardless of culture.

R.G. Franklin Jr.
Department of Behavioral Sciences, Anderson University,
316 Boulevard, Anderson, SC 29621, USA
e-mail: rfranklin@andersonuniversity.edu

M.T. Stevenson
Department of Occupational Therapy, Tufts University,
200 Boston Avenue, Suite 1725, Medford, MA 02155, USA
e-mail: michael.stevenson@tufts.edu

N. Ambady (deceased)

R.B. Adams Jr. (✉)
Department of Psychology, The Pennsylvania State University,
464 Moore Building, University Park, PA 16802-3106, USA
e-mail: radams@psu.edu

J.E. Warnick, D. Landis (eds.), *Neuroscience in Intercultural Contexts*,
International and Cultural Psychology, DOI 10.1007/978-1-4939-2260-4_5

fMRI revealed that women showed greater activation of the inferior frontal gyrus and cerebellum during the task while men showed greater activation of the superior temporal sulcus. These areas are known for reading facial cues, expressing empathy, and mirroring behavior.

Implications for how this work can affect nonverbal communication between cultures are discussed.

Throughout history and across cultures, from infancy through adulthood, the eyes hold special prominence in human social interaction. Indeed, popular folk wisdom speculates that "the eyes are the window to our souls," a presumption that remains nearly axiomatic in contemporary social interaction. This leads to two questions: Is there a language of the eyes? And, if so, to what extent is this language universal across cultures? The ability to perceive others' thoughts, intentions, and feelings, an ability referred to as mind reading or theory of mind, is regarded as a highly evolved human attribute (Allison, Puce, & McCarthy, 2000; Brune & Brune-Cohrs, 2006). Cross-cultural comparisons of mind reading, however, remain limited. To date preliminary evidence supports the presence of some biologically determined components of this ability (Avis & Harris, 1991; but see also Kobayashi, Glover, & Temple, 2006, 2007). Critically, this ability also varies as a function of whose mind is being read and by whom, which is the focus of this chapter.

The ability to process gaze information plays a pivotal role in the development of the ability to mind read (Baron-Cohen, 1995). Certain psychopathological disorders and brain damage marked by deficits in social perception are specifically linked to a failure to attend to the eyes (Adolphs et al., 2005; Baron-Cohen, Wheelwright, & Jolliffe, 1997). Thus, there does appear to be a language of the eyes. But to what extent this language is translatable across cultures and even across social groups within cultures remains unknown. These questions have been virtually neglected in the growing literature on mind reading, which has instead largely focused on our capacity to accurately infer others' mental states. It is time to extend this inquiry to include how and when this ability fails. Examining social factors that influence mental state decoding ability has the promise of furthering a basic understanding of this extraordinary ability. Given that researchers have estimated that nonverbal communication far outweighs verbal communication in our daily lives (DePaulo & Friedman, 1998), understanding cross-cultural and cross-race impairment in our mental state reasoning derived from nonverbal language could hold important insights that can be applied in interracial and international relations.

Mind reading is derived from two distinct component processes including: (1) a social-perceptual process that enables mental state decoding from nonverbal cues (e.g., the eyes) and (2) a social-cognitive process that enables abstract reasoning about another's mental state such as recognizing the potential for false beliefs in others (Sabbagh, 2004; see also Tager-Flusberg, 2001). The work that has been done to examine the social-perceptual process suggests that the eyes may hold special prominence in mental state reasoning. Besides folk wisdom, there are compelling scientific reasons for beginning an investigation on cross-group mind reading by examining the language of the eyes. For one, our nonverbal communication is achieved predominantly via the visual modality, giving the eyes a central prominence in gathering the information from which we extract our impressions of others (e.g., Baron-Cohen et al., 1997).

When reading complex messages from the eyes, they are as informative as the whole face (Baron-Cohen et al., 1996). The eyes are also particularly informative in assigning social group memberships (Zebrowitz, 2006). Thus, information about one's cultural group memberships can be readily decoded through the eyes. Baron-Cohen has developed a "reading the mind in the eyes" test (Baron-Cohen, Wheelwright, Hill, Raste, & Plumb, 2001), which has strong convergent validity with traditional tests of theory of mind and has been used now in scores of studies examining a wide range of psychopathologies known to be marked by dysfunctional social perception, including autism (e.g., Baron-Cohen et al., 1999), schizophrenia (Irani et al., 2006), and Williams syndrome (Skwerer, Verbalis, Schofield, Faja, & Tager-Flusberg, 2006).

A critical, but unresolved, question is whether mind reading from the eyes is a universal ability or one that is culturally dependent. Early work suggests that the ability to understand false beliefs in others, which represents a defining characteristic of theory of mind (Dennett, 1978), appears universal, given that it arises at about the same time in both preliterate and literate cultures (Avis & Harris, 1991). In contrast, some work suggests culture may play a critical role in how we read others' mental states. Kobayashi et al. (2006) examined the influence of cultural factors on neural processing using a false belief task and found both culture-/language-dependent and culture-/language-independent neural responsivities. Social experiences during development also impact the ability to reason about others' minds (e.g., Perner et al., 1994; Peterson & Siegal, 1997), an insight that led a number of theorists to consider potential intercultural influences on this ability (e.g., Flavell, 1999; Lillard, 1998), given that culture implies profound differences in social experience (Chiao & Ambady, 2007; see also Markus, Kitayama, & Heiman, 1996). It stands to reason then that culture could ultimately reveal itself through differences in complex mental state reasoning.

The primary question addressed by our own initial research was whether there are differences in mind reading when people read same- versus other-culture members. In order to answer this question, we developed the Cross-Cultural Reading the Mind in the Eyes Test (CCRME) and examined cultural commonalities and differences between US Caucasian and native Japanese participants while mind reading same-culture and other-culture faces (Adams et al., 2010). The CCRME consists of the Adult Reading the Mind in the Eyes Test (Baron-Cohen et al., 2001), a test initially used to distinguish between high-functioning autism (HFA) and Asperger syndrome populations and samples from the typical population in a socio-perceptual aspect of theory of mind (i.e., mental state decoding) and an Asian version of the task constructed to match the RME, but featuring Asian targets and created in conjunction with colleagues at Kyoto University. We investigated this behaviorally and

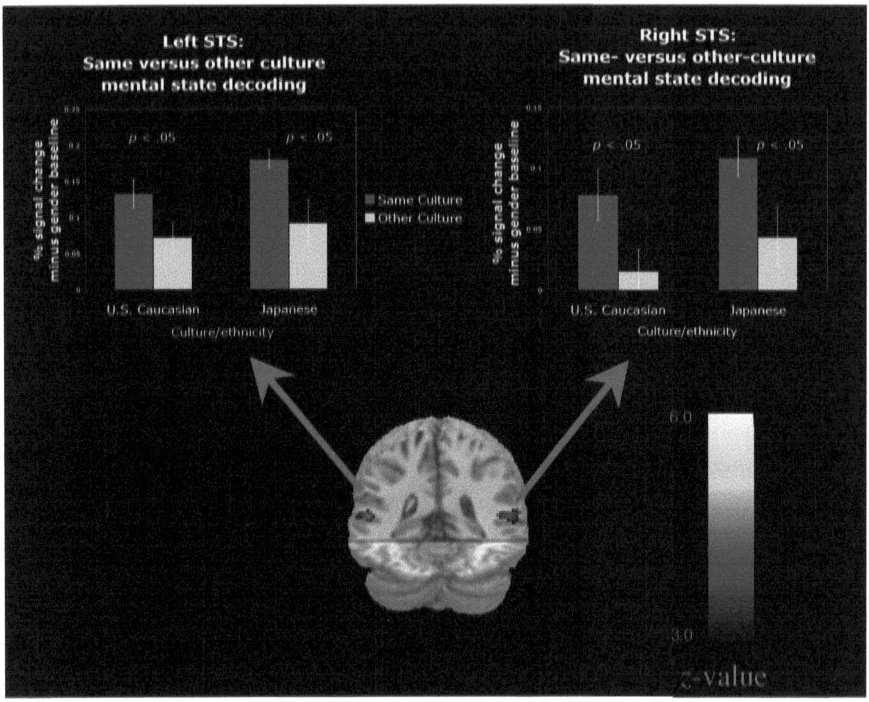

Fig. 5.1 Left and right pSTS activation for same- versus other-culture mental state decoding

using fMRI to examine the brain regions involved with mind reading same-culture and other-culture faces. We found that both groups were able to accurately recognize mental states of same-culture and other-culture individuals but both groups showed an own-group advantage, more accurately labeling faces of their own culture. This own-group advantage was paralleled by increased neural activation in the posterior superior temporal sulcus (pSTS; see Fig. 5.1), a brain region consistently implicated in understanding others' mental states.

Two theories may help explain the own-group advantage. First, nonverbal behavior may carry with it cultural dialects, or small, physical differences in the expression of emotions and mental states that vary from one culture to the next that, along with culturally defined expertise understanding those dialects, may facilitate accurate decoding of same- versus other-culture members. Second, social categorization, the tendency to allocate more attentional resources to the processing of ingroup targets than to outgroup targets, may influence the ability to properly process emotional or mental state expressions. Social categorization, in the context of visual person perception, refers to differential allocation of attentional resources to a target as a function of the perceived group membership of that target, relative to the perceiver (i.e., more attention allocated to ingroup members, less to outgroup members; see Bernstein, Young, & Hugenberg, 2007).

5.1 Perceptual Expertise

For decades, researchers have dedicated their efforts to the goal of better understanding nonverbal communication across groups. Some of the earliest work defined these groups culturally. Ekman wanted to explore the possibility of the universal nature of emotional expression and recognition, predicating this assumption on Darwin's seminal work on facial and bodily expressions (Darwin, 1872/1965). By examining expression encoding and decoding samples from several countries, Ekman (1972) provided evidence that at least some basic emotions are universally expressed and recognized (i.e., at better than chance accuracy) across cultures. From this work, Ekman put forth the notion of a *universal affect program*, or the idea that emotional communication is largely dictated and predetermined by a culturally consistent biologically determined program. As a result, many early cross-cultural emotion recognition studies have focused on identifying universality, rather than variation (Fig. 5.2).

A meta-analysis conducted by Elfenbein and Ambady (2002) supported the assertion that emotions are recognized at rates better than chance across cultures, but also the existence of an intracultural advantage for emotion recognition. Elfenbein and Ambady (2003a, 2003b) went on to posit the existence of a *specific affect program*, or adjustments to the *universal affect program*, accounting for learned differences (based on cultural membership) in the expression of emotion. Subsequent work has sought to more closely examine how culturally specific emotional expressions and their perception manifest themselves. Clearly, past research has provided ample evidence that, when examining the expression and recognition of emotions across groups, there are actual, culturally defined variations that can account for differential performance.

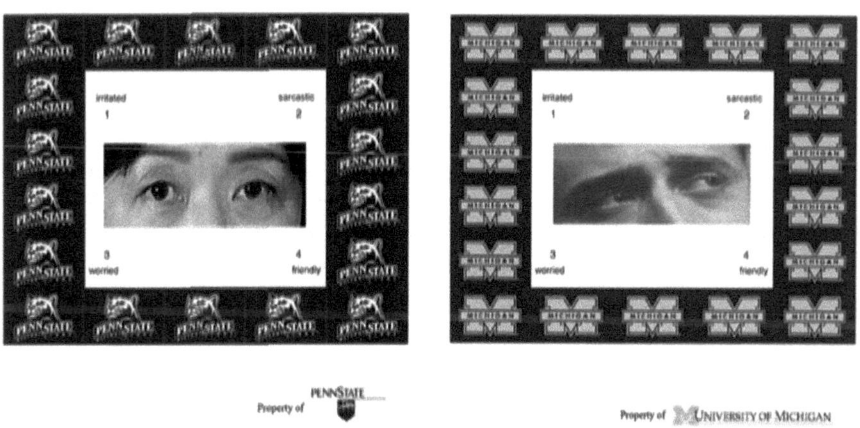

Fig. 5.2 Example stimuli of Caucasian and Asian eyes made to appear as part of same versus rival school, with same mental state depicted

Similar effects have been found in the memory literature, which reveal a pervasive own-race bias (ORB) in face memory. The ORB, generally, refers to the phenomenon of remembering faces of members of one's own race better than members of another race. Recently, the ORB debate has been focused on two explanations of this effect: perceptual expertise and social categorization (Bernstein, Young, & Hugenberg, 2007). Perceptual expertise refers to a consequence of "social segregation," such that people tend to live and interact with others of their own groups (racial, cultural, etc.). This, in turn, grants them greater exposure to, and consequent expertise in, how faces within their ingroup may vary, resulting in better memory for ingroup faces (Sporer, 2001). This explanation is very similar to the cultural dialect theory in the domain of emotion recognition in that both are based on the premise that greater ingroup (relative to outgroup) accuracy in face processing tasks depends on a level of familiarity with ingroup faces that is not present for outgroup faces. Although a seemingly promising explanation, a meta-analysis revealed that exposure to other racial groups only explains about 1 % of variance that accounts for the ORB effect (Meissner & Brigham, 2001). This finding helped give rise to an ingroup/outgroup model of the ORB (Sporer, 2001) that has recently received extensive support implicating greater allocation of resources dedicated to processing faces of ingroup relative to outgroup members and is itself sufficient to explain memory effects based on group membership (Bernstein, Young, & Hugenberg, 2007). The following section details how social categorization affects the visual perception of targets and how this may be generalizable to emotion recognition.

5.2 Social Categorization

As alluded to earlier, social categorization refers to differential allocation of attentional resources when processing ingroup (more attention) and outgroup (less attention) members. A well-documented outcome of the social categorization process is that of outgroup homogeneity. One of the basic principles of outgroup homogeneity is that outgroup members are perceived as being more similar to one another than are ingroup members (Tajfel, 1969). When we first encounter others, we make our impressions of them based on very little information. In order to make up for this lack of information, we often fill in the gaps with stereotypic information about a category available in semantic memory (Hugenberg & Sacco, 2008). When this strategy is used, it can lead to the attribution of similar information to all outgroup members. Ingroup members, on the other hand, appear to motivate a search for individuating information (i.e., information beyond stereotypic associations), leading to a more unique and often accurate perception of the person (Ostrom & Sedikides, 1992). Facial stimuli can provide sufficient information necessary to make such categorizations and to activate stereotypes associated with a category membership (e.g., Mason, Cloutier, & Macrae, 2006). Of specific interest for this proposal is the disparity in face recognition for ingroup versus outgroup members.

The well-documented recognition advantage and preference for ingroup members begin to emerge very early in life. When born, infants do not show a preference for own-race faces, but by 3 months of age, they begin to prefer faces of their own race (Kelly et al., 2005; Sangrigoli & de Schonen, 2004b). This loosely maps onto the stage of development proposed by Morton and Johnson (1991), during which children may begin to be able to extract social information from the face, as opposed to learning to pay attention to faces in general. By 3 years of age, children show the same own-race bias memory effects as do adults, indicating they are able to apply categorical labels when making judgments of others (Sangrigoli & de Schonen, 2004a). Young adults also show these effects and they persist into older age (Corenblum & Meissner, 2006).

There is also evidence that the social-cognitive mechanism driving these effects is enhanced by our basic visual machinery. For instance, Levin (1996, 2000) found that, for White participants, a Black face in an array of White faces was identified much faster than a White face in an array of Black faces. Since luminosity was controlled for in all of the stimuli, this likely illustrates a categorization effect, and not a simple visual contrast effect. Notably, the information first reaching visual awareness tends to resemble low-pass blurry visual features from which only crude distinctions can be made based on very salient cues like hairstyle (Macrae & Martin, 2007), which favor initial categorization of faces rather than individuation. Supporting this interpretation was the finding in Levin's work that the speed of categorization of Black faces predicted the subsequent degree to which an individual showed an own-race bias in memory. This suggests that people allocate fewer attentional resources to processing outgroup members once they have been initially categorized. And since visual information more quickly informs social categories than individuating features, categorizing a face as belonging to an outgroup undermines the additional processing necessary for face memory.

That social categorization can affect the depth of processing of a face (resulting in greater individuation and subsequent memory for own-race faces) speaks to the possibility that this effect may also influence the recognition of facial expressions of emotion and complex mental states. If what determines the level to which we are able to extract subtle, nuanced cues from faces (the kind of visual information that is predictive of face memory *and* facial expression) is the extent to which we engage in the early categorization of outgroup members, it stands to reason that it would influence the extent to which we extract complex social information from them. Further supporting this notion is recent social neuroscientific literature, which provides evidence for the differential processing of ingroup versus outgroup faces.

Hemispheric differences in the processing of facial stimuli based on race have been identified (Turk, Handy, & Gazzaniga, 2005). Utilizing a White, split-brain patient, White and Asian facial stimuli were presented to only one visual field at a time, and then facial recognition was assessed. An ORB for face memory was present, but only in the right hemisphere (memory performance was equal in the left hemisphere). This is particularly interesting, given previous research suggesting that the right hemisphere is where deeper, more individuated processing (the kind necessary to remember a face) takes place (Mason & Macrae, 2004).

As Levin (1996, 2000) revealed, the timecourse of social categorization can be predictive of future performance with facial stimuli, and event-related potential (ERP) research has supported this. A handful of studies have already identified effects that can distinguish the implicit categorization of ingroup versus outgroup faces (Caldara et al., 2003; Caldara, Rossion, Bovet, & Hauert, 2004; Ito, Thompson, & Caioppo, 2004; Ito & Urland, 2003, 2005). Caldara et al. (2004) found, using both behavioral and ERP studies, that Asian faces were processed faster than White faces by White participants. The ERP difference showed up at around 240 ms, in line with when Bruce and Young's (1986) functional model of face processing posits that information such as race, gender, and age are processed. While examining mostly White participants, Ito and Urland (2003) identified greater amplitude for Black relative to White facial stimuli at N100 and P200. This pattern of activation was reversed at P300, suggesting that the increased attention for Black versus White faces (N100 and P200) resulted a designation that less effort needed to be spent in order to complete the task (P300).[1] Together, these ERP findings speak toward the supposition that we are predisposed to categorizing faces as ingroup or outgroup members, and once we do so, motivation for subsequent processing of ingroup targets is increased, relative to outgroup targets. An ever-expanding body of work in the behavioral social-cognitive person perception literature provides support for this notion (e.g., Pauker et al., 2009; Rule, Ambady, Adams, & Macrae, 2008; Stevenson, Soto, & Adams, 2012).

Given that low-level visual information (e.g., gendered hairstyle) can give rise to rapid, top-down influences on the categorization of a facial stimulus, regardless of the content of the remaining features of that face (Macrae & Martin, 2007), this suggests there is an inherent, if not incidental, bias to categorizing faces quickly. Applying a similar assumption to race, Maclin and Malpass (2001) showed that racial categorization of a face stimulus is sufficient to create an ORB in memory performance when holding the actual structural makeup of the face constant. Ambiguous-race face stimuli were created, and then either stereotypical Black or Hispanic hairstyle was applied. Black and Hispanic participants displayed an ORB for the very same stimuli based solely on how the stimuli were categorized, given hairstyle information. Pauker et al. (2009), likewise, found that racial labels assigned to racially ambiguous faces predicted memory performance, based on the race of the participants.

Rule et al. (2008) identified a similar effect in the context of sexual orientation. Heterosexual and homosexual participants showed an own-group face memory bias based on how they classified the sexual orientation of the targets. That this effect can occur with ambiguously categorizable faces is especially interesting, considering that the more categorically prototypic a face is, the more quickly it is categorized and associated with stereotypes of that group (Eberhardt, Goff, Purdie, & Davies, 2004; Locke, Macrae, & Eaton, 2005; Maddox, 2004). Thus, once enough category-specific information is processed to identify a face as being an outgroup member, it appears to be processed in a qualitatively different manner than that of an ingroup member. Recently, Bernstein, Young, & Hugenberg, (2007) extended

[1]The N100 and P200 components are generally considered to be associated with early attentional effects. Specifically, increased amplitude is associated with increased attention directed at a feature of a visual stimulus. The P300 component is generally considered to be associated with working memory operations, with increased amplitude indicative of a greater degree of encoding (see Ito & Urland, 2003).

this idea to face memory, showing that the same faces (all the same race as the participants) were remembered differently, based on their belonging to the same group as the participant (either university or a personality type).

Directly relevant to examining these possible explanations (Adams et al., 2010) is our own recent evidence elaborating on our original findings of greater mindreading ability for own-culture individuals using the mind in the eyes test. We elaborated on our original study by examining the effect of combining traditional (i.e., race) and arbitrarily assigned (i.e., university affiliation) group memberships on the ability to accurately decode complex mental states. In order to do this, we used a modified version of the intergroup memory paradigm described above, which used arbitrarily assigned of faces to same versus other university affiliation (see Bernstein, Young, & Hugenberg, 2007). In this case, however, we examined the influence of group membership on a mental state decoding using the Cross-Cultural Reading the Mind in the Eyes Test (CCRME), which we used previously to examine mind reading cross-culturally (Stevenson et al., 2012).

Following the social categorization explanation, we hypothesized that the arbitrary assignment of faces to same- versus other-school affiliation would influence mental state decoding such that the eyes of those perceived to be of the same-school affiliation would be better read than those perceived to be of another school affiliation and that these effects would override previously found cross-race effects using these same stimuli. We also hypothesized that this effect would be influenced by the extent to which participants identified with their ingroup school. This is an important consideration, as it is well documented, notably in the realm of identity and stereotype threat, that group-based manipulations are most effective for those who are highly identified with the group of interest (race: Steele & Aronson, 1995; gender: Maass, Cadinu, Guarnieri, & Grasselli, 2003; academic domains: Aronson et al., 1999). Specifically, we hypothesized that the more highly identified students were with their ingroup school, the more same- versus other-school affiliation would influence performance on the RME. Further, based on the findings of Kurzban, Tooby, and Cosmides (2001), we predicted that the racial ingroup advantage would be erased for those labeled as same-school members, but would persist for other-school members. For those not highly identified with their ingroup school, we expected that the racial ingroup advantage would persist regardless of school affiliation. Thus, we collected a measure of school pride, to use as a factor in our analyses to examine these hypothesized effects (Stevenson et al., 2012).

As predicted, the racial/cultural ingroup advantage was still apparent, while school affiliation did not show an overall same-school advantage on mental state decoding. However, there was a significant 3-way interaction (school pride, perceived school affiliation, and stimulus race) that demonstrated that assignment of target stimuli into a highly meaningful, nonracial ingroup erased an otherwise robust racial advantage in mental state decoding, but only for those high in school identity. Specifically, the evaluation of other-school Asian stimuli (i.e., double outgroup) was the only instance that showed a decrease in performance in the highly identified group. In this group, same-school Asian stimuli were perceived just as accurately as White stimuli. For those low in school pride, school membership of the target face did not matter as they perform equally poorly on Asian compared to White stimuli (the previously documented racial ingroup advantage). These findings highlight the importance of considering processes such as self-identification and social categorization when considering emotional processes (e.g., mind reading).

Though several studies have examined how culture may play a role in mental state decoding, little work has examined how different social categories may interact to affect it. One such social category that has received some investigation is the role of gender. Below, we present two studies examining how gender interacts with culture in mind reading.

5.3 Exploring Gender by Culture Effects: A New Reanalysis

5.3.1 Overview

Though several studies have examined how culture may play a role in mind reading, little work has examined how different categories may interact to affect mind reading. One such social category that has received little investigation is the role of gender. Though most cultures have consistent sex stereotypes, largely as a function of division of labor into male-oriented and female-oriented roles (Wood & Eagly, 2002), socialization factors play a large role into how sex stereotypes develop and can lead to differences in different cultural settings. Though American and Japanese cultures have relatively egalitarian sex roles compared to societies across the world, the increased emphasis on hierarchy, lifelong employment, and traditional values in Japan means that women tend to be at a greater disadvantage in Japan compared to the United States (Kawaguchi, 2007; Nemoto, 2013). Specifically, compared to the United States, the rigidity of Japanese culture makes sex discrimination more likely and increases the difficulty for women to achieve leadership positions in business and politics (Toh & Leonardelli, 2012).

Given their disadvantaged role in society as compared to men, it is likely that women will show less of a cultural ingroup effect when reading the mental states of others. There are several reasons for this. First, women tend to perform better at reading emotions and understanding others' mental states, including using the RME (Baron-Cohen et al., 2001). Second, women often tend to show less intergroup conflict compared to men. Though both men and women show positive attitudes toward their own group and see other groups more negatively, men are more likely to be aggressive toward outgroups, a finding thought to be due to the increased importance of dominance and status for men (Navarrete, McDonald, Molina, & Sidanius, 2010).

However, there are several interesting questions about how two relatively developed societies with differences in gender roles may interact. One potential effect is that the increased distance that two separate categories may provide may cause increased deficits in the ability to understand others. For instance, trying to understand what a person is thinking who is of a different culture and not the same gender as one's self may be increasingly difficult for someone and thus lead to larger deficits than culture or gender alone would predict. On the other hand, it is possible that the desire to be seen as being culturally appropriate may make people perform better at understanding other-culture individuals who are of the other gender than one's self. This is especially relevant given the importance intercultural relations place on not offending members of other cultures. Specifically, Japanese individuals may tend to reduce their gender stereotypes when reading the mental states of American individuals because the United States is a more egalitarian society and Japanese participants, especially men, may be motivated to reduce their own gender stereotypes. If this is the case, we would expect better performance for reading the mental states of other gender faces in other-culture individuals, with this effect strongest for Japanese men. In either of these cases, we would expect interactions between culture and gender.

We investigated this using both behavioral and fMRI methods. First, we wanted to examine the interaction between gender and culture at mindreading performance. However, it is possible that the interaction between gender and culture may appear in neural data, but not be present in behavioral data. It could be that participants are trying harder to understand individuals who are of a different gender and race than themselves, but that this is not reflected in performance. Therefore, we examined these differences using both methods.

5.4 Study 1

5.4.1 Method

5.4.1.1 Participants

Sixty-one (38 female) native Japanese students at Kyoto University (mean age = 23 years) and sixty (30 female) US Caucasian students from the Pennsylvania State University participated in this study (mean age = 20 years). Participants received research credit for an introductory psychology course.

5.4.1.2 Stimuli

The "Reading the Mind in the Eyes Test" (Baron-Cohen et al., 2001) as well as a new, Asian stimuli, Japanese language version of the test were used to create our Cross-Cultural Reading the Mind in the Eyes Test. The Asian version of the test was created at Kyoto University, following closely the procedure outlined by Baron-Cohen et al. (2001). Each of the mental state words was translated into Japanese. A large number of Asian eyes stimuli were created, and a group of raters chose stimuli from this to match the original target and foil mental state words, as well as the same target gender distribution. A total of seventy-two stimuli (36 Asian and 36 White) comprise our final test.

The White version of the eyes test consists of 36 photographs depicting just the eye region of individual faces. The eye region was delineated by a rectangular area approximately 15 × 10 cm, running the width of the face, from midway up the nose to right above the brow. All photographs were collected from magazines. Four mental state terms accompany each stimulus (one target word and three foils) and are presented at each corner of the photograph.

To generate the Asian version of the eyes test, we employed similar procedures used to generate the White version of the eyes test. Rather than generating new mental state words, however, the mental state words used in the White version of the test were first translated into Japanese, and then Asian eye stimuli were collected from magazines, the Internet, a database of amateur models, and commercial DVDs to match each of the 36 sets of words. The Japanese test was subsequently back-translated by a Japanese student at Tufts University to check for agreement with the original test. The back-translation revealed exact agreement for all but six English terms. Importantly, in each case where an alternate English translation was offered, the alternate word was a close synonym of the original word (e.g., daydreaming instead of fantasizing, scared instead of terrified, etc.). The resultant test was piloted on nine students at Kyoto University until each test item reached criterion levels of consensus (i.e., at least five of the nine judges picked the target word on each test item). Four pilot tests of this kind were required to reach criterion on all items. In two of these four female and five male judges were used and in the other two five female and four male judges. A preliminary study revealed overall test performance exceeding 70 % correct responses for a nonclinical sample of Kyoto University undergraduate students, performance that is comparable to that previously reported for the White version of the eyes test using a nonclinical sample of Cambridge University undergraduate students.

5.4.1.3 Procedure

US participants entered a room individually or in groups ranging in size from one to four per session. Each participant was seated in their own cubicle containing a computer with 17-in. CRT displays. The Cross-Cultural Reading the Mind in the Eyes Test was presented and responses recorded using Cedrus' SuperLab Pro 2.0. Stimuli, size was approximately 10 cm × 4 cm, and participants sat approximately 45 cm from the display. Mental state words were presented surrounding the eyes stimuli and were numbered 1 through 4. Participants were instructed to take as much time as necessary to determine which word best described what the person in each photograph was thinking or feeling. A response was made by pressing the number key on a standard keyboard that corresponded with their word choice. Stimuli were presented randomly in two blocks (one Asian, one White), the order of which was counterbalanced across participants. Japanese participants completed a paper and pencil version of the Cross-Cultural Reading the Mind in the Eyes Test. On each page of the test booklet, one stimulus was presented surrounded by four mental state words. Participants were instructed to take as much time as necessary to choose which word best described what the person in each photograph is thinking or feeling. Participants indicated their choice by circling one of the four words.

5.4.2 *Results*

In order to address the hypothesis of interest, we computed a 2 (culture: Japanese, US Caucasian) by 2 (gender: male, female) by 2 (stimulus race: Asian, Caucasian) mixed effects ANOVA. The ANOVA revealed no main effects of culture, $F(1,117)=0.939$, $p=0.335$, suggesting that neither culture has a particular advantage in performing the RME task. Additionally, there was no main effect of stimulus race, $F(1,117)=0.061$, $p=0.805$. There was, however, a main effect of gender, such that females ($M=0.696$) performed more accurately than males ($M=0.667$) on the RME task, $F(1,117)=3.915$, $p=0.05$. There was also the predicted culture × stimulus race interaction, which proved to be significant, $F(1,117)=50.659$, $p<0.0001$. Simple effects analyses revealed that Japanese participants performed more accurately on the Asian stimuli ($M=0.735$) than on Caucasian stimuli ($M=0.654$), $t(60)=5.831$, $p<0.0001$. Likewise, US Caucasian participants performed more accurately on the Caucasian stimuli ($M=0.711$) than on Asian stimuli ($M=0.638$), $t(59)=4.694$, $p<0.0001$, thus providing evidence for an intracultural advantage for mental state decoding. Gender did not interact with either culture or stimulus race.

5.4.3 *Discussion*

In Study 1, we replicated previous findings for better performance in mind reading for same-culture individuals (Adams et al., 2010). However, we found no effects of stimulus gender or any stimulus gender by participant gender effects. This finding indicates that though participants had trouble recognizing the mental states of other-culture individuals, this effect was not moderated by gender. We did find, however, that females performed better than males overall, a finding consistent with Baron-Cohen and colleagues' (2001) original Reading the Mind in the Eyes Test, showing that this effect generalizes to other cultures.

5.5 Study 2: fMRI

In Study 2, we wanted to investigate whether there were any neural differences in mindreading ability. Study 1 found no behavioral differences due to the gender of stimulus or any interactions between participant and stimulus gender, or gender and culture. However, it is possible that there are critical differences in the interaction between gender and culture if we use different brain regions to examine the mental states of males and females of different cultures despite the lack of behavioral differences.

In order to investigate this, we reanalyzed neural data from our previous study (Adams et al., 2010) and investigated gender by culture effects. Due to the nature of our design, where we mixed male and female faces within blocks, we were unable to examine effects due to the gender of the stimulus. Therefore, we restricted our analysis to examining effects of the gender of the participant and same- versus other-culture faces. Below, we summarize the methods we used in that study and present the results of our new analyses.

5.5.1 *Method*

5.5.1.1 Participants

Fourteen White American participants (9 women) and 14 native Japanese participants (9 women) completed this study. The Japanese participants were recruited from a summer English language program and had only minimal experience with visiting the United States prior to the study.

5.5.1.2 Stimuli

The same stimuli as used in Study 1 were used in this study. As in previous brain imaging studies using the eyes test, a modified two-choice task was used in the magnet portion of this study, including one target and one foil word presented in the top corners of the photograph. We chose foil words using pilot data to determine which of the original three foil words was chosen the least often for each item by both Japanese and US White participants on both Japanese and White versions of the test. The resultant foils were thus identical across both the Asian and White versions of the test.

5.5.1.3 Procedure

Participants viewed 72 photographs, depicting 36 White American and 36 Asian eye stimuli twice during our study, for a total of 144 stimulus presentations. The stimuli were presented once with the mental state labels and once with gender labels as a control task. The photos were presented within each run in approximately 35 s blocks in a periodic ABA design, with task A being the mental state judgment task and task B being the gender judgment task. The gender task was chosen to be a control for low-level visual and motor processing. Participants saw the pictures using a back-projection system with individuals viewing the screen using a mounted mirror. Functional data were collected using two runs, one with all Asian stimuli and the other with all Caucasian stimuli. In each block, stimuli were presented on the screen for 5 s, with randomized interstimulus intervals between 0.5 and 1 s between stimuli. Participants chose between the labels by pressing one of two buttons. Responses were collected as long as the stimulus remained on the screen. The order of the blocks was counterbalanced across participants. After scanning, participants completed a self-paced behavioral task using the same methods as described in Study 1.

5.5.1.4 fMRI Data Analysis

Participants were scanned in a supine position using a 1.5 Tesla Siemens Avanto scanner. Data were collected using echoplanar imaging and we also collected anatomical images to coregister functional data and normalize data to standard Montreal Neurological Institute space. Data were processed using SPM5 software. In order to examine specific contrasts, we computed first-level contrasts by examining neural activation for mentalizing minus the gender task for male and female faces separately, separate for same-culture and other-culture faces. For more details, see Adams et al. (2010).

For this specific study, we examined hypotheses of the interaction between stimulus gender and culture using a 2 (participant gender: male versus female) × 2 (culture: ingroup versus outgroup) mixed factors ANOVA with gender being a between-subject factor and culture being a within-subject factor. In this manuscript, we report the main effect of gender and the interaction between gender and culture. We do not examine the main effect of culture because those results have previously been reported in Adams et al. (2010) utilizing an analysis designed specifically to examine the main effect of culture in contrast than the analysis reported here, which included a between-subject factor. Our data was thresholded to only allow clusters significant at a $p < 0.005$ level, with a cluster size larger than 80 mm^3 (10 voxels).

5.5.2 *Results*

5.5.2.1 Behavioral

In order to examine sex differences in performance, we replicated the analyses in Study 1 examining the influence of gender on culture. We found no behavioral differences between males and females, which is likely due to the reduced power in this smaller sample.

5.5.2.2 Neural

The neural results are reported in Table 5.1. To summarize our findings, we found that men had greater activation of the superior temporal sulcus (STS) in contrast to women, which is interesting, given that men showed less accuracy in Study 1 and in other variations of the mindreading task. Our own previous work found greater STS activation for same-culture individuals, which paralleled better performance for same-culture individuals. However, in this case, increased STS activation did not reflect better performance. In contrast, however, women showed greater activation in regions involved with the inferior frontal gyrus (IFG) and cerebellum than men did. These regions are known to be involved in mind reading, specifically using facial cues. The IFG is involved in understanding intentions and empathy for

Table 5.1 Regions active for the main effect of sex and interaction between sex and culture in Study 2

Main effect of sex						Ingroup beta		Outgroup beta	
Brain region	X	Y	Z	F	Extent	Male	Female	Male	Female
R. STS	58	−50	−2	14.17	59	0.178	−0.042	0.097	−0.069
L. Pons	−4	−32	−30	12.12	39	−0.048	0.042	−0.044	0.103
R. precentral gyrus	32	0	46	12.04	12	0.080	−0.004	0.076	−0.019
R. cerebellum	8	−38	−26	12.04	10	−0.055	0.023	−0.053	0.107
L. posterior cingulate	−14	−42	24	10.85	12	0.042	−0.018	0.044	−0.047
R. inferior frontal gyrus	42	36	−8	10.3	13	−0.037	0.051	−0.023	0.076
L. lingual gyrus	−46	−68	−18	10.15	12	0.127	0.280	−0.126	0.313
Interaction									
L. mPFC	−12	54	12	19.18	85	0.001	−0.113	−0.112	−0.003
R. caudate	2	20	−2	18.08	117	0.059	−0.049	−0.195	0.041
R. anterior insula	20	24	−10	17.98	38	0.043	−0.040	−0.088	0.021
L. IFG	−26	26	8	14.6	31	−0.009	0.052	0.106	0.005
R. OFC	34	54	−16	13.87	16	0.085	−0.097	−0.071	0.046
R. precentral gyrus	42	−4	38	13.82	19	−0.029	0.092	0.102	0.002
R. anterior PFC	12	66	−2	12.6	16	0.034	−0.089	−0.135	0.009
R. vmPFC	6	50	−18	12.49	68	0.097	0.008	−0.097	0.072
L. vmPFC	−12	50	−12	12.31	68	0.019	−0.022	−0.093	0.063
L. temporal pole	−58	0	−28	12.47	64	0.070	−0.001	−0.066	0.088
L. cerebellum	−8	−88	−20	12.18	42	0.380	0.095	−0.265	0.382
R. ACC	12	34	−8	11.55	36	0.076	−0.025	−0.066	0.018
L. parietooccipital sulcus	−34	−58	12	11.2	12	0.068	−0.002	−0.048	0.070
L. dmPFC	−2	62	32	11.08	23	0.202	0.064	−0.109	0.207
L. posterior cingulate	−2	−56	28	10.18	21	0.090	−0.107	−0.245	0.016

Beta values reflect the mean level of activation for mental state decoding minus the baseline task
ACC anterior cingulate, *dmPFC* dorsomedial prefrontal cortex, *IFG* inferior frontal gyrus, *mPFC* medial prefrontal cortex, *OFC* orbitofrontal cortex, *STS* superior temporal sulcus, *vmPFC* ventromedial prefrontal cortex

emotional states (e.g., Beeney, Franklin, Levy, & Adams, 2011; Franklin et al., 2013; Shamay-Tsoory, Aharon-Peretz, & Perry, 2009), and activation may reflect greater empathy for emotional states in women than men. The cerebellum has been posited to have several roles in mentalizing, including mirroring others' behaviors or examining the impact of traits on others' behaviors (Van Overwalle, Baetens, Mariën, & Vandekerckhove, 2014).

We also found many regions active for the interaction between gender and emotion. Most of these regions were more responsive for males viewing ingroup faces and females viewing outgroup faces. These regions include the medial prefrontal cortex, cerebellum, and temporal poles. Each of these regions is implicated in mindreading affective states. The temporal poles are involved in reading others' affective states (Shamay-Tsoory, Tomer, Berger, Goldsher, & Aharon-Peretz, 2005). The medial prefrontal cortex, especially including ventral regions, is involved in understanding what similar others are thinking, largely due to self-referential thought (Mitchell, Banaji, & Macrae, 2005; Mitchell, Macrae, & Banaji, 2006). These findings suggest that men tend to use more self-referential processing of affective states for ingroup members versus outgroup members, but that this pattern is reduced in women, who may even tend to show the reverse pattern in these brain regions.

5.5.3 *Discussion*

In contrast to finding only a main effect of gender in performance on the mindreading task in Study 1, we found several neural regions which responded differently as a function of gender and for the interaction between gender and culture. This activation reflects the pattern that males showed more mentalizing-related neural activation to same-culture individuals as opposed to females, who showed less of a difference between ingroup and outgroup individuals or even may have shown higher levels of activation for outgroup faces. These findings suggest women show less of a cultural ingroup effect than men in their neural activation in these regions. Though we found no behavioral evidence of this effect, our neural evidence reveals that men might be more prone, in both cultures, to differentially processing same- versus other-culture/race faces, which deserves extended examination in future research. Such differential processing could have implications for approaches to remediation that could have implications for cross-group mental state decoding, with particular consequence for cross-race, and international relations.

5.6 General Discussion

Study 1 revealed the predicted interaction between culture of participant and race of stimulus eyes, replicating that found in our previous fMRI study (Adams et al., 2010). No main effects were found and direct comparisons showed that Japanese participants were both better at reading mental states from Asian than Caucasian eyes and better at reading Asian eyes than were Caucasian participants, with the opposite pattern of effects evident for Caucasian participants. The only main effect to reach significance was that for sex of participant, such that women were better than men at decoding mental states overall, replicating previous findings for sex differences in the Reading the Mind in the Eyes Test (Baron-Cohen et al., 2001; Hall, 1978). However, no interactions were found between sex of participant and other variables of interest.

Previous work examining these effects has focused on individual differences, particularly those associated with clinical dysfunction such as autism that is marked, and to some extent defined by impairments to social perception. The current work demonstrates that this ability varies not only on the individual level but as a function of who is being read and their social group membership in relation to who is doing the reading. Study 1 demonstrated an intracultural advantage for mental state reasoning and main effect of participant gender. Study 2 examined, through reanalysis (see Adams et al., 2010), whether these effects extend to neural responsivity as well. This reanalysis of existing data replicated those previously found for culture of participant by race of stimulus face. As in Study 1, a main effect of gender was also found, with female participants showing greater activation in several regions related to socioemotional processing (see Adams et al., 2010). Critically, this reanalysis also revealed an interaction between gender and culture of participant in several regions involved with mind reading. These findings are consistent with men showing greater neural activations for ingroups whereas women show less of an ingroup bias, or even greater neural activation for outgroups.

These findings have clear implications for better understanding cross-race and international relations by underscoring the profound consequences impairments in cross-race mental state decoding can have. For instance, these findings support recent conjectures in the popular media that a breakdown in the otherwise universal ability to read the mental states of others must be at least partially responsible for incidents such as the Amadou Diallo killing, in which four White police officers shot 41 rounds at an unarmed Black man. Diallo must have been expressing sheer terror, yet the police officers apparently read violent motive into his expression. Indeed, in the laboratory, White participants are more prone to shooting Black than White targets in simulated trials (e.g., Correll, Park, Wittenbrink, & Judd, 2002). Future research efforts are necessary to determine if individual differences in the intra-racial advantage for mental state decoding may predict the extent to which individuals are prone to such errors.

The question that remains is whether the mechanism of influence for this effect is predominantly perceptual or cognitive in nature. A lack of exposure to other-race faces may explain intracultural advantages. Empirical tests have supported this theory by demonstrating that increased contact with individuals of other racial groups yielded increased cross-race memory ability as well (e.g., Chiroro & Valentine, 1995; Cross, Cross, & Daly, 1971; Hancock & Rhodes, 2008; Valentine & Endo, 1992) and that own-race faces are processed holistically or configurally as compared to other-race faces (e.g., McKone, Brewer, MacPherson, Rhodes, & Hayward, 2007; Michel, Rossion, Han, Chung, & Caldara, 2006; Tanaka, Kiefer, & Bukach, 2004; Turk, Handy, & Gazzaniga, 2005). Similar effects may be apparent for perceptual exposure to expression in the face as well (see Elfenbein & Ambady, 2002). As already noted, simple social categorization can affect the depth of processing of a face (resulting in greater individuation and subsequent memory for own-race faces). This speaks to the possibility that this effect may extend to other social groups as well. If what determines the level to which we are able to extract subtle, nuanced cues from faces (the kind of visual information that is predictive of face memory) are how we categorize others as being an ingroup member relative to an outgroup member; it stands to reason that it also determines the extent to which we are able to extract complex social information (such as emotional expressions).

Taken together, these studies offer the first comprehensive evidence we are aware of that mental state decoding ability varies as a function of either cultural or racial group memberships. Notably Japanese and US Caucasian participants (Study 1) showed a clear detriment in the ability to read meaning from other-racial groups' eyes. The implications for this are clear for international relations, both corporate and political, and for racial relations within the same culture. The extent to which we depend on nonverbal cues in cross-cultural and cross-racial exchange can have profound consequences. How generalizable these effects might be to regional differences or even to smaller social cliques, however, remains an open question. What is clear is that at least part of phenomenon in question is driven by more generalizable intergroup process.

5.7 Implications for International Relations

The evidence we present here shows the importance of cultural group membership in mental state reasoning. Regardless of the cause of these effects, the findings show that cultural group membership affects how we read others' emotional messages, which is an important factor to consider given how much we use nonverbal cues to communicate with those in other cultures. When this type of communication breaks down, consequences can be severe. As situations become increasingly ambiguous (which can be a result of unclear face-to-face interpersonal communication), so too does the opportunity for group biases to enter the picture, as people are likely to employ heuristics (which include bias) to manage their social lives and make decisions. Thus, decreased accuracy in emotion recognition capacity would allow for those participating in international relations to fall back on their racial and/or cultural prejudices (if they are indeed present) and to perhaps suspect deceit during meaningful discourse. This, clearly, is not a recipe for successful relations, especially when breakdowns in communication can result in tragedy, including war, genocide, and economic turmoil (as some extreme examples). In this section of the paper, we consider some of these issues and discuss possible ways of addressing cross-cultural mind reading in international relations.

One important explanation to consider is a phenomenon that can be considered a special type of own-group bias, that of infrahumanization. Infrahumanization refers to the belief that the human nature (or essence, encompassing such qualities as intelligence, language, and emotionality) of ingroup members is qualitatively different than that of outgroup members (Leyens et al., 2001). In a number of studies, (Paladino et al., 2002; Vaes, Paladino, & Leyens, 2006), participants have reported associated primary (or nonuniquely human) emotions more with outgroup members and complex (or uniquely human) emotions with ingroup members. This may also lead to detriments in how complex emotions are decoded from facial expressions in outgroup members, since an associative link between emotional complexity and outgroup membership does not exist, creating a much more difficult process. Infrahumanization may help explain why we struggle in reasoning about the mental states of those from other cultures, but it provides a potential mechanism to help address those deficits. If people are able to understand the uniquely human experiences within other cultures, then they may be more likely to attribute more complex emotions to other cultures and thus be better at mind reading those in other cultures. This suggests that learning about specifically human attributes of other cultures may help reduce deficits in intergroup mind reading.

The present findings can also be considered through the logic of the common ingroup identity model (Dovidio, Gaertner, & Saguy, 2009; Gaertner & Dovidio, 2000). This model is based on social identity theory (Tajfel & Turner, 1979) and self-categorization theory (Turner et al., 1987) and focuses on the recategorization process. Since social categorization is not a fixed process, increasing the relevance of a more inclusive social group (one that allows more members) can result in fewer of the negative consequences of exclusive categorization. The manipulation of the relevance and inclusiveness of a social category has yet to be examined in the context of visual person perception and thus seems like a prudent avenue for future work in the study of intergroup communication. Specifically, it provides hope that one's web of social inclusion may be malleable in the sense that it can be cast wider, thus helping to improve accurate intergroup communication of emotions and thus intergroup relations.

The present work took advantage of participants' exclusive social categorization (i.e., using outgroups for comparison) as a part of its design. The common ingroup identity model would suggest that, by utilizing a sufficiently expansive ingroup definition, similar effects could be found without the need for an outgroup to suffer the negative consequences of categorization processes. The most expansive ingroup in the context of person perception would, of course, be "human." That infrahumanization of at least some outgroup members appears to a consequence of social categorization suggests that expanding ingroup inclusivity to its most beneficial level may hinge on the ability to exorcise the tendency to *deny humanity* to perceived outgroup members.

Our findings in regard to gender also have important implications for international relations. The findings here, along with other research, suggest that women show less of a cultural ingroup effect compared to men. One possible reason for this is that men tend to be more competitive while completing ingroup tasks whereas women tend to value cooperation (e.g., Eckel & Grossman, 2001). The increased value that women reportedly tend to place on cooperation may make it more important to understand what others are thinking, especially for other-culture individuals, and lead them to allocate additional mental resources to understanding other-culture individuals. This increased value that women place on reasoning about the mental states of other-culture individuals may be a function of socially learned gender roles (e.g., an emphasis on maintaining communal harmony), rather than a biologically based ability. If true, this suggests that potential international relation professionals could be trained to overcome difficulties in decoding complex emotional states of others of different cultures. This specifically could be done by manipulations designed to increase the importance of cooperation by focusing on working together to accomplish common goals (e.g. Gaertner et al., 2000; Sherif, 1958). This is a very encouraging proposition, as it could prove to be a feasible form of training, especially considering the profound international problems that could be avoided.

To summarize, this research shows the importance of cultural in considering the language of the eyes. Like with verbal language, important meaning may be "lost in translation" when reading the nonverbal cues from those from other cultures. However, the research reviewed here offers potential solutions to addressing these problems. Understanding the cultural background of others as well as allocating additional effort may provide a way to address these misunderstandings and help ensure smooth international relations.

Acknowledgements While writing this paper, we lost our dear friend and co-author Nalini Ambady. It is with our deepest sadness and greatest respect that we dedicate this paper to her memory.

We thank Sakiko Yoshikawa of Kyoto University for helping us generate the Asian Reading the Mind in the Eyes Test and to which we owe our data from Japanese participants (reported in Study 1).

This research was supported in part by the Social Science Research Institute, the Pennsylvania State University to RBA, Jr.

References

Adams, Jr., R. B., Rule, N. O., Franklin, R. G., Wang, E., Stevenson, M. T., Yoshikawa, S., et al. (2010). Cross-cultural reading the mind in the eyes: An fMRI investigation. *Journal of Cognitive Neuroscience, 22*, 97–108.

Adolphs, R., Gosselin, F., Buchanan, T. W., Tranel, D., Schyns, P., & Damasio, A. R. (2005). A mechanism for impaired fear recognition after amygdala damage. *Nature, 433*(7021), 68–72.

Allison, T., Puce, A., & McCarthy, G. (2000). Social perception from visual cues: role of the STS region. *Trends in cognitive sciences, 4*, 267–278.

Aronson, J., Lustina, M. J., Good, C., Keough, K., Steele, C. M., & Brown, J. (1999). When white men can't do math: Necessary and sufficient factors in stereotype threat. *Journal of Experimental Social Psychology, 35*, 29–46.

Avis, J., & Harris, P. L. (1991). Belief-desire reasoning among Baka children: Evidence for a universal conception of mind. *Child Development, 62*(3), 460–467.

Baron-Cohen, S. (1995). *Mindblindness: An essay on autism and theory of mind*. Boston: MIT Press/Bradford Books.

Baron-Cohen, S., Ring, H. A., Wheelwright, S., Bullmore, E. T., Brammer, M. J., Simmons, A., et al. (1999). Social intelligence in the normal and autistic brain: An fMRI study. *European Journal of Neuroscience, 11*, 1891–1898.

Baron-Cohen, S., Riviere, A., Fukushima, M., French, D., Hadwin, J., Cross, P., et al. (1996). Reading the mind in the face: A cross-cultural and developmental study. *Visual Cognition, 3*, 39–59.

Baron-Cohen, S., Wheelwright, S., Hill, J., Raste, Y., & Plumb, I. (2001). The "reading the mind in the eyes" test revised version: A study with normal adults, and adults with asperger syndrome or high-functioning autism. *Journal of Child Psychology and Psychiatry, 42*, 241–251.

Baron-Cohen, S., Wheelwright, S., & Jolliffe, T. (1997). Is there a "language of the eyes?" Evidence from normal adults, and adults with autism or asperger syndrome. *Visual Cognition, 4*, 311–331.

Bernstein, M. J., Young, S. G., & Hugenberg, K. (2007). The cross-category effect mere social categorization is sufficient to elicit an own-group bias in face recognition. *Psychological Science, 18*, 706–712.

Bruce, V., & Young, A. (1986). Understanding face recognition. *British Journal of Psychology, 77*, 305–327.

Brune, M., & Brune-Cohrs, U. (2006). Theory of mind: Evolution, ontogeny, brain mechanisms and psychopathology. *Neuroscience and Biobehavioral Reviews, 30*, 437–455.
Caldara, R., Rossion, B., Bovet, P., & Hauert, C. (2004). Event-related potentials and time course of the “other-race” face classification advantage. *Neuroreport, 15*(5), 905–910.
Caldara, R., Thut, G., Servoir, P., Michel, C. M., Bovet, P., & Renault, B. (2003). Face versus non-face object perception and the “other-race” effect: A spatio-temporal event-related potential study. *Clinical Neurophysiology, 114*, 515–528.
Chiao, J. Y., & Ambady, N. (2007). Cultural neuroscience: Parsing universality and diversity across levels of analysis. In D. Cohen & S. Kitayama (Eds.), *Handbook of cultural psychology* (pp. 237–254). New York, NY: Guilford Press.
Chiroro, P., & Valentine, T. (1995). An investigation of the contact hypothesis of the own-race bias in face recognition. *The Quarterly Journal of Experimental Psychology, 48*, 879–894.
Corenblum, B., & Meissner, C. A. (2006). Recognition of faces of ingroup and outgroup children and adults. *Journal of Experimental Child Psychology, 93*, 187–206.
Correll, J., Park, B., Wittenbrink, B., & Judd, C. M. (2002). The police officer’s dilemma: Using ethnicity to disambiguate potentially threatening individuals. *Journal of Personality and Social Psychology, 83*, 1314–1329.
Cross, J. F., Cross, J., & Daly, J. (1971). Sex, race, age, and beauty as factors in recognition of faces. *Perception & Psychophysics, 10*, 393–396.
Darwin, C. (1872/1965). *The expression of the emotions in man and animals*. Chicago: University of Chicago Press.
Dennett, D. C. (1978). Beliefs about beliefs. *Behavioral and Brain Sciences, 1*, 568–570.
DePaulo, B. M., & Friedman, H. S. (1998). Nonverbal communication. In D. Gilbert, S. T. Fiske, & G. Lindzey (Eds.), *Handbook of social psychology* (4th ed., Vol. 2, pp. 3–40). New York: McGraw-Hill.
Dovidio, J. F., Gaertner, S. L., & Saguy, T. (2009). Commonality and the complexity of “we”: Social attitudes and social change. *Personality and Social Psychology Review, 13*, 3–20.
Eberhardt, J. L., Goff, P. A., Purdie, V. J., & Davies, P. G. (2004). Seeing black: Race, crime, and visual processing. *Journal of Personality and Social Psychology, 87*(6), 876–893.
Eckel, C. C., & Grossman, P. J. (2001). Chivalry and solidarity in ultimatum games. *Economic Inquiry, 39*(2), 171–188.
Ekman, P. (1972). Universals and cultural differences in facial expressions of emotion. In J. Cole (Ed.), *Nebraska symposium on motivation; 1971* (Vol. 19, pp. 207–282). Lincoln: University of Nebraska Press.
Elfenbein, H. A., & Ambady, N. (2002). On the universality and cultural specificity of emotion recognition: A meta-analysis. *Psychological Bulletin, 128*(2), 203–235.
Elfenbein, H. A., & Ambady, N. (2003a). Universals and cultural differences in recognizing emotions. *Current Directions in Psychological Science, 12*(5), 159–164.
Elfenbein, H. A., & Ambady, N. (2003b). When familiarity breeds accuracy: Cultural exposure and facial emotion recognition. *Journal of Personality and Social Psychology, 85*, 276–290.
Flavell, J. H. (1999). Cognitive development: Children’s knowledge about the mind. *Annual Review of Psychology, 50*, 21–45.
Gaertner, S. L., & Dovidio, J. F. (2000). *Reducing intergroup bias: The common ingroup identity model*. Philadelphia: Psychology Press.
Gaertner, S. L., Dovidio, J. F., Banker, B. S., Houlette, M., Johnson, K. M., & McGlynn, E. A. (2000). Reducing intergroup conflict: From superordinate goals to decategorization, recategorization, and mutual differentiation. *Group Dynamics: Theory, Research, and Practice, 4*(1), 98–114.
Hall, J. (1978). Gender effects in decoding nonverbal cues. *Psychological Bulletin, 85*, 845–858.
Hancock, K. J., & Rhodes, G. (2008). Contact, configural coding and the other-race effect in face recognition. *British Journal of Psychology, 99*(1), 45–56.
Hugenberg, K., & Sacco, D. F. (2008). Social categorization and stereotyping: How social categorization biases person perception and face memory. *Social and Personality Psychology Compass, 2*, 1052–1072.

Irani, F., Platek, S. M., Panyavin, I. S., Calkins, M. E., Kohler, C., Siegel, S. J., et al. (2006). Self-face recognition and theory of mind in patients with schizophrenia and first-degree relatives. *Schizophrenia Research, 88*, 151–160.

Ito, T. A., Thompson, E., & Caioppo, J. T. (2004). Tracking the timecourse of social perception: The effects of racial cues on event-related brain potentials. *Personality and Social Psychology Bulletin, 30*, 1267–1280.

Ito, T. A., & Urland, G. R. (2003). Race and gender on the brain: Electrocortical measures of attention to the race and gender of multiply categorizable individuals. *Journal of Personality and Social Psychology, 85*(4), 616–626.

Ito, T. A., & Urland, G. R. (2005). The influence of processing objectives on the perception of faces: An ERP study of race and gender perception. *Cognitive, Affective, & Behavioral Neuroscience, 5*(1), 21–36.

Kawaguchi, D. (2007). A market test for sex discrimination: Evidence from Japanese firm-level panel data. *International Journal of Industrial Organization, 25*, 441–460.

Kelly, D. J., Quinn, P. C., Slater, A. M., Lee, K., Gibson, A., Smith, M., & Pascalis, O. (2005). Three-month-olds, but not newborns, prefer own-race faces. *Developmental Science, 8*, F31–F36.

Kobayashi, C., Glover, G. H., & Temple, E. (2006). Cultural and linguistic influence on neural bases of "theory of mind": An fMRI study with Japanese bilinguals. *Brain and Language, 98*, 210–220.

Kobayashi, C., Glover, G. H., & Temple, E. (2007). Children's and adults' neural bases of verbal and nonverbal 'theory of mind'. *Neuropsychologia, 45*, 1522–1532.

Levin, D. T. (1996). Classifying faces by race: The structure of face categories. *Journal of Experimental Psychology, 22*, 1364–1382.

Levin, D. T. (2000). Race as a visual feature: Using visual search and perceptual discrimination tasks to understand face categories and the cross-race recognition deficit. *Journal of Experimental Psychology: General, 128*(4), 559–574.

Leyens, J. P., Rodriguez-Perez, A., Rodriguez-Torres, R., Gaunt, R., Paladino, M. P., Vaes, J., et al. (2001). Psychological essentialism and the differential attribution of uniquely human emotions to ingroups and outgroups. *European Journal of Social Psychology, 31*, 395–411.

Lillard, A. S. (1998). Ethnopsychologies: Cultural variations in theories of mind. *Psychological Bulletin, 123*, 3–32.

Locke, V., Macrae, C. N., & Eaton, J. L. (2005). Is person categorization modulated by exemplar typicality? *Social Cognition, 23*(5), 417–428.

Maass, A., Cadinu, M., Guarnieri, G., & Grasselli, A. (2003). Sexual harassment under social identity threat: The computer harassment paradigm. *Journal of Personality and Social Psychology, 85*, 853–870.

Maclin, O. H., & Malpass, R. S. (2001). Racial categorization of faces: The ambiguous race face effect. *Psychology, Public Policy, and Law, 7*(1), 98–118.

Macrae, C. N., & Martin, D. (2007). A boy primed sue: Feature-based processing and person construal. *European Journal of Social Psychology, 37*, 793–805.

Maddox, K. B. (2004). Perspectives on racial phenotypicality bias. *Personality and Social Psychology Review, 8*(4), 383–401.

Markus, H., Kitayama, S., & Heiman, R. (1996). Culture and "basic" psychological principles. In E. T. Higgins & A. W. Kruglanski (Eds.), *Social psychology: Handbook of basic principles*. New York: Guilford.

Mason, M. F., Cloutier, J., & Macrae, C. N. (2006). On construing others: Category and stereotype activation from facial cues. *Social Cognition, 24*, 540–562.

Mason, M. F., & Macrae, C. N. (2004). Categorizing and individuating others: The neural substrates of person perception. *Journal of Cognitive Neuroscience, 16*, 1785–1795.

McKone, E., Brewer, J. L., MacPherson, S., Rhodes, G., & Hayward, W. G. (2007). Familiar other-race faces show normal holistic processing and are robust to perceptual stress. *Perception, 36*(2), 224.

Meissner, C. A., & Brigham, J. C. (2001). Thirty years of investigating the own-race bias in memory for faces. *Psychology, Public Policy, and Law, 7*, 3–35.
Michel, C., Rossion, B., Han, J., Chung, C., & Caldara, R. (2006). Holistic processing is finely tuned for faces of one's own race. *Psychological Science, 17*, 608–615.
Mitchell, J. P., Banaji, M. R., & Macrae, C. N. (2005). The link between social cognition and self-referential thought in the medial prefrontal cortex. *Journal of cognitive neuroscience, 17*(8), 1306–1315.
Mitchell, J. P., Macrae, C. N., & Banaji, M. R. (2006). Dissociable medial prefrontal contributions to judgments of similar and dissimilar others. *Neuron, 50*, 655–663.
Morton, J., & Johnson, M. H. (1991). CONSPEC and CONLERN: a two-process theory of infant face recognition. *Psychological Review, 98*, 164–181.
Navarrete, C. D., McDonald, M. M., Molina, L. E., & Sidanius, J. (2010). Prejudice at the nexus of race and gender: An outgroup male target hypothesis. *Journal of Personality and Social Psychology, 98*(6), 933–945.
Nemoto, K. (2013). When culture resists progress: Masculine organizational culture and its impacts on the vertical segregation of women in Japanese companies. *Work, Employment & Society, 27*, 153–169.
Ostrom, T. M., & Sedikides, C. (1992). Out-group homogeneity effects in natural and minimal groups. *Psychological Bulletin, 112*, 536–552.
Paladino, P. M., Leyens, J. P., Rodriguez, R. T., Rodriguez, A. P., Gaunt, R., & Demoulin, S. (2002). Differential association of uniquely and nonuniquely human emotions to the ingroup and the outgroups. *Group Processes and Intergroup Relations, 5*, 105–117.
Pauker, K., Ambady, N., Weisbuch, M., Sommers, S. R., Adams, R. B., Jr., & Ivcevic, Z. (2009). Not so black and white: Memory for ambiguous group members. *Journal of Personality and Social Psychology, 96*, 795–810.
Perner, J., Ruffman, T., & Leekam, S. R. (1994). Theory of mind is contagious: You catch it from your sibs. *Child Development, 65*, 1228–1238.
Peterson, C. C. & Siegal M. (1997). Psychological, physical, and biological thinking in normal, autistic, and deaf children. In T*he Emergence of Core Domains of Thought: Children's Reasoning About Physical, Psychological, and Biological Phenomena* (*New Directions for Child Development*, No. 75) (pp. 55–70). San Francisco: Jossey-Bass.
Rule, N. O., Ambady, N., Adams, R. B., Jr., & Macrae, C. N. (2008). Accuracy and awareness in the perception and categorization of male sexual orientation. *Journal of Personality and Social Psychology, 95*, 1019–1028.
Sabbagh, M. A. (2004). Understanding orbitofrontal contributions to theory-of-mind reasoning: Implications for autism. *Brain and Cognition, 55*, 209–219.
Sangrigoli, S., & De Schonen, S. (2004a). Effect of visual experience on face processing: a developmental study of inversion and non-native effects. *Developmental Science, 7*, 74–87.
Sangrigoli, S., & De Schonen, S. (2004b). Recognition of own-race and other-race faces by three-month-old infants. *Journal of Child Psychology and Psychiatry, 45*, 1219–1227.
Shamay-Tsoory, S. G., Aharon-Peretz, J., & Perry, D. (2009). Two systems for empathy: A double dissociation between emotional and cognitive empathy in inferior frontal gyrus versus ventromedial prefrontal lesions. *Brain, 132*, 617–627.
Shamay-Tsoory, S. G., Tomer, R., Berger, B. D., Goldsher, D., & Aharon-Peretz, J. (2005). Impaired "affective theory of mind" is associated with right ventromedial prefrontal damage. *Cognitive and Behavioral Neurology, 18*, 55–67.
Sherif, M. (1958). Superordinate goals in the reduction of intergroup conflict. *American Journal of Sociology, 63*, 349–356.
Skwerer, D. P., Verbalis, A., Schofield, C., Faja, S., & Tager-Flusberg, H. (2006). Social-perceptual abilities in adolescents and adults with Williams syndrome. *Cognitive Neuropsychology, 23*(2), 338–349.
Sporer, S. L. (2001). Recognizing faces of other ethnic groups: An integration of theories. *Psychology, Public Policy, and Law, 7*, 36–97.

Steele, C. M., & Aronson, J. (1995). Stereotype threat and the intellectual test performance of African Americans. *Journal of Personality and Social Psychology, 69*, 797–811.
Stevenson, M. T., Soto, J. A., & Adams, R. B., Jr. (2012). More than meets the eye: The role of self-identity in decoding complex emotional states. *Emotion, 12*, 882–886.
Tager-Flusberg, H. (2001). A reexamination of the theory of mind hypothesis of autism. In J. A. Burack, T. Charman, N. Yimiya, & P. R. Zelazo (Eds.), *The development of autism: Perspectives from development and theory* (pp. 173–193). Mahwah, NJ: Erlbaum.
Tajfel, H. (1969). Cognitive aspects of prejudice. *Journal of Social Issues, 25*(4), 79–97.
Tajfel, H., & Turner, J. C. (1979). An integrative theory of intergroup conflict. In W. G. Austin & S. Worchel (Eds.), *The social psychology of intergroup relations* (pp. 33–48). Monterey, CA: Brooks/Cole.
Tanaka, J. W., Kiefer, M., & Bukach, C. M. (2004). A holistic account of the own-race effect in face recognition: Evidence from a cross-cultural study. *Cognition, 93*, B1–B9.
Toh, S. M., & Leonardelli, G. J. (2012). Cultural constraints on the emergence of women as leaders. *Journal of World Business, 47*(4), 604–611.
Turk, D. J., Handy, T. C., & Gazzaniga, M. S. (2005). Can perceptual expertise account for the own-race bias in face recognition? A split-brain study. *Cognitive Neuropsychology, 22*, 877–883.
Turner, J. C., Hogg, M. A., Oakes, P. J., Reicher, S. D., & Wetherell, M. S. (1987). *Rediscovering the social group: A self-categorization theory*. Oxford, UK: Basil Blackwell.
Vaes, J., Paladino, M. P., & Leyens, J. (2006). Priming uniquely human emotion and the in-group(but not the out-group) activates humanity concepts. *European Journal of Social Psychology, 36*, 169–181.
Valentine, T., & Endo, M. (1992). Towards an exemplar model of face processing: The effects of race and distinctiveness. *The Quarterly Journal of Experimental Psychology, 44*(4), 671–703.
Van Overwalle, F., Baetens, K., Mariën, P., & Vandekerckhove, M. (2014). Social cognition and the cerebellum: A meta-analysis of over 350 fMRI studies. *NeuroImage, 86*, 554–572.
Wood, W., & Eagly, A. H. (2002). A cross-cultural analysis of the behavior of women and men: Implications for the origins of sex differences. *Psychological Bulletin, 128*(5), 699–729.
Zebrowitz, L. A. (2006). Finally, faces find favor. *Social Cognition, 24*, 657–701.

Chapter 6
Brain-As-Predictor Approach: An Alternative Way to Explore Acculturation Processes

Pin-Hao A. Chen, Todd F. Heatherton, and Jonathan B. Freeman

Abstract Chen, Heatherton, and Freeman review the unique strategy of longitudinally assessing the neural changes that occur during the acculturation process. This process differs from the typical research strategy of comparing individuals from two unique cultures or using bicultural individuals and priming them with different cultural cues. The authors use this strategy to determine whether longitudinal study of neural responses can predict acculturation better than typical behavioral measures.

Neuroimaging studies indicate that the ventromedial prefrontal cortex is activated in response to self-judgments in both Western and Eastern cultures and in response to judgments about close others (e.g., one's mother) in Eastern cultures. Longitudinal analysis showed that Chinese immigrants to the USA who displayed greater acculturation had more activation in this brain region for self-judgments than for judgments about close others.

Studies of the reaction to positive emotion expressions of in-group and out-group faces have shown activation of the ventral striatum, a structure in the brain's reward circuit. Longitudinal study of Chinese graduate students who were in school in the USA showed that higher ventral striatum activity to in-group facial expressions showed more in-group friends on a social networking website. Lower ventral striatum activity to in-group facial expressions showed lower number of in-group friends on a social networking website.

The authors conclude by offering several ideas for future research, including using diffuse tensor imaging to investigate white matter integrity changes over time and using more cultural specific cues in experimental protocol.

P.-H.A. Chen (✉) • T.F. Heatherton
Department of Psychological and Brain Sciences, Dartmouth College,
6207 Moore Hall, Hanover, NH 03755, USA
e-mail: pin-hao.chen.gr@dartmouth.edu; todd.f.heatherton@dartmouth.edu

J.B. Freeman
Department of Psychology, New York University,
6 Washington Place, New York, NY 10003, USA
e-mail: jon.freeman@nyu.edu

J.E. Warnick, D. Landis (eds.), *Neuroscience in Intercultural Contexts*,
International and Cultural Psychology, DOI 10.1007/978-1-4939-2260-4_6

Due to increasing globalization within the past decade, immigration has become both more large-scale and frequent. According to a recent report from the International Organization for Migration (2008), more than 200 million immigrants now exist worldwide. Their top destination is Europe, followed by North America and Asia. Among migrant populations, immigrants from Asian countries form the largest majority. An enormous number of intercultural contacts occur between these millions of immigrants. These cross-cultural exchanges result in changes at both group (e.g., cultural) and individual (e.g., psychological) levels (Berry, 1997). At the group level, both the migrant and native group influence changes in their respective cultures. Indeed, intercultural contact inevitably compels incorporation and evolution for every cultural group. However, analysis at the group level is beyond the scope of this review. In this chapter, we will instead focus mainly on the individual level of intercultural impact.

What happens when people move from their native countries and familiar cultures to settle in an entirely new environment? Among a vast array of possible adaptations, they may need to adjust to an unfamiliar language, to understand locals' unfamiliar sayings and jokes, and to change their behaviors in order to be accepted by the new culture. These changes are usually depicted as a progressive process, which is called acculturation (Berry, 2003). Acculturation involves the ways in which people change following immersion within a new culture. It can be defined as a process of cultural and psychological adjustment following intercultural contact. Adaptation is another term frequently used in acculturation studies, which can be defined as the extent to which immigrants are able to conform to the acculturation process (Sam & Berry, 2010). It can be psychological (for example, well-being) or sociocultural (for example, learning a new language), and often is regarded as the consequence of acculturation.

Previous studies have found that individual differences in levels of acculturation can be enormous, even between immigrants who share a cultural origin and settle in the same region (Nauck, 2008). Four major types of acculturation strategies or states contribute to the acculturation outcome. These have been identified as integration, assimilation, separation, and marginalization (Sam & Berry, 2010). Integration can be described as a strategy by which immigrants maintain their own culture while also integrating the parameters of the host culture into their cultural value. Immigrants who use this strategy not only interact with their ethnic group but also show interest in interacting with out-group members. Use of this strategy generally correlates with better psychological and sociocultural adaption than that achieved using other strategies. Assimilation, on the other hand, carries the integrative aspect of the first strategy to an extreme. In assimilating, individuals abandon their original cultural values and try to accept all aspects of the host culture. Under this circumstance, immigrants will interact for the most part with the out-group members and avoid interacting with people from the same culture. Unlike integration or assimilation, separation is defined as a strategy that strongly emphasizes the maintenance of immigrants' own cultural values and identities, with less interest in the host culture. Individuals who use this strategy will interact predominantly with their own ethnic group and avoid interacting with out-group members. Although this seems like a less adaptive strategy, it actually produces some positive signs of adaptation (Berry, Phinney, Sam, & Vedder, 2006). The least adaptive strategy is marginalization. Marginalization describes immigrants who lack interest in either maintaining their own culture or accepting a new culture. In the end, they become socially isolated and potentially face exclusion by both in-group and out-group members.

Although researchers have clearly identified four distinct types of acculturation, the attempt to use behavioral assessment tools to predict immigrants' acculturation styles has not been successful (Sam & Berry, 2010). Developing such predictive tools is important because they may enable governments to provide adequate intervention programs for at-risk immigrants at the early stage of acculturation. Previous studies have tried to use the Big Five personality traits (Ward, Leong, & Low, 2004), motivation (Kosic, Kruglanski, Pierro, & Mannetti, 2004), or attachment styles (Bakker, van Oudenhoven, & van der Zee, 2004) to make predictions in acculturation, but the results from these studies were inconsistent and failed to make clear predictions. It seems that sticking to behavioral tools as a means of predicting acculturation simply does not produce accurate results. Using another method, such as the brain-as-predictor approach (Berkman & Falk, 2013), might be an alternative solution to form more precise predictions of immigrants' acculturation outcomes. Hopefully, this could lead to the development of sufficient resources to help immigrants adapt more healthfully to a new culture soon after arrival.

Since the rebirth of cultural psychology in the early 1990s, the number of studies investigating the impact of cultural differences on affective or cognitive processes has grown dramatically (Kitayama & Uskul, 2011). Following in the steps of cultural psychology, cultural neuroscience has become an interdisciplinary subfield in cognitive neuroscience. This new field has enabled researchers to explore how cultural values influence the neural mechanisms underlying different affective and cognitive processes (Han et al., 2013; Kitayama & Uskul, 2011).

Among studies in the field of cultural neuroscience, two major approaches exist. The first approach is to compare one group of individuals who live in one type of culture, such as an interdependent culture (e.g., Japan), to another group of individuals who live in another type, such as a collectivist culture (e.g., the USA) (Chiao et al., 2008, 2009b). This can be regarded as the monocultural comparison approach. Findings from such studies support the notion that neural mechanisms subserving the same psychological processes can be modulated by different cultural processes. The second approach is to recruit individuals who grew up under exposure to two distinct cultures, and to prime them with different cultural cues (Chiao et al., 2009a; Ng, Han, Mao, & Lai, 2010). Because bicultural individuals have been influenced by processes from two distinct cultures and have integrated them into their cognitive systems, they tend to show flexibility in neural responses according to which cultural values are primed at that moment. This is the bicultural priming approach, which is also another common approach in this field.

On the other hand, longitudinal designs, which can be used to answer the key questions surrounding acculturation processes, have rarely been used in cultural neuroscience studies. Based on a recent review (Kitayama & Uskul, 2011), acculturation is revealed to be not only a learning process but also an adaption process.

Longitudinal designs are therefore required in order to form a better understanding of what happens in each subprocess. In addition to this advantage, the implementation of longitudinal designs can test whether using neural responses is more accurate than using behavioral measurements to predict immigrants' acculturation outcomes over time. This approach can provide an alternative avenue for cultural psychologists to gain a deeper understanding of dynamic cultural-learning processes.

It has long been known that there are prominent cultural differences in both the concept of the self as well as in thinking about the relationship between self and others (Markus & Kitayama, 1991). Furthermore, these cultural differences in self-construal influence almost every aspect of affective and cognitive processing. Thus, understanding how self-construal modulates the neural representations of the self and others may be the first step towards understanding the processes and consequences of acculturation (Kitayama & Park, 2010). In the first part of this chapter, we review what brain regions are involved in representations of abstract knowledge of self and others, followed by a review on how cultural differences in self-construal modulate the neural representations of the self and close others. This first section concludes with the findings from our own longitudinal studies on immigrants, which may provide more insight into their acculturating neural systems. In the second part of this chapter, we review the neural mechanisms involved in social affiliative and avoidant behaviors, and the brain regions subserving in-group favoritism. Since acculturation is always involved in affiliating with or avoiding in-group or out-group members, understanding these mechanisms will enhance our understanding of acculturation processes. Next, we focus on reward-related processing, which may play an important role in in-group favoritism. At the end of this section, we present our recent findings exploring how individual differences in reward reactivity can predict immigrants' friendship patterns. These findings might provide more information about a determining force in the acculturation process. In the final section of this chapter, we propose an integrated brain-as-predictor approach, which brings together self-referential processing and reward processing at the same time, and uses the data to predict immigrants' acculturation outcomes. We also propose several possible research lines that intercultural neuroscientists could use to generate and test hypotheses in their future studies. We hope that this chapter will provide an alternative approach for cultural neuroscientists who seek a deeper understanding of acculturation processes, filling in this critical part of cultural neuroscience.

6.1 Cultural Differences Modulate Neural Representations of the Self and Close Others

If you have ever travelled in China or in Chinatown in the USA, you may have noticed the different shape of dining tables at restaurants. The tables are round, which may enable a group of up to ten people to talk to each other clearly, but hardly takes personal space into account. By contrast, the shape of tables in an American restaurant is usually rectangular, which allows for more personal space for individuals. Under this circumstance, the people whom you interact with most are the ones who sit beside or in front of you. This slight difference in table shape echoes more fundamental differences in concepts of the self and the relationship with others across cultures.

According to a comprehensive review on cultural differences in self-construal, two distinct types of self-construals, namely, independent and interdependent self-construal, have been identified (Markus & Kitayama, 1991). Independent self-construal involves viewing the self as an entity, which is defined by dispositional attributes and is relatively detached from context. As a consequence, individuals with independent self-construal emphasize their autonomy from others and their uniqueness. Their self-esteem is, in large part, dependent upon being better than others. Furthermore, individuals' behaviors are often attributed to personal internal factors rather than contextual factors. The self is regarded as being independent from others, even close others. The boundary between the self and others is solid and clear. By contrast, interdependent self-construal involves viewing the self as enmeshed within the social network. The self is partially defined by the social context and is less differentiated from others. As a consequence, interdependent individuals emphasize the control of personal desires in order to pursue harmonious social interactions, and self-esteem is regarded as synonymous with the capability to exert control over one's own needs. Instead of being attributed to one's internal factors, individuals' behaviors are more readily attributed to contextual factors. Thus, the self is embedded in the social network, and the boundary between the self and others is fuzzier than that of individuals with independent self-construals.

6.1.1 *Neural Substrates for Self-Knowledge*

Before reviewing how cultural differences in self-construal modulate neural representations of self and close others, a more general review of studies exploring cognitive and neural representations of self and others is necessary. One of the well-known arguments in social psychology is whether there is a "superordinate schema" for the self in the cognitive system (Rogers, Kuiper, & Kirker, 1977; Symons & Johnson, 1997). The earliest exploration of this issue started with a series of studies examining the mnemonic advantage effect when participants take their selves as a reference point in a memory task. The common finding in these studies was that words associated with one's self would be more easily encoded in the memory system, which showed up in the following memory recall task. In short, using one's self as the reference increases the likelihood of those associated words being encoded in the memory system, which is the mnemonic advantage effect in a nutshell. Although this mnemonic advantage effect is quite consistent across studies, this effect can be perfectly explained by two conflicting interpretations. The first interpretation is that the self is special and has its own superordinate schema in the cognitive system (Rogers et al., 1977). Because of this superordinate schema, "self" has more immediate access to the memory system. Thus, trait words associated with the self would show subsequently mnemonic advantage effect. By contrast, the other interpretation argues that self is not special at all and there is no special space for the self (Klein & Kihlstrom, 1986; Klein & Loftus, 1988). Instead of having its own schema in the cognitive system, this mnemonic advantage effect only reflects a deeper "depth-of-processing."

After decades of attempts to solve this puzzle using behavioral findings, researchers finally realized that this tactic would be insufficient. It turned out that behavioral results could support either theory perfectly (Symons & Johnson, 1997). Yet where the behavioral tools failed, advancements in neuroscience afforded a different avenue for exploration. Neuroimaging techniques allow researchers to directly explore the engagement of particular brain regions in specific tasks. Thus, these techniques enabled researchers to map cognitive processes to brain regions and directly test these two conflicting theories. Based on what we know about the functional neuroanatomy of social cognition, the medial prefrontal cortex (MPFC) plays a central role in neural representations of self and others (Amodio & Frith, 2006; Mitchell, Macrae, & Banaji, 2006; Wagner, Haxby, & Heatherton, 2012). The MPFC is located at the medial wall of two cerebral hemispheres, and can be divided into a ventral and a dorsal part. The ventral portion of the MPFC, which is also known as Brodmann's area 10, is named as the ventral MPFC (VMPFC). Another part of the MPFC, which is located at more dorsal portion of the MPFC (including Brodmann's areas 8 and 9), is named as the dorsal MPFC (DMPFC). Based on a recent review (Wagner et al., 2012), these two regions function differently in thinking about one's self and others. In short, the VMPFC subserves primarily self-referential processing, whereas the DMPFC subserves primarily other-referential processing.

The search for the neural basis of the self started with a positron emission tomography (PET) study (Craik et al., 1999), which investigated which regions engage in self-referential memory encoding. The task contained three conditions. Participants had to judge whether different trait words could describe themselves or a familiar figure, or had to think about the semantic meaning of these words. Not surprisingly, the MPFC activity was greater for the self-referential condition than the semantic-judgment condition. However, there were no differences between self- and other-referential conditions, which was inconsistent with the classical self-referential memory advantage effect. Due to this inconsistency and the relatively small sample size used in this study, another group of researchers (Kelley et al., 2002) used another neuroimaging technique, functional magnetic resonance imaging, to re-explore whether self-referential processing is distinct from other semantic processing. Similarly to Craik et al.'s study, three conditions (self-judgment, other-judgment, and case-judgment) were used in this study. As expected, regions involved in semantic processing, such as the left inferior frontal cortex, showed greater activity for self- and other-judgments than case-judgment. However, the only region that showed selective activity for self-referential processing was the VMPFC. Unfortunately, this study had no memory task after the scanning, which makes the correlational test between VMPFC magnitudes during self-referential conditions and subsequent memory performances impossible. Some researchers (Macrae, Moran, Heatherton, Banfield, & Kelley, 2004) revisited this issue and found supportive evidence for this correlation, however. Their findings revealed that the MPFC magnitude during self-referential encoding predicts the subsequent memory effect for the trait words being associated with the self during the task. As a consequence, it is obvious that self-referential is not just a deeper type of semantic processing. Instead, self-referential processing is special and one specific brain region, the VMPFC, is engaged in this processing.

6.1.2 Cultural Differences in Neural Representations of Self and Mother

Since the VMPFC primarily engages in thinking about the self, researchers started to ask whether this region also engages in the moment when individuals think about close others, such as one's mother or best friend. This speculation stems from what psychologists found decades ago when the self-referential processing task was conducted with the reference target as one's close friend or a family member (Bower & Gilligan, 1979; Ferguson, Rule, & Carlson, 1983; Kuiper & Rogers, 1979). A similar mnemonic advantage effect was found for words paired with one's close others, but not a familiar other (e.g., a political figure, like Bush). This finding suggested that it is intimacy, but not familiarity, that contributes to the mnemonic advantage effect. As a result, researchers (A. Aron, Aron, & Smollan, 1992; A. Aron, Aron, Tudor, & Nelson, 1991) proposed that individuals might incorporate close others, such as best friends or family members, into their self-schemas. In order to test the hypothesis that the VMPFC is also involved in representing the abstract knowledge of close others, Heatherton et al. (Heatherton et al., 2006) conducted a pioneer fMRI study by using a classical trait-word judgment task. Participants were asked to judge trait words in three conditions, including self-judgment, best friend-judgment, and a semantic judgment condition. The brain signal changes were extracted from an a-prior region-of-interest (ROI) from Kelley et al.'s VMPFC region. The results revealed that the activity in the VMPFC was greatest for self-judgment, but there was little difference in the activity level for the other two judgments. This evidence revealed that although thinking about close others might show a similar memory advantage effect behaviorally, the VMPFC is only specifically subserved for representing abstract knowledge about one's self.

It seems clear, according to the studies reviewed above, that the VMPFC uniquely subserves representations of abstract knowledge of one's self. Notably, however, the majority of these studies recruited participants solely from independent cultures. Would different patterns be observed in participants from interdependent cultures? Interdependent self-representations feature fuzzier boundaries between one's concept of self and of others. Therefore, it is plausible that the neural representations between one's self and one's closest other (e.g., one's mother) may likewise overlap, specifically in the VMPFC. To test this hypothesis, Zhu and colleagues (Zhu, Zhang, Fan, & Han, 2007) conducted an fMRI study recruiting participants from China, an example of an interdependent culture, and participants from independent Western cultures.

Both groups of participants were asked to judge trait words in three conditions, comprising self-judgment, mother-judgment, or font-judgment. Not surprisingly, as Heatherton et al. found, VMPFC activity was significantly higher for judgments of self than for those of one's mother in Western participants. However, the VMPFC showed no self–mother differentiation in Chinese participants. This finding provided supportive evidence that cultural differences in self-construal might not only be reflected at an abstract level but also in neural representations of the self and of close others. Although the sample size was quite small ($N=13$ for each group) in Zhu et al.'s study, a later study (Wang et al., 2012) replicated these findings and also included judgments for one's father and best friends. They found that while no statistical difference in VMPFC activation was observed between self and mother, there was significantly less relative activation for the father judgments and even less for the judgments of a best friend. This finding suggests that mothers may be uniquely incorporated into their children's self-schema, and play an important role in the intimate relationships within Chinese culture. Based on the above findings, it is plausible that differential VMPFC patterns between one's self and mother may be modulated by cultural differences in self-construal.

This hypothesized relationship has received support from a recent study (Chiao et al., 2009b) that considered only individual differences in self-construal styles, regardless of participants' country of origin. Participants who possessed independent self-construal styles showed greater MPFC activity for general than for contextual self-judgments, whereas participants who held interdependent self-construal styles showed the reverse pattern. This finding demonstrated the modulatory effect of self-construal style upon neural representations of self within the MPFC. Although the focus of this study was not on the neural differentiations between concepts of self and mother, the results provide supportive evidence that individual differences in self-construal might modulate these neural dissociations. Behavioral studies have also found that substantial heterogeneity in self-construal styles can occur within the same culture, even within a culture traditionally defined as more interdependent like that of China (Green, Deschamps, & Paez, 2005). For example, the observed degree of independence and interdependence is quite variable among Chinese participants (H. Li, Zhang, Bhatt, & Yum, 2006). This phenomenon has been characterized as within-cultural variance (Freeman, 2013).

One hypothesis related to within-cultural variance is the voluntary-settlement hypothesis, which explains the underlying motives behind immigration (Kitayama, Ishii, Imada, Takemura, & Ramaswamy, 2006). According to this hypothesis, voluntary immigrants moving to frontiers, such as wilderness regions, foreign countries, or metropolitan cities, have higher independence or lower interdependence than those who stay in their place of origin. For example, individuals who grew up in Japan are generally highly interdependent in self-construal. However, findings have revealed that Hokkaido, the northern territory and the last frontier in Japan, contains more highly independent individuals than other places in Japan. Interestingly, these geographic alterations in autonomy can be observed in a very intimate metric—the names parents give their children (Varnum & Kitayama, 2011). Specifically, children in more recently settled regions of Japan are less likely to have been given popular baby names than are their counterparts in long-settled regions. This trend also holds true in the USA and Canada. In short, it is clear that immigrants generally hold more independent self-construals than do individuals who permanently settle within their place of origin. Thus, immigrants from interdependent cultures, such as China, would likely still show significant self–mother differentiations in the VMPFC.

6.1.3 Neural Differentiations Between Self and Mother in Chinese Immigrants

In order to test this possibility, our first study (Chen, Wagner, Kelley, Powers, & Heatherton, 2013) recruited 19 newly arrived Chinese immigrants within the first 2 months of their arrival in the USA. None of these immigrants had ever previously lived in a foreign country for more than 2 months. This criterion ensured that these individuals would be free from pre-exposure cultural effects. During four functional imaging runs, participants were asked to judge trait words presented in their native language or in their secondary language in three different conditions (self-judgment, mother-judgment, or font-judgment). Surprisingly, the findings from this study were quite different from what had been found in previous studies done in China. The whole brain analysis revealed significantly greater VMPFC and posterior cingulate cortex (PCC) activity for self-judgments than mother-judgments (Fig. 6.1) regardless of the language used. In order to further confirm our findings in an unbiased way, two independent VMPFC regions-of-interest (ROI) from two previous studies (Kelley et al., 2002; Wang et al., 2012) were used for the ROI analysis.

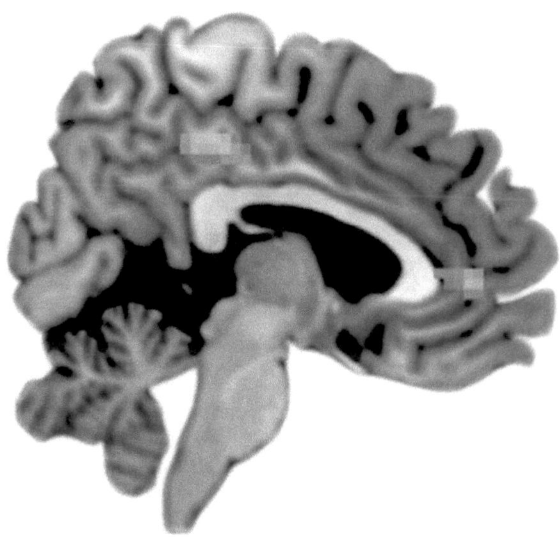

Fig. 6.1 Results from a whole-brain analysis of SELF versus MOTHER-judgments contrast across two languages in newly arrived Chinese immigrants ($P<0.05$, corrected). Results showed that VMPFC and PCC were more engaged for SELF-judgments than MOTHER-judgments across two languages (adapted from Chen et al., 2013)

The VMPFC activity extracted from these two VMPFC ROIs were both greater for self-judgments than mother-judgments, confirming the findings from the whole brain analysis (Fig. 6.2). Both the whole brain analysis and the ROI analysis showed consistently greater VMPFC activity for self-judgments than for mother-judgments

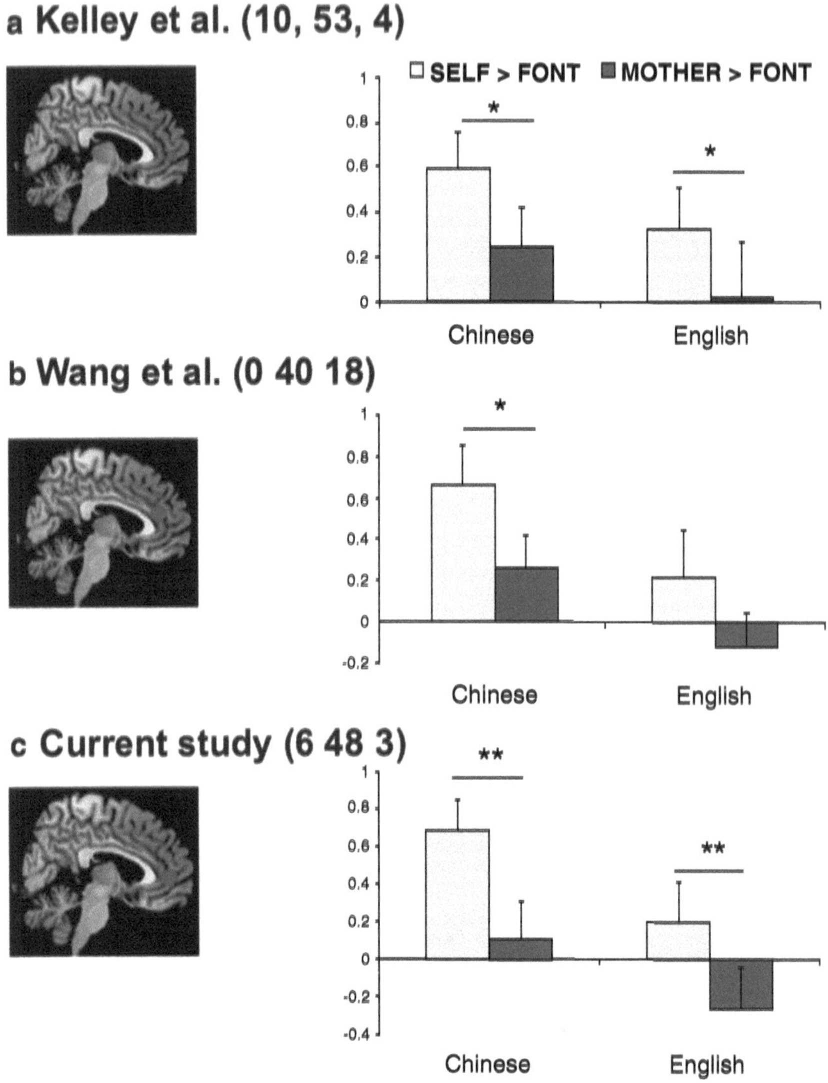

Fig. 6.2 Analysis of differences scores from parameter estimates in three different VMPFC ROIs for SELF>MOTHER contrast in Chinese and in English, respectively. VMPFC activity was significantly greater for SELF- than MOTHER-judgments in both Chinese and English based on the results from three different ROIs. (**a**) The MPFC ROI defined by Kelley et al. (2002). (**b**) The MPFC ROI defined by Wang et al. (2012). (**c**) The MPFC defined in the current study. $^{*}P<0.05$, $^{**}P<0.01$. *Bars* indicated standard error of the mean (adapted from Chen et al., 2013)

across different ROIs and languages. This finding suggests that these immigrants might possess a more distinct and autonomous self-construal compared to those who choose to stay in China. This speculation was supported by participants' self-construal scores, which were equal in independence and interdependence. Essentially, moving to the USA may have been comparable to moving to a frontier area for most of these immigrants. Therefore, we think it quite possible to generalize that immigrants who choose of their own free will to move can be found to hold more independent or less interdependent self-construals than those who choose to stay in their motherlands. Most importantly, their distinct self-construals are reflected not only at the behavioral level but also in the activity of the VMPFC.

Although the VMPFC self–mother differentiation pattern seems to be a reliable index for assessing individual differences in self-construal, little is known about whether behavioral and psychological changes incurred by acculturation processes manifest in the VMPFC self–mother differentiation pattern. Most of the previous studies employ either the monocultural comparison approach (Wang et al., 2012; Zhu et al., 2007) or the bicultural priming approach (Chiao et al., 2009a; Kitayama & Uskul, 2011; Ng et al., 2010). Further, cultural neuroscientists seldom take a longitudinal approach, which is critical for answering the key questions surrounding acculturation processes (Chen et al., 2013; Kitayama & Uskul, 2011).

6.1.4 Changes in the Interdependent Self-Construal Modulate Self–Mother Differentiation in the VMPFC

According to what we know about acculturation strategies, some immigrants tend to integrate host cultural values into their identities, whereas others show less interest in taking on host cultural values and instead maintain or enhance their original cultural values. For example, immigrants from China, who are equally high in interdependence and independence (Chen et al., 2013; Kringelbach & Berridge, 2009; Smith, Berridge, & Aldridge, 2011), may show divergent changes in self-construal styles across time. Some of these Chinese immigrants may become less interdependent in self-construal and more like Americans, whereas others may become even more interdependent in self-construal and more like Chinese. Our second study (Chen et al., 2015) was designed to examine whether these two different groups of immigrants would show distinct self vs. mother differentiations in the MPFC. We hypothesized that individuals who become less interdependent would show significant self vs. mother differentiation, whereas those who become more interdependent would show no such differentiation 6 months after their arrival. Twenty-seven newly arrived native Chinese-speaking participants were recruited for this study. The same trait-judgment task (self, mother, or font judgments in English or Chinese) from our first study was conducted within the first 2 months of their arrival in the USA (Time 1). After the scanning, the self-construal scale (Singelis, 1994) was administrated and their interdependence scores were calculated. Six months after the initial scan (Time 2), the same participants performed the same self-referential task again, and their interdependence scores were also collected. Their difference scores in the

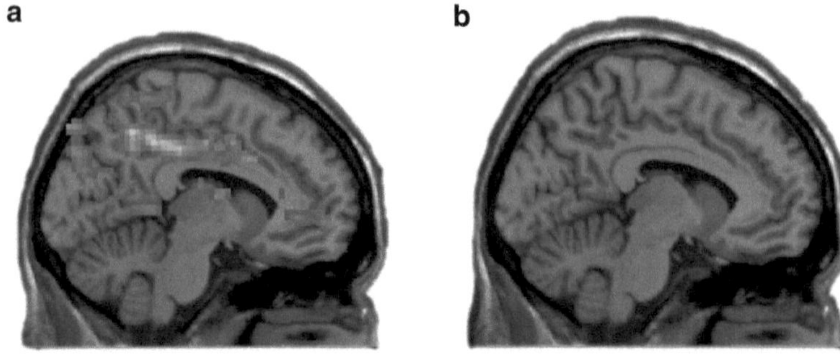

Fig. 6.3 Whole-brain analyses of SELF versus MOTHER-judgments in two groups of immigrants who showed divergent changes in the interdependent self-construal six moths later of their arrival ($P<0.05$, corrected). (**a**) The immigrants who decreased in interdependent self-construal scores showed prominent SELF–MOTHER differentiations in the VMPFC. (**b**) The immigrants who increased in interdependent self-construal scores showed no such differentiation pattern in the VMPFC (adapted from Chen et al., 2015)

interdependence measure were computed based on the difference in their interdependent scores at Time 2 versus scores at Time 1. We divided participants into two groups based on whether they showed a decrease or increase in these difference scores. To test our hypothesis, we analyzed the fMRI data collected at Time 2 only. The whole brain analysis for the individuals in the decrease group, who become more like Americans, still showed greater VMPFC activity for self-judgments than mother-judgment. By contrast, those in the increase group, who become more like Chinese, showed no such self–mother differentiation pattern in the VMPFC (Fig. 6.3). The ROI analysis using Chen et al.'s ROI (2013) supported these correlations and our hypothesis (Fig. 6.4). In sum, it seems that changes in the self-construal can be reflected by changes in this VMPFC self–mother differentiation pattern during acculturation processes. These findings also suggest that the neural differentiation patterns between self and mother could potentially yield deeper understandings of acculturation processes.

In the first section, we begin by reviewing two conflicting theories that explain the self-mnemonic advantage effect, and follow up with an exploration of how functional imaging methods can be used to solve this puzzle. Then, we review which brain regions are involved in representations of abstract knowledge of self and close others, and how cultural differences in self-construal modulate these neural representations. Lastly, we present findings from our first study, showing that immigrants may be inherently more independent in self-construal than those who stay in their motherlands. Their distinct self-construals may be reflected in the self–mother differentiations in the VMPFC. Furthermore, in our second longitudinal study, we demonstrate that immigrants' changes in self-construal could be reflected by changes in this VMPFC self–mother differentiation pattern during acculturation processes. These findings suggest that this differentiation pattern could be a potential tool for understanding the divergent acculturation processes experienced by immigrants, and for predicting immigrants' acculturation outcomes. We come back to this topic in the last section of this chapter.

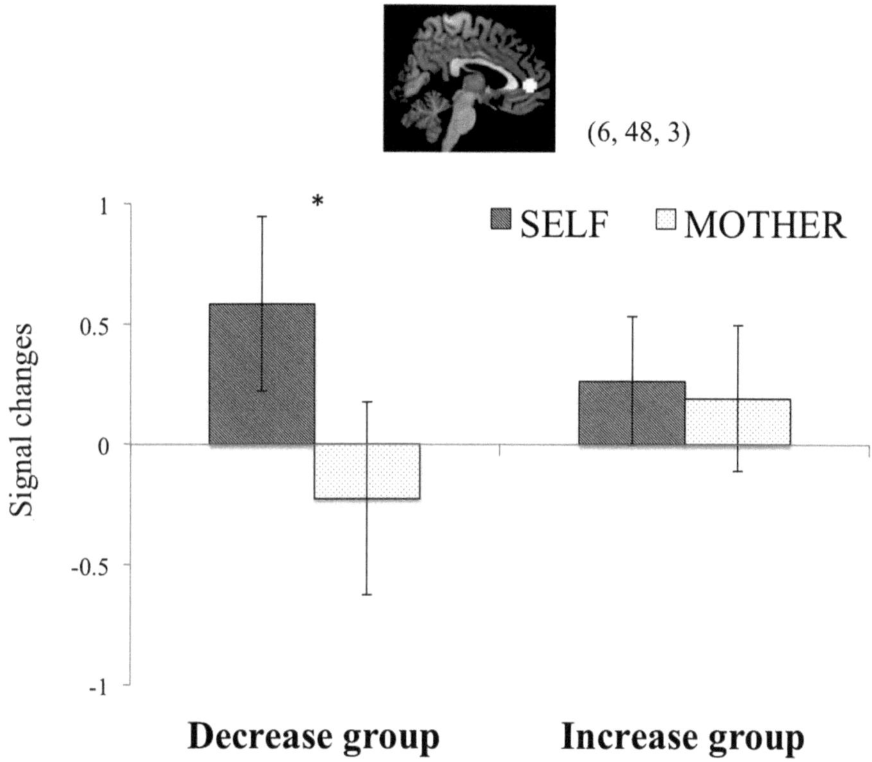

Fig. 6.4 Analysis of differences scores from parameter estimates in Chen et al.'s VMPFC ROI for SELF > MOTHER contrast. The VMPFC activity was significantly greater for SELF- than MOTHER-judgments for immigrants who showed decrease in the interdependent self-construal scores, whereas the VMPFC activity showed no such SELF–MOTHER differentiation for immigrants who showed increase in the interdependent self-construal scores (adapted from Chen et al., 2015)

6.2 Neural Mechanisms for Reward Processing and In-Group Reward Reactivity

6.2.1 Neural Mechanisms for Reward Processing

Before moving on to the reward processes underlying in-group favoritism, we have to review the fundamental neural mechanisms underlying these processes. Reward is critical for motivating incentive-based learning, shaping reactions to different stimuli, and triggering goal-directed behaviors. Researchers have found three distinct aspects within the broad concept of reward, namely, liking, wanting, and learning (Kringelbach & Berridge, 2009; Smith et al., 2011). Among these three aspects, "liking" is a hedonic aspect, and reflects the pleasure component of reward. "Wanting" is a motivational aspect, which reflects the incentive salience component of reward. The last component is "earning," which reflects the processes of associating, representing, and predicting future outcomes based on past experiences. Although these three aspects sometimes can be processed consciously, most of the time they are processed without subjective awareness.

Among all of the brain regions, the ventral striatum (VS) is primarily in charge of the liking and wanting components (Delgado, 2007; Kringelbach & Berridge, 2009), and the wanting component is closely related to cue-association learning. Moreover, the VS's anatomical connections with other reward-related brain regions and the neurotransmitters it receives (Delgado, 2007) make cue-association learning possible in this region. The VS is a compound structure, which consists of the ventral putamen (VP), ventral caudate nucleus, and nucleus accumbens (NAcc). The VS not only receives inputs from the amygdala, dorsal striatum, and ventral prefrontal cortex but also receives dopaminergic (DA) inputs originating from mesolimbic dopamine regions, such as the ventral tegmental area (VTA). During the early stage of reward learning, the activation of these DA neurons from the VTA is synchronized at the point of time at which the reward is delivered. However, if some specific cues always appear before the reward delivery, after several trials of cues and reward association, the activation of these DA neurons shifts to the point of time at which the cues are presented (Glimcher, 2011). As the result, the activation of these DA neurons synchronizes at the presence of cues instead of the delivery of rewards. Due to the fact that the activation of these DA neurons from the VTA results in DA release in the VS, this cue-association activity is also established in the VS. Most importantly, this cue-association characteristic plays an important role not only in the formation of addiction but also in different kinds of social learning. The VS is part of the broader cortico-basal ganglia circuitry, which has been described as integrating the motivation and action components into reward processes (Haber & Knutson, 2010). Within this circuitry, the VS primarily subserves for the motivation component (liking and wanting), whereas the dorsal caudate and putamen subserve for moving from motivation to action. By integrating these subcomponents in this circuitry, individuals can execute goal-directed behaviors through incentive-based learning processes.

Another line of studies (Demos, Heatherton, & Kelley, 2012) suggests that the neural signals in the VS, especially the NAcc, can predict real-world behaviors, such as eating behaviors and daily sexual desires. By using a classic cue-reactivity paradigm, participants were asked to see different kinds of photographs, including food, people in sexual activity, and scenery, in the MRI scanner, and these participants were asked to judge whether these photographs were shot indoors or outdoors. Because of the indoors and outdoors judgment, participants were unaware of the purpose of this task, which was to measure their incidental brain reward activity toward cues in specific categories. The NAcc activity for food and sexual images were extracted from the NAcc ROI. Individual differences in this NAcc activity for food cues were found to predict participants' weight gain 6 months later, whereas individual differences in this NAcc activity for sexual scenes predicted participants' daily sexual desires 6 months later (Demos et al., 2012). This finding reveals that NAcc activity for specific cues may have a determinant role in real-world behaviors that are linked with these specific cues. Extending from the above finding, it also suggests that brain-as-predictor approach can be applied to acculturation studies when appropriate cues are used in the paradigm.

6.2.2 Self-Categorization and In-Group Favoritism

A growing number of imaging studies has shown that regions within this cortico-basal ganglia circuitry not only responds to primary rewards, such as food and sex, but also to other secondary rewards, such as money, attractive faces, symbols of status, and social cues (Aharon et al., 2001; Schultz, 2000). Thus, it is not surprising that regions within this circuitry may also play an important role in social interaction. To be more specific, reward processes may motivate in-group or out-group social affiliation behaviors (Powers & Heatherton, 2012).

Differentiating people into in-group or out-group members is an essential ability for human survival, as proven by numerous social psychology studies (Correll & Park, 2005). When there is no other prominent cue, the default cue for this in-and-out-group differentiation is often race (Sporer, 2001; Van Bavel & Cunningham, 2008). Research has shown that people usually show more positive or stronger responses to in-group members than out-group members (Sporer, 2001; Van Bavel & Cunningham, 2008). For example, Elfenbein and Ambady (Elfenbein & Ambady, 2003) have reported in-group advantage in emotional recognition. In this study, Chinese participants showed a shorter reaction time and a better accuracy rate to Chinese emotional expressions than the Caucasian ones. In another study investigating the effect of group membership on the automatic impact of emotional expressions (Weisbuch & Ambady, 2008), participants automatically showed negative responses to in-group fearful expressions, but showed positive responses to out-group ones. However, participants showed a reversed pattern for happy expressions. This in-group vs. out-group difference was not only found at the behavioral level but also at the neural level. Chiao et al. (2008) found that both Japanese and Americans showed greater amygdala responses to in-group fearful expressions than out-group ones, but this effect was not found for either happy or angry expressions. This finding was partially in line with the findings from Weisbuch and Ambady's study, in which participants showed more negative responses to in-group fearful expressions. This response might be reflected in the heightened amygdala responses to the in-group fearful expressions. However, in-group positive responses to in-group happy expressions were not found at the neural level. There are at least two possible explanations. First, the target region in Chiao et al.'s study was the amygdala, which might be more sensitive to negative emotions than positive ones. Instead of the amygdala, regions within the reward circuitry, such as the VS and orbitofrontal cortex (OFC) might subserve for processing this rewarding stimuli (Berridge, Robinson, & Aldridge, 2009; O'Doherty et al., 2003). Second, Chiao et al. used an explicit face-rating paradigm whereas an affective priming paradigm was used in Weisbuch and Ambady's study, which is an implicit priming paradigm. It is possible that the findings in an implicit paradigm might be distinct from the findings in an explicit paradigm.

The other line of research highly relevant to the in-group advantage is in-group bias. In-group bias is not only observed at the behavioral level (Van Bavel & Cunningham, 2008) but also at the neural level (Van Bavel, Packer, & Cunningham, 2008). This novel neuroimaging study found that by assigning Caucasian participants to a classic minimal-group paradigm, several brain regions, especially amygdala and

OFC, responded more highly to new in-group faces than to new out-group faces regardless of the race. This finding suggests that the role of amygdala is to process salient and motivationally related stimuli, not just the automatically negative responses to the stigmatized racial group. In this study, the liking rating for the in-group Black faces was much higher than the out-group Black faces, but there was no rating difference in the in-group Caucasian and Black faces. Moreover, this self-reported liking was mediated by the OFC, which was part of the reward circuitry. The researchers concluded that the key process for this in-group bias was self-categorization as a member of a particular group.

The effect of self-categorization with a particular group might turn into in-group favoritism. In a recent study (Cikara, Botvinick, & Fiske, 2011) exploring the neural mechanism of intergroup competition, the VS activity was higher when participants saw the positive outcomes in a baseball game (their supporting team won the game or their rival team lost the game) than negative outcomes. The participants' pleasure ratings were also positively correlated with the magnitude of activities in the VS. This finding indicates that regions subserving for primary rewards, such as the VS, might engage in in-group favoritism when in-group members have positive outcomes. It is very possible that the underlying driving force is whether participants consciously self-categorized themselves belonging to a particular group (Balcetis & Dunning, 2006; Sporer, 2001).

6.2.3 Individual Differences in In-Group Favoritism Predicts In-Group Friendship Patterns

This self-categorization process might play an important role in acculturation processes. According to the four major types of acculturation strategies, integration can be described as a strategy by which immigrants maintain their own culture while also integrating the host cultural value into their identities. On the contrary, assimilation describes the state when individuals abandon their original cultural values, and try to accept all aspects of the host culture. Although these two strategies seem different from each other, these two strategies have one common feature, which is that immigrants who use these two strategies are inclined to incorporating host cultural values into their identities. This feature implies that these immigrants may either have no specific tendency to self-categorize into the original or host cultural group, or have stronger tendency to self-categorize into the host cultural group. Compared to these two strategies, immigrants who use the separation strategy are those who emphasize only maintaining one's own cultural values and have less interest in integrating host cultural values into their identities. As the result, these immigrants may have a strong tendency to self-categorize themselves as members in the original cultural group, and will still show extreme in-group favoritism toward their original group members. Based on the above speculation, one possibility is that the determination of whether a group belongs to the in-group does not depend on individual's original cultural group. Rather, it may depend on this self-categorization process. Once immigrants self-categorize themselves to a particular group, they will automatically show in-group favoritism toward that target group. Thus, this in-group favoritism may lead immigrants into divergent acculturation outcomes.

Our longitudinal study (Chen et al., submitted) was designed to test this possibility, which is that individual differences in in-group favoritism toward the original cultural group may drive immigrants to different acculturation outcomes, specifically friendship patterns. As noted, an early challenge that newly arrived immigrants encounter is determining with whom they should interact. Indeed, new arrivals vary greatly in their friendship formation patterns. Some individuals make more friends among out-group members, whereas others isolate themselves from out-group members and associate mainly with other newly arrived in-group members (Sam & Berry, 2010). Because social affiliation occurs automatically (Powers & Heatherton, 2012), individuals may lack insight into their affiliative processes and behaviors. Therefore, self-reported motives for friendship formations may be biased. Under this circumstance, functional imaging can be used as an alternative means to test the hypothesis that distinct patterns of neural reward reactivity may underlie differential in-group vs. out-group interaction patterns.

Prior imaging studies investigating reward activity provide supportive evidence for the hypothesis that affiliative behaviors are associated with activity in brain reward regions. For example, studies examining facial expressions of happiness as experimental stimuli found activation in the VS (Phan, Wager, Taylor, & Liberzon, 2002). Additionally, a go/no go study that used different emotional expressions as targets found high VS activation in response to happy facial expressions, which also made it more difficult for participants to inhibit responses (Hare, Tottenham, Davidson, Glover, & Casey, 2005). The VS activation induced by happy expressions is likely to motivate approach behaviors. Based on happiness' approach-provoking characteristics, we hypothesized that new arrivals who showed higher reward reactivity for in-group compared to out-group happy expressions would show a preference for approaching in-group members in daily life. In the long run, their percentage of in-group friends might increase. By contrast, those who showed equivalent reactivity for in-group and out-group facial expressions of happiness would be expected to show more balance in their in-group and out-group friendship patterns.

To test this hypothesis, 27 newly arrived Chinese international graduate students were recruited within the first month of their arrivals in the USA. One was excluded from analysis due to excessive movement during scanning. In order to assess how new arrivals expand their social networks with in-group or out-group members in a new culture, percentages of in-group and out-group friends from social networking services (e.g., Facebook) were recorded. Importantly, as Facebook is blocked in China, these Chinese new arrivals had to create a whole new online friendship network in the USA, making it possible to track their changing friendship patterns without contamination from their previous friendship networks. The percentages of in-group friends for each of the participants were recorded immediately after the scanning and again 6 months later, and subsequently changes in the percentage of in-group friends were computed.

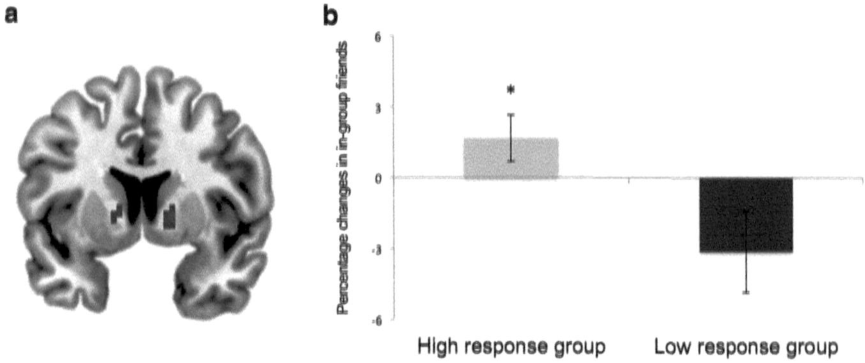

Fig. 6.5 Individual differences in in-group reward reactivity predict changes in the percentage of in-group friends after 6 months. (**a**) The VS reactivity was greater for in-group than out-group masked happy expressions. (**b**) New arrivals who were higher in in-group reward reactivity had a significant increase in the percentage of in-group friends than those who were lower in in-group reward reactivity (adapted from Chen, Whalen, Freeman, Taylor, & Heatherton, in press)

During three runs of functional imaging, the participants were asked to passively view blocked presentations of masked faces. In each masked-face block, one of three in-group and out-group emotional faces (fearful, happy, and surprised) was presented. Based on previous findings of regional brain activity to emotional stimuli presented without subjective awareness (Whalen et al., 1998), the current study presented emotional expressions in a backward masking paradigm. The goal was to examine whether VS activity in response to in-group and out-group happy expressions in the absence of explicit knowledge of the stimuli could predict new arrivals' friendship patterns over 6 months. Following the standard procedures of prior work (Whalen et al., 1998), four participants who indicated subjective awareness of the masked faces were excluded from further analysis. Based on the above-mentioned hypotheses concerning VS responses representing reward reactivity to positive affiliative stimuli, whole brain analysis focused on the comparison between in-group and out-group masked-happy conditions in the VS. To examine whether the VS activity could predict changes in the percentage of in-group friends over 6 month, the parameter estimates were extracted from this VS ROI, difference scores were computed, and then participants were divided into two groups based on the median-split (high and low VS responses to in-group expressions). The whole brain analysis revealed higher activity for in-group than out-group masked happy expressions in the bilateral VS, left medial prefrontal gyrus, right inferior occipital gyrus, left middle occipital gyrus, and bilateral cerebellum. Interestingly, participants with lower ventral striatal activity in response to in-group masked happy expressions showed a reduction in the percentage of in-group friends, whereas those with greater ventral striatal activity showed an increased in the percentage of in-group friends 6 months later (Fig. 6.5).

Our findings support the hypothesis that those with the greatest reward reactivity to in-group happy expressions subsequently developed more friendships with in-group members. Importantly, these affiliation processes may occur without subjective awareness, and this possibility has been proven by post-scan measurements of objective awareness. The increase group and decrease group showed no difference in d-prime difference scores in discriminating in-group and out-group happy expressions, suggesting that this in-group favoritism occurs without subjective awareness and is reflected at neural responses within reward circuitry. Moreover, the new arrivals generally showed greater reward reactivity for in-group than for out-group masked happy expressions, which supports the general idea of in-group favoritism (Cikara et al., 2011; Van Bavel et al., 2008). Individual differences in this in-group reward reactivity, however, predicted divergent friendship patterns 6 months later. This brain-as-predictor approach (Berkman & Falk, 2013) has been used to successfully predict the successful cassation of smoking behaviors (Berkman, Falk, & Lieberman, 2011). Findings from the present study further suggest that this approach might be a useful tool for exploring the acculturation process of immigrants, particularly the dynamics of friendship network formation in a new country.

6.2.4 *Linking Individual Differences in Self–Mother Differentiation to In-Group Reward Reactivity*

The first section of this chapter reviews how cultural differences in self-construal modulate the neural differentiation between self and mother observed in the VMPFC. Individuals from independent cultures showed a prominent self–mother differentiation in the VMPFC, whereas those from interdependent cultures showed no such differentiation (Heatherton et al., 2006; Wang et al., 2012; Zhu et al., 2007). Moreover, this self-construal modulation effect occurred not only at the cultural level but also at the individual level. We found that newly arrived Chinese immigrants showed this prominent self–mother differentiation, which reflected their unique self-construal, characterized by equivalent independence and interdependence. This finding further supports the voluntary-settlement hypothesis. (As a reminder, this hypothesis states that those who move to the frontiers are inherently different in self-construal from those who stay in their motherlands.) Subsequently, we also found that these self-motivated immigrants showed divergent changes in their interdependent self-construal 6 months later, suggesting that differing degrees of acculturation occurred within a few months. The immigrants who became closer to Americans in their self-construal style showed significant self–mother differentiation in the VMPFC. By contrast, those whose activation patterns became even closer to the Chinese self-construal style showed no such differentiation. This finding suggested that changes in self-construal could be reflected by changes in this VMPFC self–mother differentiation pattern during acculturation processes. Moreover, based on our preliminary findings (Chen et al., 2015), those immigrants who became even more like Chinese revealed fundamental differences during the first fMRI scanning, which was collected within the first months of their arrival. Although they still showed self–mother differentiation in the VMPFC, their differentiation patterns were less prominent than those who became more like Americans.

However, their self-construal scores were not significantly different from those who became more like Americans. This discrepancy suggests that behavioral measurements may not be sensitive enough to detect slight individual differences in self-construal, perhaps the reason why previous studies have failed to make successful acculturation predictions. In contrast, neuroimaging techniques are sensitive enough to detect these subtle differences in the neural mechanisms underlying automatic processes.

In the second section of this chapter, we describe a study that employed a brain-as-predictor approach to predict immigrants' friendship patterns. Although immigrants generally showed greater reward reactivity to in-group than to out-group happy expressions presented under subjective awareness, there were substantial individual differences in this in-group reward reactivity. Immigrants with relatively less ventral striatal activity in response to in-group masked happy expressions showed a reduction over time in their percentage of in-group friends, whereas those with greater ventral striatal activity showed an increase in the percentage of in-group friends 6 months later.

Based on the above findings, it seems that individual differences in the VMPFC self–mother differentiation and in-group reward reactivity can independently predict different aspects of acculturation outcomes. Is it possible that individual differences in one domain are correlated with individual differences in the other domain? A recent review (Northoff & Hayes, 2011) proposed that three possible models could be used to explain the relationships between self-processing and reward processing: integration, segregation, and parallel model. The integration model suggests that self and reward processes are highly overlapping, whereas the segregation model suggests that these two processes are distinct. The parallel model presents a compromise between these polar theories, and suggests that certain degrees of overlap exist between self and reward processing. Of these three models, the authors conclude that the parallel model is the best substantiated by existing evidence.

Since there are likely certain degrees of overlap existing between self and reward processes, it is rational to hypothesize that individual differences in the VMPFC self–mother differentiation may be correlated with individual differences in in-group reward reactivity. Based on findings from our studies, immigrants who show a greater self–mother differentiation are those who possess a more independent self-construal. This is observed upon their initial arrival in the USA, and also holds true 6 months later. Because these immigrants become more like Americans in self-construal style, their acculturation processes may foster the development of a state of integration or assimilation. Thus, they may also show less in-group reward reactivity when they first arrive in the USA. As a result, their percentage of out-group friends will increase over time. In contrast to these immigrants, immigrants who show no prominent self–mother differentiation are those who have a more interdependent self-construal style. They generally become even more like Chinese in self-construal style 6 months later. Thus, it is possible that they are on the path to the state of separation during acculturation processes. These immigrants may show higher in-group reward reactivity when they first arrive in the USA. In the long run, their percentage of out-group friends will decrease, while they will show a prominent increase in in-group friendships. Unfortunately, the participants in the self–mother differentiation study and the in-group reward study are recruited from different group of participants, making it impossible to directly test the above hypothesis. Future studies are needed to test this hypothesis, and to develop a better model to understand the complex acculturation processes.

Several limitations exist within our studies. First of all, these studies lack a more comprehensive acculturation battery. We only administered a self-construal scale in these studies, and suggest that other acculturation questionnaires, such as the Suinn–Lew Asian Self Identity Acculturation Rating Scale (Suinn, Rickard-Figueroa, Lew, & Vigil, 1987) and General Ethnicity Questionnaire (Tsai, Ying, & Lee, 2000), should be incorporated in future studies. Secondly, more detailed data regarding friendship patterns should be collected in future studies. In our research, we use a binary approach (in-group vs. out-group) to analyze the friendship patterns. However, this binary assignment might fail to detect subtle affiliation patterns with different ethnic out-groups. Moreover, this study also failed to measure the strength of friendships within their friendship grids. This strength information might be quite meaningful when researchers start doing more detailed analyses. Lastly, we only recruited immigrants who had moved from China to the USA, rather than recruiting from other "Western" populations. Thus, our findings may not be generalizable to immigrants moving from one independent culture to another. In this case, the self–mother differentiation pattern may not have the same predictive power, as the self–mother differentiation pattern is more prominent in immigrants from interdependent cultures. As an alternative, the in-group reactivity paradigm may be used. Future studies must address all of the limitations above.

6.3 Future Directions

6.3.1 Incorporating Diffusion Tensor Imaging into the Brain-As-Predictor Approach

A recent study (Chavez & Heatherton, in press) using diffusion tensor imaging found that the integrity of white matter tracts between brain regions subserving self-referential processing (i.e., the VMPFC) and regions subserving reward processing (i.e., the VS) predicts individual differences in trait self-esteem. This finding implies that individual differences in traits, which are stable across time, may not reflect differences in brain activation patterns. Rather, they may represent differences in structural integrity between brain regions. According to what we found in one of our studies, immigrants who became more like Americans 6 months later still showed prominent self–mother differentiation not only in the VMPFC but also in one of the reward regions, the VS. Is it possible that those who show an increase in independent self-construal style are those who have more white matter integrity between the VMPFC and the VS? One previous review (Johansen-Berg, 2010) concludes that individual differences in the white matter structures play an important role in individual differences in behavioral consequences. Since immigrants generally show huge individual differences during the acculturation process, some of which are stable across time, it is possible that inherent differences in white matter integrity motivate immigrants to acculturate. Also, due to the fact that learning leads to neural plasticity in the brain (Kitayama & Uskul, 2011), it is possible that white matter integrity will also show dynamic changes during acculturation. Based on our previous findings, the regions likely incurring these changes are the VMPFC, which subserves neural representations of self and close others, and the VS, which subserves in-group reward processing. Future studies are needed to test whether white matter integrity between these two regions can predict individual differences in acculturation outcomes, as well as to test whether any changes in white matter integrity are affected by the acculturation process.

6.3.2 *Individual Differences in Reward Responses to Culture-Specific Cues*

Different cultures reinforce different behaviors, and this relationship extends to affect the neural level. For example, American culture reinforces more dominant behaviors, whereas Japanese culture reinforces more subordinate behaviors. Freeman and colleagues (Freeman, Rule, Adams, & Ambady, 2009) conducted a study by recruiting Americans and Japanese to passively view pictorial cues relevant to dominant or subordinate behaviors while in the MRI scanner. The findings revealed that the bilateral caudate nucleus, part of the cortico-basal ganglia circuitry, showed greater activity for dominant than for subordinate cues in Americans, whereas the same regions showed greater activity for subordinate than dominant cues in Japanese. Moreover, individual differences in activity in the right caudate nucleus were correlated with individual differences in the self-reported behavioral tendency toward dominant or subordinate behaviors. These findings suggest that reward activity to distinct behavioral cues can be shaped not only by individuals' cultural backgrounds but also by individuals' behavioral preferences. Based on this implication, it will be interesting to test whether individual differences in reward activity to culture-specific cues can predict immigrants' acculturation states. It seems that greater reward activity to one's own culture-specific cues may indicate stronger in-group favoritism and a higher tendency toward the acculturation state of separation. Thus, Japanese who show no reward activity to subordinate cues or show even greater reward activity to dominant cues are those who may move toward the state of integration or assimilation. Conversely, those who show greater reward activity to subordinate cues may move toward the state of separation. The same hypothesis can also be made in relation to Americans and other cultural groups once the appropriate cultural cues are selected. Future acculturation studies may use this paradigm to test the above hypothesis.

6.3.3 Linking Research Using Mouse-Tracking to In-Group Reward Reactivity

Findings from one of our studies indicate that immigrants have greater reward reactivity to in-group than out-group happy expressions. This heightened in-group reward reactivity represents an approach motivation toward in-group members. However, the avoidance motivation away from out-group members was not tested in this study. To test whether avoidance motivation is also an important factor, mouse-tracking methods could be applied in future work. For example, the MouseTracker software package enables researchers to record and analyze the real-time dynamics underlying a variety of perceptual and cognitive decisions using participants' hand movements en route to specific responses on the screen (Freeman & Ambady, 2010, 2011). MouseTracker makes it possible to visualize the time course of an evolving behavioral response by analyzing how participants' hand movements (mouse trajectories) settle into a given response over time—and how they may be partially "pulled" toward alternative responses in parallel (Freeman, Dale, & Farmer, 2011). In the context of approach/avoidance behavior, MouseTracker would allow the recording and analysis of how participants' hand movements may be particularly direct and facilitated toward hypothetical approach-related response options, and particularly indirect and "pushed away" from hypothetical avoidance-related response options.

The setup of this approach/avoidance to in-/out-group members experiment starts with the selection of appropriate emotional expressions for triggering approach or avoidance tendencies. In our previous in-group reward reactivity study (Chen et al., in press), only the approach tendency was tested. The ideal target for eliciting an approach tendency would be happy expressions (as was done in the previous study), whereas the ideal target for eliciting an avoidance tendency would be angry expressions (Phan et al., 2002). Mouse trajectories that are particularly direct would suggest an approach tendency toward a given face, whereas those that are particularly indirect would suggest an avoidance tendency toward a given face. It would therefore be interesting to test whether approach and avoidance tendencies to angry and happy expressions are at all moderated by a face's in-group/out-group status. MouseTracker in this case would allow us to use motor behavior to gain insight into spontaneous approach and avoidance tendencies toward in-group and out-group faces varying in emotion expressions.

By looking at both of these motivations, particularly the approach tendency toward in-group happy expressions and the avoidance tendency elicited by out-group angry expressions, we may better capture the underlying mechanisms that determine immigrants' acculturation outcomes. Based on a previous meta-analysis of the functional anatomy of emotions (Phan et al., 2002), it is known that angry expressions consistently induce activity in the anterior cingulate cortex (ACC), whereas happy expressions consistently induce VS activity. By combining the mouse tracking data and neural response measures from the VS and ACC, cultural psychologists may be able to form a more comprehensive picture of the acculturation process and make more accurate acculturation predictions for immigrants.

6.4 Conclusion

Acculturation is a complex and multidimensional process. Since previous attempts to use behavioral tools to predict immigrants' acculturation outcomes have not been fully successful (Sam & Berry, 2010), using a brain-as-predictor approach may be a better alternative. Combining neuroimaging with behavioral observations allows us to understand the acculturation process more deeply, to make more precise acculturation predictions. This is the brain-as-predictor approach as discussed in a recent review (Berkman & Falk, 2013), a means of linking brain imaging data to outcomes beyond the immediate experimental session in the lab setting. By using this approach, neural responses that encode information can be used to make predictions for subsequent real-world behaviors.

Although this approach is a promising tool for future studies, it has several flaws that we need to take into consideration. First is the reliability of neuroimaging data. In general, neuroimaging data is easily affected by hardware, the interval length between two imaging scans, and the complexity of the underlying cognitive processes. Secondly, researchers must choose brain regions carefully based on the results of meta-analyses or prior research. Lastly, discrimination analysis should be used in this approach in order to make more precise predictions before conducting more advanced analyses. Without taking these issues into account, researchers may not be able to make precise and accurate predictions.

In this chapter, we first review how cultural differences in self-construal modulate the neural representations of the self and close others. This is followed by findings from our longitudinal studies on immigrants, which may provide more insight into changes in their VMPFC self–mother differentiation during acculturation processes. In the second part, we review the neural mechanisms subserving in-group favoritism. We also report that individual differences in this in-group reward reactivity could predict immigrants' friendship patterns. In the last part of this chapter, we propose an integrated brain-as-predictor approach, which brings together self-referential processing and reward processing to predict immigrants' acculturation outcomes. In this part, we also propose several future directions researchers may want to take in order to gain a better understanding of the acculturation process and foster more accurate predictions. We hope that this exploration into the underlying neural mechanisms involved in the acculturation process will lead to interventions that guide immigrants toward the integration state of acculturation, which allows immigrants from different cultures to gain a better understanding of each other and better communicate across cultures.

References

Aharon, I., Etcoff, N., Ariely, D., Chabris, C. F., O'Connor, E., & Breiter, H. C. (2001). Beautiful faces have variable reward value: fMRI and behavioral evidence. *Neuron, 32*, 537–551.

Amodio, D. M., & Frith, C. D. (2006). Meeting of minds: The medial frontal cortex and social cognition. *Nature Reviews Neuroscience, 7*(4), 268–277. doi:10.1038/nrn1884.

Aron, A., Aron, E. N., Tudor, M., & Nelson, G. (1991). Close relationships as including other in the self. *Journal of Personality and Social Psychology, 60*(2), 241–253.

Aron, A., Aron, E. N., & Smollan, D. (1992). Inclusion of other in the self scale and the structure of interpersonal closeness. *Journal of Personality and Social Psychology, 63*(4), 596–612.

Bakker, W., van Oudenhoven, J. P., & van der Zee, K. I. (2004). Attachment styles, personality, and Dutch emigrants' intercultural adjustment. *European Journal of Personality, 18*(5), 387–404. doi:10.1002/per.515.

Balcetis, E., & Dunning, D. (2006). See what you want to see: Motivational influences on visual perception. *Journal of Personality and Social Psychology, 91*(4), 612–625. doi:10.1037/0022-3514.91.4.612.

Berkman, E. T., & Falk, E. B. (2013). Beyond brain mapping: Using neural measures to predict real-world outcomes. *Current Directions in Psychological Science, 22*(1), 45–50. doi:10.1177/0963721412469394.

Berkman, E. T., Falk, E. B., & Lieberman, M. D. (2011). In the trenches of real-world self-control: Neural correlates of breaking the link between craving and smoking. *Psychological Science, 22*(4), 498–506. doi:10.1177/0956797611400918.

Berridge, K. C., Robinson, T. E., & Aldridge, J. W. (2009). Dissecting components of reward: 'liking', 'wanting', and learning. *Current Opinion in Pharmacology, 9*(1), 65–73. doi:10.1016/j.coph.2008.12.014.

Berry, J. W. (1997). Immigration, acculturation and adaptation. *Applied Psychology, 46*, 5–68.

Berry, J. W. (2003). Conceptual approaches to acculturation. In K. Chun, P. Balls-Organista, & G. Marin (Eds.), *Acculturation: Advances in theory, measurement and applied research* (pp. 17–37). Washington, DC: American Psychological Association.

Berry, J. W., Phinney, J. S., Sam, D. L., & Vedder, P. (2006). Immigrant youth: Acculturation, identity and adaptation. *Applied Psychology, 55*(3), 303–332.

Bower, G., & Gilligan, S. (1979). Remembering information related to one's self. *Journal of Research in Personality, 13*, 420–432.

Chavez R. S., & Heatherton T. F. (in press). Multi-modal frontostriatal connectivity underlies individual differences in self-esteem. *Social Cognitive and Affective Neuroscience*.

Chen, P.-H. A., Wagner, D. D., Kelley, W. M., Powers, K. E., & Heatherton, T. F. (2013). Medial prefrontal cortex differentiates self from mother in Chinese: Evidence from self-motivated immigrants. *Culture and Brain, 1*(1), 3–15. doi:10.1007/s40167-013-0001-5.

Chen, P.-H. A., Wagner, D. D., Kelley, W. M., & Heatherton, T. F. (2015). Activity in the cortical midline structures is modulated by self-construal changes during acculturation. *Culture and Brain, 3*(1), 39–52. doi:10.1007/s40167-015-0026-z.

Chen, P. H. A., Whalen, P. J., Freeman, J. B., Taylor, J. M., & Heatherton, T. F. (in press). Brain reward activity to masked in-group smiling faces predicts friendship development. Social Psychological and Personality Science. doi:10.1177/1948550614566093.

Chiao, J. Y., Harada, T., Komeda, H., Li, Z., Mano, Y., Saito, D., et al. (2009a). Dynamic cultural influences on neural representations of the self. *Journal of Cognitive Neuroscience, 22*(1), 1–11.

Chiao, J. Y., Harada, T., Komeda, H., Li, Z., Mano, Y., Saito, D., et al. (2009b). Neural basis of individualistic and collectivistic views of self. *Human Brain Mapping, 30*(9), 2813–2820. doi:10.1002/hbm.20707.

Chiao, J. Y., Iidaka, T., Gordon, H. L., Nogawa, J., Bar, M., Aminoff, E., et al. (2008). Cultural specificity in amygdala response to fear faces. *Journal of Cognitive Neuroscience, 20*(12), 2167–2174.

Cikara, M., Botvinick, M. M., & Fiske, S. T. (2011). Us versus them: Social identity shapes neural responses to intergroup competition and harm. *Psychological Science, 22*(3), 306–313. doi:10.1177/0956797610397667.

Correll, J., & Park, B. (2005). A model of the ingroup as a social resource. *Personality and Social Psychology Review, 9*(4), 341–359. doi:10.1207/s15327957pspr0904_4.

Craik, F. I. M., Moroz, T. M., Moscovitch, M., Stuss, D. T., Winocur, G., Tulving, E., et al. (1999). In search of the self: A positron emission tomography study. *Psychological Science, 10*(1), 26–34. doi:10.1111/1467-9280.00102.

Delgado, M. R. (2007). Reward-related responses in the human striatum. *Annals of the New York Academy of Sciences, 1104*(1), 70–88. doi:10.1196/annals.1390.002.

Demos, K. E., Heatherton, T. F., & Kelley, W. M. (2012). Individual differences in nucleus accumbens activity to food and sexual images predict weight gain and sexual behavior. *Journal of Neuroscience, 32*(16), 5549–5552. doi:10.1523/JNEUROSCI. 5958-11.2012.

Elfenbein, H. A., & Ambady, N. (2003). When familiarity breeds accuracy: Cultural exposure and facial emotion recognition. *Journal of Personality and Social Psychology, 85*(2), 276–290. doi:10.1037/0022-3514.85.2.276.

Ferguson, T. J., Rule, B. G., & Carlson, D. (1983). Memory for personally relevant information. *Journal of Personality and Social Psychology, 44*(2), 251–261.

Freeman, J. B. (2013). Within-cultural variation and the scope of cultural neuroscience. *Psychological Inquiry, 24*(1), 26–30. doi:10.1080/1047840X.2013.767069.

Freeman, J. B., & Ambady, N. (2010). MouseTracker: Software for studying real-time mental processing using a computer mouse-tracking method. *Behavior Research Methods, 42*(1), 226–241. doi:10.3758/BRM.42.1.226.

Freeman, J. B., & Ambady, N. (2011). Hand movements reveal the time-course of shape and pigmentation processing in face categorization. *Psychonomic Bulletin & Review, 18*(4), 705–712. doi:10.3758/s13423-011-0097-6.

Freeman, J. B., Dale, R., & Farmer, T. A. (2011). Hand in motion reveals mind in motion. *Frontiers in Psychology, 2*, 1–6. doi:10.3389/fpsyg.2011.00059/abstract.

Freeman, J. B., Rule, N. O., Adams, R. B., Jr., & Ambady, N. (2009). Culture shapes a mesolimbic response to signals of dominance and subordination that associates with behavior. *NeuroImage, 47*(1), 353–359. doi:10.1016/j.neuroimage.2009.04.038.

Glimcher, P. W. (2011). Understanding dopamine and reinforcement learning: The dopamine reward prediction error hypothesis. *Proceedings of the National Academy of Sciences of the United States of America, 108*(Suppl 3), 15647–15654. doi:10.1073/pnas.1014269108.

Green, E. G. T., Deschamps, J.-C., & Paez, D. (2005). Variation of individualism and collectivism within and between 20 countries: A typological analysis. *Journal of Cross-Cultural Psychology, 36*(3), 321–339. doi:10.1177/0022022104273654.

Haber, S. N., & Knutson, B. (2010). The reward circuit: Linking primate anatomy and human imaging. *Neuropsychopharmacology, 35*(1), 4–26. doi:10.1038/npp.2009.129.

Han, S., Northoff, G., Vogeley, K., Wexler, B. E., Kitayama, S., & Varnum, M. E. W. (2013). A cultural neuroscience approach to the biosocial nature of the human brain. *Annual Review of Psychology, 64*(1), 335–359. doi:10.1146/annurev-psych-071112-054629.

Hare, T. A., Tottenham, N., Davidson, M. C., Glover, G. H., & Casey, B. J. (2005). Contributions of amygdala and striatal activity in emotion regulation. *Biological Psychiatry, 57*(6), 624–632. doi:10.1016/j.biopsych.2004.12.038.

Heatherton, T. F., Wyland, C. L., Macrae, C. N., Demos, K. E., Denny, B. T., & Kelley, W. M. (2006). Medial prefrontal activity differentiates self from close others. *Social Cognitive and Affective Neuroscience, 1*(1), 18–25. doi:10.1093/scan/nsl001.

International Organization for Migration. (2008). *Global estimates and trends*. Geneva: International Organization for Migration.

Johansen-Berg, H. (2010). Behavioural relevance of variation in white matter microstructure. *Current Opinion in Neurology, 23*, 351. doi:10.1097/WCO.0b013e32833b7631.

Kelley, W. M., Macrae, C. N., Wyland, C. L., Caglar, S., Inati, S., & Heatherton, T. F. (2002). Finding the self? An event-related fMRI study. *Journal of Cognitive Neuroscience, 14*(5), 785–794.

Kitayama, S., Ishii, K., Imada, T., Takemura, K., & Ramaswamy, J. (2006). Voluntary settlement and the spirit of independence: Evidence from Japan's 'northern frontier'. *Journal of Personality and Social Psychology, 91*(3), 369–384. doi:10.1037/0022-3514.91.3.369.

Kitayama, S., & Park, J. (2010). Cultural neuroscience of the self: Understanding the social grounding of the brain. *Social Cognitive and Affective Neuroscience, 5*(2–3), 111–129. doi:10.1093/scan/nsq052.

Kitayama, S., & Uskul, A. K. (2011). Culture, mind, and the brain: Current evidence and future directions. *Annual Review of Psychology, 62*(1), 419–449. doi:10.1146/annurev-psych-120709-145357.

Klein, S. B., & Kihlstrom, J. F. (1986). Elaboration, organization, and the self-reference effect in memory. *Journal of Personality and Social Psychology, 115*(1), 26–38.

Klein, S. B., & Loftus, J. (1988). The nature of self-referent encoding: The contributions of elaborative and organizational processes. *Journal of Personality and Social Psychology, 55*(1), 5–11.

Kosic, A., Kruglanski, A. W., Pierro, A., & Mannetti, L. (2004). The social cognition of immigrants' acculturation: Effects of the need for closure and the reference group at entry. *Journal of Personality and Social Psychology, 86*(6), 796–813. doi:10.1037/0022-3514.86.6.796.

Kringelbach, M. L., & Berridge, K. C. (2009). Towards a functional neuroanatomy of pleasure and happiness. *Trends in Cognitive Sciences, 13*(11), 479–487. doi:10.1016/j.tics.2009.08.006.

Kuiper, N. A., & Rogers, T. B. (1979). Encoding of personal information: Self-other differences. *Journal of Personality and Social Psychology, 37*(4), 499–514.

Li, H., Zhang, Z., Bhatt, G., & Yum, Y.-O. (2006). Rethinking culture and self-construal: China as a middle land. *The Journal of Social Psychology, 146*(5), 591–610. doi:10.3200/SOCP.146.5.591-610.

Macrae, C. N., Moran, J. M., Heatherton, T. F., Banfield, J. F., & Kelley, W. M. (2004). Medial prefrontal activity predicts memory for self. *Cerebral Cortex, 14*(6), 647–654. doi:10.1093/cercor/bhh025.

Markus, H. R., & Kitayama, S. (1991). Culture and the self: Implications for cognition, emotion, and motivation. *Psychological Review, 98*(2), 224–253.

Mitchell, J. P., Macrae, C. N., & Banaji, M. R. (2006). Dissociable medial prefrontal contributions to judgments of similar and dissimilar others. *Neuron, 50*(4), 655–663. doi:10.1016/j.neuron.2006.03.040.

Nauck, B. (2008). Acculturation. In F. J. Van de Vijver, D. A. van Hemert, & Y. H. Poortinga (Eds.), *Multilevel analysis of individuals and cultures* (pp. 379–410). Mahwah, NJ: Erlbaum.

Ng, S. H., Han, S., Mao, L., & Lai, J. C. L. (2010). Dynamic bicultural brains: fMRI study of their flexible neural representation of self and significant others in response to culture primes. *Asian Journal of Social Psychology, 13*(2), 83–91. doi:10.1111/j.1467-839X.2010.01303.x.

Northoff, G., & Hayes, D. J. (2011). Is our self nothing but reward? *Biological Psychiatry, 69*(11), 1019–1025. doi:10.1016/j.biopsych.2010.12.014.

O'Doherty, J., Winston, J., Critchley, H., Perrett, D., Burt, D. M., & Dolan, R. J. (2003). Beauty in a smile: The role of medial orbitofrontal cortex in facial attractiveness. *Neuropsychologia, 41*, 147–155.

Phan, K. L., Wager, T., Taylor, S. F., & Liberzon, I. (2002). Functional neuroanatomy of emotion: A meta-analysis of emotion activation studies in PET and fMRI. *NeuroImage, 16*(2), 331–348. doi:10.1006/nimg.2002.1087.

Powers, K. E., & Heatherton, T. F. (2012). Characterizing socially avoidant and affiliative responses to social exclusion. *Frontiers in Integrative Neuroscience, 6*(46), 1–5. doi:10.3389/fnint.2012.00046/abstract.

Rogers, T. B., Kuiper, N. A., & Kirker, W. S. (1977). Self-reference and the encoding of personal information. *Journal of Personality and Social Psychology, 35*(9), 677–688.

Sam, D. L., & Berry, J. W. (2010). Acculturation: When individuals and groups of different cultural backgrounds meet. *Perspectives on Psychological Science, 5*(4), 472–481. doi:10.1177/1745691610373075.

Schultz, W. (2000). Multiple reward signals in the brain. *Nature Review Neuroscience, 1*, 199–207.

Singelis, T. M. (1994). The measurements of independent and interdependent self-construals. *Personality and Social Psychological Bulletin, 20*, 580–591.

Smith, K. S., Berridge, K. C., & Aldridge, J. W. (2011). Disentangling pleasure from incentive salience and learning signals in brain reward circuitry. *Proceedings of the National Academy of Sciences of the United States of America, 108*(27), E255–E264. doi:10.1073/pnas.1101920108/-/DCSupplemental.

Sporer, S. L. (2001). Recognizing faces of other ethnic groups: An integration of theories. *Psychology, Public Policy, and Law, 7*(1), 36–97. doi:10.1037//1076-8971.7.1.36.

Suinn, R. M., Rickard-Figueroa, K., Lew, S., & Vigil, P. (1987). The Suinn-Lew Asian self-identity acculturation scale: An initial report. *Educational and Psychological Measurement, 47*(2), 401–407.
Symons, C. S., & Johnson, B. T. (1997). The self-reference effect in memory: A meta-analysis. *Psychological Bulletin, 121*(3), 371–394.
Tsai, J. L., Ying, Y. W., & Lee, P. A. (2000). The meaning of "Being Chinese" and "Being American": Variation among Chinese American young adults. *Journal of Cross-Cultural Psychology, 31*(3), 302–332. doi:10.1177/0022022100031003002.
Van Bavel, J. J., & Cunningham, W. A. (2008). Self-categorization with a novel mixed-race group moderates automatic social and racial biases. *Personality and Social Psychology Bulletin, 35*(3), 321–335. doi:10.1177/0146167208327743.
Van Bavel, J. J., Packer, D. J., & Cunningham, W. A. (2008). The neural substrates of in-group bias: A functional magnetic resonance imaging investigation. *Psychological Science, 19*(11), 1131–1139. doi:10.1111/j.1467-9280.2008.02214.x.
Varnum, M. E. W., & Kitayama, S. (2011). What's in a name?: Popular names are less common on frontiers. *Psychological Science, 22*(2), 176–183. doi:10.1177/0956797610395396.
Wagner, D. D., Haxby, J. V., & Heatherton, T. F. (2012). The representation of self and person knowledge in the medial prefrontal cortex. *Wiley Interdisciplinary Reviews: Cognitive Science, 3*(4), 451–470. doi:10.1002/wcs.1183.
Wang, G., Mao, L., Ma, Y., Yang, X., Cao, J., Liu, X., et al. (2012). Neural representations of close others in collectivistic brains. *Social Cognitive and Affective Neuroscience, 7*(2), 222–229. doi:10.1093/scan/nsr002.
Ward, C., Leong, C.-H., & Low, M. (2004). Personality and sojourner adjustment: An exploration of the "Big Five" and the "Cultural Fit" proposition. *Journal of Cross-Cultural Psychology, 35*(2), 137–151. doi:10.1177/0022022103260719.
Weisbuch, M., & Ambady, N. (2008). Affective divergence: Automatic responses to others' emotions depend on group membership. *Journal of Personality and Social Psychology, 95*(5), 1063–1079. doi:10.1037/a0011993.
Whalen, P. J., Rauch, S. L., Etcoff, N. L., McInerney, S. C., Lee, M. B., & Jenike, M. A. (1998). Masked presentations of emotional facial expressions modulate amygdala activity without explicit knowledge. *Journal of Neuroscience, 18*(1), 411–418.
Zhu, Y., Zhang, L., Fan, J., & Han, S. (2007). Neural basis of cultural influence on self-representation. *NeuroImage, 34*(3), 1310–1316. doi:10.1016/j.neuroimage.2006.08.047.

Chapter 7
Implications of Behavioral and Neuroscience Research for Cross-Cultural Training

Sharon Glazer, Sergey Blok, Alissa J. Mrazek, and Andrew M. Mathis

Abstract Glazer, Blok, Mrazek, and Mathis first review the literature on priming of cultural syndromes as a gateway to targeting brain mechanisms underlying cultural differences. Most of this research, they point out, involves the priming of individualism or collectivism. They criticize this work as being too narrow, and ignoring the fact that most intercultural interactions involve relationships. They propose using the relational models of Fiske and colleagues as more representative of the reality of these relationships. They present data that show that these models can be successfully primed. They then suggest specific brain areas that would be activated when each model is primed. The implications of behavioral and neuroscience findings for research and training in the areas of intercultural relations are discussed.

S. Glazer (✉)
University of Maryland Center for Advanced Study of Language, College Park, MD, USA

Division of Applied Behavioral Sciences, University of Baltimore,
1420 N. Charles St., Baltimore, MD 21201, USA
e-mail: sglazer@ubalt.edu

S. Blok
University of Maryland Center for Advanced Study of Language, College Park, MD, USA

Booz Allen Hamilton Inc., One Preserve Parkway, Rockville, MD 20852, USA
e-mail: blok_sergey@bah.com; svb2@cornell.edu

A.J. Mrazek
Department of Psychology, Northwestern University, 2029 Sheridan Rd,
Evanston, IL 60208, USA
e-mail: alissamrazek2016@u.northwestern.edu

A.M. Mathis
University of Maryland Center for Advanced Study of Language, College Park, MD, USA

Booz Allen Hamilton Inc., 901 15th St NW Suite 400, Washington, DC 20005, USA
e-mail: andrew.m.mathis@gmail.com; mathis_andrew@bah.com

J.E. Warnick, D. Landis (eds.), *Neuroscience in Intercultural Contexts*,
International and Cultural Psychology, DOI 10.1007/978-1-4939-2260-4_7

7.1 Introduction

An important goal for employees working with communication content from people of different cultural backgrounds is to understand the cultural subtext of *what* they say. A prominent dimensional theory of culture invokes the concepts of individualism and collectivism to account for differences and explain the subtext. However, recent theoretical and empirical work has pointed to serious limitations of this view (e.g., Oyserman, Coon, & Kemmelmeier, 2002). In particular, this view is likely overly coarse and carries too much noise in which to truly detect the cultural influences on communication, decisions, affects, and actions. For example, the concept of collectivism confounds multiple distinct kinds of social interaction, including interpersonal connectivity, group belonging, duty, harmony, advice-seeking, hierarchy, and preference for group interaction (Brewer & Gardner, 1996; Fiske, 2002). Moreover, these types of social interactions are not clear manifestations of a collectivistic culture, nor are not necessarily incompatible with an individualistic one (Oyserman et al., 2002). They are ways in which relationships, at the dyadic (group) level of analysis, manifest; they are not necessarily labels of a culture itself.

In this chapter, we draw upon priming research, research on relational models, and a few prominent published neuroscientific findings that show a link between (1) how people think about others and (2) brain activity to argue that more priming research is needed to demonstrate how preferred relationship structures differ across cultures. Before we delve into relational perspectives, we describe culture and cultural priming studies as a foundation for our call for more research focused on dyadic interactions. We then highlight Fiske's (1992) Relational Models Theory (RMT) as a possible framework that would enable understanding of the dynamic interplay of communication exchange and capture more nuanced understanding of interpersonal interactions. Finally, we believe that neuroimaging studies would enhance people's understanding of priming, and therefore we review neuroscientific findings to theorize the underlying mechanisms for the priming of relational models. Neuroscientific studies of individualism and collectivism, as well as hierarchy-related constructs, also show distinct neural activation patterns related to processing of those culture cues (Chiao et al., 2009; Han, 2010; Iacoboni et al., 2004). Ultimately, results from priming research (as tested through behavioral data and survey responses, as well as functional magnetic resonance imaging or fMRI) would aid cross-cultural training designers and instructors in tailoring training programs that would include modules other than the typical ones on macro conceptualization of culture, such as individualistic vs. collectivistic or high vs. low power distance (Hofstede, Pedersen, & Hofstede, 2002). We anticipate that an enhanced training program that folds in a dyadic level of cultural understanding would broaden people's skills in considering alternative relational perspectives (vs. different cultural perspectives or personalities).

Advances in cultural priming coupled with cultural neuroscience are expected to provide objective evidence to substantiate development and implementation of culture training. Completion of these studies might also suggest that priming could be a quick aid for individuals to shift their attention toward different cultural and relationship perspectives and, thus, produce predictable changes in location and strength of brain activity. For these reasons, we contend cultural priming is potentially relevant to training that is geared toward improving cross-cultural communication. A major aim of the current chapter is to advance propositions about the link between cultural priming of relationship structures and their functional presence in the brain. Such relationships, we believe, are at the heart of intercultural interactions.

In this chapter, findings from social psychology experiments and neuroscientific studies are the basis from which we propose that priming different social relational orientations can be a promising approach for testing cultural theories and a practical method of improving intercultural perspective taking. We focus on Fiske's (1992) RMT, because it provides a framework for culturally universal modes of social relations, as well as cultural differences with respect to saliency of these modes. In particular, the RMT addresses four relational models (RMs): Communal Sharing, Market Pricing, Equality Matching, and Authority Ranking. Each of the RMs is described in Sect. 7.5 of this chapter. We also believe that RMs might be functionally represented in the brain. If this assertion is correct, then priming may be a viable tool for cuing alternative perspective-taking. In the next sections, we describe culture, cultural priming, priming's consequences, and finally attitudinal shifts and neural pathways implicated by RMs. We conclude this chapter by describing the promising role for neuroscience research in the development and validation of theoretical and practical approaches to stimulating cultural perspectives. If relational models can be cued, as evidenced by changes in brain function and human judgment, then practical methods may be designed for improving intercultural communication in the workplace.

7.2 Culture

Culture provides a framework for interpreting and explaining why and how individuals think, feel, and act (Bond & Leung, 2009; Kashima, 2009). Although people function within cultures and cultures shape individuals' affects (emotions), behaviors, and cognitions (thoughts), not all individuals uphold the culture's defining characteristics. One category of defining characteristics is cultural values. Cultural values are the most often studied defining characteristics in cross-cultural psychology research (e.g., individualism and collectivism, see Sect. 7.2.1). As general principles of what is desired, good, and important, cultural values serve as motivating and guiding factors that aid sense-making of social systems and influence people, though they are not necessarily principles held by all individual members of a social system. A social system is a group in which there are meaningful interactions among individuals (Schwartz, 2009). Examples of social systems are nations, regions, organizations, ethnic groups, gender groups, and families. All social systems have cultures. Thus we can refer to peer group cultures, corporate cultures, and, as we do in this chapter, national cultures.

In order to describe cultures, we need to refer back to the definition of culture, which includes generalized beliefs, values, and norms. These factors explain how and why observed features manifest in a social system. Cultural values are the most often studied cultural factors. They influence individuals' personal (or human) values, and these human values serve as abstract social cognitions of desired end states (Schwartz, 1992). According to Kim and Markus (1999), cultural values can

Fig. 7.1 Situated meaning as a mediator of the relationship between cultural context and individuals' affect, behavior, and/or cognition

explain observable features, such as behaviors, in a social system, because "actions are expressions of cultural values" (p. 792). However, it is the psychological processes of human values and individual belief systems that help situate the meaning of the features and, thus mediate the relationships between cultural values and affect, behavior, and cognition (see Fig. 7.1). Most research describing why certain features differed across cultures have used the individualism vs. collectivism dimension (Oyserman et al., 2002), even though the studies do not present a clear process through which affects, behaviors, and cognitions manifest.

7.2.1 *Individualism and Collectivism*

Individualism–collectivism refers to how individuals' cultural identities are formed relative to a collective. In individualistic cultures, individuals' identities are based on their uniqueness from others. In contrast, in collectivistic cultures, individuals' identities are based on group belongingness. This identity construct is a cultural level descriptor and is important for understanding overall social contexts. In fact, when people seek to adopt a different cultural mindset, it is vital to understand the cultural values, beliefs, and norms that people of a social system are socialized (i.e., taught and reinforced) to endorse. This is because culture serves as a vehicle for sense-making or giving meaning to situations that arise within a social system (Oyserman & Sorensen, 2009). Oyserman and Sorensen (2009) further suggest that the meaning spun on to a situation yields a culturally influenced effect. For example, in a social system that values individualism a worker might interpret equal distribution of an unexpected financial reward among members of a project team as unfair (Chen, 1995; Leung & Bond, 1984), which might then lead the worker who thinks s/he contributed more than his/her teammates to file a grievance against the company. In contrast, in China, where collectivism is valued, one might conjecture that equal distribution of an unexpected reward among team members is generous and the person or entity responsible for its distribution would be graciously thanked (Leung & Bond, 1984), because the expectation is differential reward distribution (Chen, 1995; Fischer & Smith, 2003).

Still, not all individuals in a social system would subscribe to "individualism or collectivism uniformly;" asserting so would be an ecological fallacy (Hofstede, 2001).

For example, assuming all US citizens are individualistic based on the nation's individualistic proclivity would be akin to assuming all US citizens are wealthy, by virtue of the nation being considered among the world's wealthiest. Moreover, at the culture level of analysis, the individualism–collectivism constructs are multifaceted, and it is difficult to determine which aspect of the conceptual definitions is being captured in a cross-cultural comparative study. Specifically, Oyserman (2007) and colleagues (2002) depicted individualism at the individual level of analysis (i.e., idiocentrism) in terms of

1. The importance placed on one's autonomy (*independence*), but also in terms of
2. *Personal goals* one wants to achieve,
3. Competing for status or resources (*competition*),
4. Asserting one's individuality (*uniqueness*),
5. Need and appreciation for separation from others (*privacy*),
6. Acquiring knowledge of oneself (*self-knowledge*), and
7. *Direct*, unambiguous *communication*.

Oyserman and colleagues also depicted collectivism at the individual level of analysis (i.e., allocentrism) in terms of

1. How individuals relate with one another (*relationality*),
2. The extent to which belonging to a group is important (*group belonging*),
3. One's sense of obligation to others in one's group (*duty*),
4. Importance of maintaining a balanced life with others and nature (*harmony*),
5. Social support network in which information is sought from elders (*advice-seeking*),
6. Indirect communication that requires a great deal of attention to relational cues (*contextualization*),
7. Clear demarcation of power, roles, and status (*hierarchy*), and
8. *Preference for group interaction*.

A problem with the facets comprising individualism is that they do not necessarily oppose facets comprising collectivism. In fact, some are complementary. For example, evoking competition (an individualism facet) cannot also evoke harmony (a collectivism facet), but invoking competition can still evoke belonging to a group (a collectivism facet). In fact, it is quite possible that as a result of belonging to a group, individuals will compete with other groups for resources to sustain their own group. Likewise, evoking hierarchy (a collectivism facet) cannot also evoke personal goals (an individualism facet), but it can evoke direct communication (an individualism facet).

7.2.2 *Culture as Situated Cognition*

Given the major shortcoming such a multifaceted conceptual definition of individualism and collectivism has in drawing causal conclusions about culture from observed differences, researchers have been investigating culture as situated cognition. The situational view assumes that schemas (i.e., patterned views of how thoughts and actions ought to be organized) governing social interaction are widely shared across all humans (Baumeister & Leary, 1995; Burris & Rempel, 2004) and that culture determines the relative accessibility of a particular schema (Hong, 2009). For example, a North American undergraduate student is capable of thinking of him or herself as an independent or an interdependent being, but would be more likely to endorse values consistent with individualism. Importantly, the situational view emphasizes the role of context in shifting self-construal. The same student may endorse collectivist values upon joining a social organization or while engaging in projects organized by the group (e.g., Gardner, Gabriel, & Lee, 1999). Taken as a whole, the situational view can explain cross-national patterns of differences by appealing to the chronic salience of particular cultural syndromes (i.e., a social characterization or descriptor of culture, e.g., individualism). In addition, it can also account for within-culture and cross-situational differences in behavior (e.g., Uskul, Kitayama, & Nisbett, 2008) by assuming context-appropriate schema activation.

Oyserman and Lee (2007) highlight that the goal in cultural and cross-cultural research is to understand not only culture's influence on what people think, but also *how* culture influences cognition, by identifying cultural characteristics that influence human cognitive processes. Given that we cannot manipulate macro characteristics of a social system, but believing that culture is a manifestation that resides within an individual and that can be manipulated, cultural psychologists (Gardner et al., 1999; also see review by Oyserman, 2011) have been testing whether cultural factors (individualism or collectivism) can be cued to yield different measurement outcomes (e.g., values) and a shift in assumed cultural identity. Findings from these research programs validate the vital role culture plays in understanding situations and the importance of deep cultural knowledge.

Moreover, if culture exerts its influence through differential access to knowledge structures, then enhancing access to general cultural syndromes should cause behavioral consequences similar to those arising from cross-national studies (Gardner et al., 1999). A related claim is that for bicultural individuals, enhancing access to specific cultural knowledge of one culture (e.g., artifacts, symbols) should shift behavior towards consistency with that culture by virtue of the connection between specific cultural knowledge and general syndromes (Hong, Morris, Chiu, & Benet-Martinez, 2000). Nonetheless, if someone does not have real cultural experiences in another social system, one can still enact stereotypical behaviors if (1) the person has been exposed to cognitive processes prominent in another culture and (2) the person is primed (or cued) to evaluate the situation from another's viewpoint.

7.3 Review of Cultural Priming Experiments

7.3.1 What Is Priming?

Priming refers to psychological conditions or stimuli that change an organism's readiness to make a response. Encountering a word, face, or object may speed up processing of, or enhance decision making about, the same or related object in the near future. For example, reading a particular sign while driving along a highway is likely to speed up the processing of the same sign if it is observed again. In this case, it can be said that the first sign "primes" the processing of the subsequent sign. In a laboratory version of this scenario, participants were faster at identifying briefly flashed words if they were previously exposed to (i.e., "primed with") the same words in an earlier task (Jacoby & Dallas, 1981).

The theoretical basis for priming is rooted in the "new look in perception," where researchers such as Gardner and Long (1962) showed that individuals have selective attention, and Bruner and Goodman (1947) demonstrated that a condition (poverty or wealth) coupled with an individual's values and needs, causes a person to select certain behaviors. Later researchers began to discover that at any given time, some concepts in the brain are more active or accessible than others. An idea or a concept is active if it can be applied to the task at hand. What determines which concepts are active include the task or the situation (Bruner, 1957; Posner & Snyder, 1975; Solley & Murphy, 1960), as well as previously active knowledge that carries over to the current task (Neely, 1977). The extent to which recent knowledge influences current processing depends on whether recent knowledge is *related* to the current task (Srull, 2005; Wyer & Carlston, 1979). For example, knowledge activated by playing the guitar is more likely to carry over to playing the harp than to playing baseball. Thus, tasks that are related in terms of their goals and procedures tend to draw on similar sets of cognitive resources and activate the same kinds of knowledge (Bargh, 1989; Srull & Wyer, 1978).

Taken together, earlier findings have shaped today's studies in which researchers are creating conditions by manipulating cultural values and needs to motivate alternative selected behaviors. Priming effects have been found broadly across levels of processing, including low-level repetition effects, semantic relations, and relatively abstract processing, including reasoning and social judgment. We focus on priming of higher-level processes, such as social judgment and decision-making, because these processes have greater relevance to the workplace context that motivates this research.

7.3.2 Construal Priming

Construal priming operates broadly at the level of situational understanding (e.g., Srull & Wyer, 1979), and therefore goes beyond the simple judgments used in many well-known priming examples (e.g., deciding whether a string of characters is a word). Some examples of construal priming include the facilitation of insight problem solving by cueing (Higgins & Chaires, 1980), or the priming of photographs of environments (e.g. "library") in order to facilitate the endorsement of relevant social norms ("silence") (Aarts & Dijksterhuis, 2003). In addition, priming social power makes responsibility goals more salient (Lee-Chai & Bargh, 2001) while priming emotional states can carry over to unrelated decision-making tasks (Lerner, Small, & Loewenstein, 2004). Finally, a number of researchers have examined the priming of self-construal (Gardner et al., 1999; Solomon, Greenberg, & Pyszczynski, 1991). In a seminal study, Gardner et al. (1999) asked participants to search for first person singular or plural pronouns ("I" vs. "we") in a paragraph, and found that those primed with "I" were more likely to exhibit an independent self-construal than those primed with "we." These authors also found that this shift in self-construal had a similar effect on endorsement of values. Taken together, this result provides support for the "situated" view of culture that members of a group can flexibly shift cultural orientation, depending on the context or task. In other words, the way in which one perceives him/her-self, as either autonomous from or embedded in a group, provides an interpretative frame from which to view the world.

7.3.3 Overview of Culture Priming Research

Experiments using the culture priming techniques can be divided into two types: those that prime general cultural *syndromes* (typically individualism and collectivism) and those that prime cultural *symbols*. The first paradigm has received the bulk of attention in the literature. The culture syndrome priming paradigm typically involves studying a single group (e.g., U.S. undergraduates) and testing the extent to which priming cultural syndromes of individualism or collectivism changes behavior on tests of social or cognitive performance (Gardner et al., 1999; Oyserman & Lee, 2008a). Such experiments have been used to test a variety of interesting hypotheses about the effects of rendering cultural syndromes differentially salient, including whether priming affects endorsement of values, changes perceived closeness with others, changes well-being and happiness, and changes a variety of cognitive outcomes. To preview, the literature consistently shows that priming independence leads people to respond in ways consistent with an individualistic mindset, while priming interdependence leads people to behave in ways consistent with collectivism (Oyserman & Lee, 2008b). As mentioned in the previous section, these results have been interpreted as support for the situational view of culture because they purport to emulate cross-national differences in chronic levels of syndrome accessibility.

The second type of paradigm, cultural symbol priming, looks at individuals who are members of two different cultures. An ideal participant in such a study has sufficient experience in both cultures to function fluently in either environment (Benet-Martínez, Leu, Lee, & Morris, 2002; Hong et al., 2000). Bicultural participants are typically bilingual (Marian & Kaushanskaya, 2004) and have the subjective experience of switching between cultural modes depending on the situation (Bond & Yang, 1982). An often-studied population of bicultural people is Hong Kong Chinese, a group that has been substantially influenced by both the Chinese mainland and British culture. Priming studies with bicultural individuals examine differences in social and cognitive behavior as a function of the cultural self that is made temporarily salient in a given person. In a typical priming manipulation, bicultural Chinese American participants are presented with either a Chinese Dragon or an American Flag icon (Hong, Chiu, & Kung, 1997, Hong et al., 1997). Afterwards, participants are asked to perform a social judgment task that is well known to exhibit Chinese-Western cross-national differences. Findings typically indicate that priming one side of a person's bicultural identity tends to cause him or her to behave in a manner that is congruent with that culture's norms (Hong, 2009).

Similar to syndrome priming studies, studies of symbol priming with bicultural individuals provide support for the situational view of culture. First, bicultural participants, who are typically bilingual, can be influenced to switch between identities using relatively subtle priming manipulations, supporting the idea that situational cues automatically activate cultural procedures and norms that guide behavior in context (Bargh & Chartrand, 2000; Brewer & Gardner, 1996; Chartrand & Bargh, 1996; Srull & Wyer, 1979; Trafimow, Triandis, & Goto, 1991). Second, echoing the general conclusions from the syndrome priming studies, studies of bicultural individuals show that cultural effects arise not only out of differences between members of varying ethnic or national groups, but can also manifest themselves across situations within individuals' minds.

From a practical point of view, it is useful to think of syndrome and symbol priming approaches as looking at two sides of a continuum of cultural knowledge. On the one side, individuals with very low knowledge of another culture will nonetheless be susceptible to priming of abstract syndromes. As we have argued, members of all groups possess the basic modes of thought, such as being able to see separation and connection. These are opposite but necessary mental frameworks (Oyserman & Lee, 2008a). On the other side of the continuum are individuals with richer cultural knowledge, including knowledge of language, customs, norms, artifacts, and symbols. These individuals will be effectively primed by cultural syndromes, but will furthermore be susceptible to priming using more specific cultural knowledge.

7.4 Cultural Priming's Consequences

7.4.1 Shifting Values

To demonstrate that personal values can shift depending upon one's cultural mindset, researchers (e.g., Briley & Wyer, 2001, 2002; Gardner et al., 1999) used primes indicative of an independent mindset, which would influence greater endorsement of "freedom" values, in addition to primes that would trigger an interdependent mindset, which would influence "friendship" values. In Gardner et al.'s (1999) study, participants were tasked with circling either the singular or plural pronouns in order to activate independent or interdependent self-construal. They were then asked to provide ratings on Schwartz's (1992) Values Survey. Participants in the interdependent ("we") prime condition compared to those in the independent ("I") prime condition were more likely to endorse those values that Chinese respondents tended to endorse (see also Briley & Wyer, 2001 Study 4; Gardner, Gabriel, & Dean, 2004; Kemmelmeier, 2003 for other studies using this priming method). Several studies also examined effects of primes on judgments about specific scenarios, such as acceptance of euthanasia or affirmative action (Kemmelmeier, Wieczorkowska, Erb, & Burnstein, 2002).

Other methods employed to prime independent vs. interdependent self-construal included a story prime (a general had to choose a warrior to send to the king, either on the basis of individual qualifications for the job and personal gains or on the basis of family obligations and gains) (Gardner et al., 1999), informing participants that they would be working alone or in a group (Briley & Wyer, 2002 Study 1), and using cultural iconic symbols, such as the U.S. flag and a Chinese dragon (Briley & Wyer, 2001). In all three studies, in-group stimuli and Chinese cultural icons increased preference for "collectivistic" type values. Even language can prime bilinguals. In one study, Hong Kong Chinese bilingual participants were more likely to endorse "modern" values and reject traditional Chinese proverbs when completing a survey written in English than one written in Chinese (Bond & Yang, 1982). Likewise, bilingual-Chinese participants responded in a more Western way to measures presented in English than one presented in Chinese (Kemmelmeier & Cheng, 2004; Ralston, Cunniff, & Gustafson, 1995; Ross, Xun, & Wilson, 2002).

Taken together, these studies show that personal values can move towards the primed cultural norm. However, it is not clear still if such shifts can influence decision-making (Briley & Wyer, 2001). If the independent vs. interdependent primes cannot predict judgments and cognitions beyond scale responses, its utility to model culture is, at best, limited. We suggest that neuroscience can provide a promising alternative to looking at scale responses in surveys by systematically uncovering brain areas that may play a role in culture-specific processing. We expand on this point in later sections of this chapter.

7.4.2 Changing Attitudes Toward Obligation and Cooperation

Social obligation and cooperation is often operationalized in terms of prosocial behavior, perceived social support from others, as well as automatic behaviors, such as how closely one sits next to others (Oyserman & Lee, 2008a). Obligation and cooperation are likely to have implications on message encoding. For example, a subordinate might feel an obligation not to confirm or deny any contractual requirements in order to not overstep his or her role in relation to the supervisor. Thus, a sense of obligation, if sufficiently strong, can serve as a powerful motive for action. For people in the USA, this ambiguous messaging might be misconstrued as laziness or apathy, whereas for a person in India it is construed as respecting one's position and obligations to the welfare of his or her work group. This kind of cultural perspective can begin to occur after one is made aware and learns about normative practices regarding interpersonal interactions in the workplace and experiences interacting with people from the other culture. However, the current chapter's researchers suggest that this alternative interpretation might also be more readily retrieved and considered if the knowledge is cued by a cultural prime. Likewise, feeling a need to cooperate with others might be due to a strong sense of harmony, even if the consequences of cooperation yield negative consequences. For example, Space Shuttle Challenger's 1986 accident soon after liftoff is thought to have occurred out of NASA's culture to comply with purported "expert" opinions (Rossow, 2012).

Several studies have looked at obligation and cooperation as dependent variables. Miller, Bersoff, and Harwood (1990) found that adults and children in India (vs. the USA) had a stricter sense of social responsibility across a variety of need situations, including life-threatening, moderately serious, or minor with parent, best friend, or stranger. Of those situations, U.S. participants were more likely to aid in the case of life- 384 threatening need or parents responding to children's serious needs. Later priming studies have shown a similar pattern of results. Gardner et al. (1999) used a social judgment task describing a person who refused to help a friend needing directions to a store. Using the pronoun circling task and the story-based priming task, they found that participants in the interdependent prime condition were more likely to desire punishing the hypothetical person who refused to meet an obligation than participants who received an independent manipulation. These results suggest that close relationships with others increased sensitivity to social obligations in a hypothetical actor (see also Gardner et al., 2004).

Oyserman et al. (2002) reviewed cross-national differences in cooperation styles in conflict resolution scenarios and found that the majority of study participants supported the idea that Americans adapt a relatively self-oriented conflict resolution style. Similarly, Wong and Hong (2005), as well as Utz (2004), found that when bicultural Chinese Americans were primed with a Chinese cultural identity, their level of cooperation in the Prisoner's Dilemma game increased. Interestingly, the effect of priming held only for pairs of individuals who were friends; pairs of strangers were unaffected by the prime. Although a full explanation of this finding is necessary, several cross-national studies point to greater reliance on equality-based interaction norms for in-group members in East Asian societies (Leung & Bond, 1984).

In summary, existing laboratory research shows that priming can shift social processing to increase or decrease sensitivity to social interaction norms, depending on prime content. Priming of an interdependent self-construal tends to increase sensitivity to social obligations and to facilitate cooperation in a conflict resolution scenario. These priming findings mirror those in the cross-national literature.

7.4.3 Improving the Accuracy of Causal Explanation

Understanding others' behaviors means generating accurate explanations for them. Questions of causal explanation and the difficulty of such judgments lie at the heart of dyadic interactions. For example, how accurate are we at assigning the right explanation for another person's communications and behaviors? A long line of research on this topic shows that U.S. respondents' explanations are biased towards attributing behavior to personality and other internal (or dispositional) forces of the actor (Ross & Nisbett, 1991), whereas in East-Asian and Indian cultures, the dispositional bias is much weaker or nonexistent (e.g., Menon, Morris, Chiu, & Hong, 2005; Miller, 1984; Morris & Peng, 1994). Thus, it is clear that explanation styles differ across cultures (Choi, Nisbett, & Norenzayan, 1999; Oyserman & Lee, 2007). The tendency to assign explanations for behavior to individual dispositions rather than situational factors is referred to as the Fundamental Attribution Error

(Gilbert & Malone, 1995; Ross, 1977). Explanations that are based on temporary situational factors are called "situational inferences." The dispositional bias leads people not only to over-attribute observed behavior to internal personality factors but also to overestimate the consistency of behavior for a particular individual across situations (Kunda & Nisbett, 1986). Finally, the bias leads people to be overly confident in making predictions about the behaviors of others (Dunning, Griffin, Milojkovic, & Ross, 1990).

Differences in attribution between Western and non-Western cultural groups have been documented for a variety of paradigms and stimulus materials (Al-Zahrani & Kaplowitz, 1993; Miller, 1984, 1986; Morris & Peng, 1994). Importantly, these differences persist even when situational information is made salient and is in conflict with dispositional factors (Morris & Peng, 1994). The bottom line is that European-Americans tend to be less accurate than Asians in producing social explanations because they tend to disregard situational factors that play a role in behavior.

How can the attribution error be reduced, especially in European-American observers who are more prone to the Fundamental Attribution Error? Hong et al. (2000) conducted a study of self-construal by displaying an animated ambiguous interaction among fish that is consistent with either an internal or an external/situational disposition. They found that Hong Kong Chinese bicultural participants were more likely to assign a situational attribution to the target fish ("escaping") if primed with Chinese primes than American or neutral primes. The findings were also replicated with Chinese-Americans living in the USA (Hong et al., 2000; Hong, Benet-Martinez, Chiu, & Morris, 2003).

It is possible that Asians have a reduced dispositional bias because of the group's general tendency to pay more attention to the contextual factors in a situation (Nisbett & Miyamoto, 2005). This difference in information processing style has been demonstrated using very basic perceptual stimuli (Ji, Peng, & Nisbett, 2000; Masuda & Nisbett, 2006). Employing a priming paradigm, Kühnen and Oyserman (2002) showed that participants primed with interdependent self-construals responded faster than those primed with independent self-concepts to a local vs. a global feature of a Navon (1977) figure[1] (see also Lin & Han, 2009). These results indicate that priming can direct attention toward local or global features.

We believe that cultural priming for understanding social communications can help people better interpret interpersonal communications. Given the tendency for Americans to bias attributions towards internal characteristics of actors, priming on social relational factors might alleviate this bias and reduce the chances of a fundamental attribution error.

[1] The type of figure used in Navon (1977) consists of repetitions of a particular letter of the alphabet (the local-level features), arranged to form the larger shape of another, possibly different letter of the alphabet (the global-level feature). For example, one such figure consists of many repetitions of the letter "S" that form a larger letter "H."

7.4.4 Tracking Common Ground in Conversation

Effective understanding of a conversation requires going beyond the literal meaning of spoken words toward inferences about meaning (Clark, 1985). Speakers must monitor and update a representation of shared knowledge, or common ground, as they progress through the conversation. In this sense, conversation can be thought of as a game of meanings that follows particular rules (Anderson, 1983; Sperber & Wilson, 1995). Haberstroh, Oyserman, Schwarz, Kühnen, and Ji (2002) examined the relationship between self-construal and sensitivity to pragmatic rules in a conversation. Specifically, the authors manipulated whether two very similar survey questions were written by the same or different researchers. In the "same" condition, the pragmatic cues point to the need for different responses, because it is unlikely that the same researcher would add a truly redundant question to the survey. In the "different" condition, it is more plausible that the questions are truly redundant, given the different authors. In addition, a self-construal priming manipulation was employed. In the independent prime condition, there was no significant difference between the question frames. By contrast, when primed with an interdependent self-construal, participants tended to give the same answer in the "different" condition and the different answers in the "same" condition supporting the idea that the interdependent prime increased people's sensitivity to pragmatic context.

7.4.5 Relationship Between the Situation or Task and Priming Effectiveness

Hong and colleagues (2003) suggest that priming manipulations will be effective to the extent that the primed self-construal is *applicable* to the task characteristics (see also Choi et al., 1999). Drawing on earlier research on automatic behavior (Higgins, 1996; Higgins & Brendl, 1995; Strack & Hannover, 1996), Hong et al. (2003) suggest that a primed construal will become maximally applicable if the primed cultural orientation helps to resolve competing features of the stimuli. For example, when Hong et al. (1997) visually reinforced the difference between the individual and the group (vis a vis a visual cue of one fish swimming in relation to other fish), there was a cultural priming effect on social explanation. Chinese bicultural participants in the Chinese prime condition were more likely to select a situational interpretation (fish "escaping" from the group) than those in the American-prime condition. Similarly, Wong and Hong (2005) found a positive effect of priming collectivism on cooperation, but only if the game was played with a friend. This finding is consonant with the idea that playing against a friend created a condition that highlighted the tension between personal gain (winning the game) and social obligation to cooperate.

7.4.6 Primes That Activate Default Orientations

Gardner and colleagues (1999) suggest that self-construal primes are only effective if the prime goes against the default or dominant construal in the culture. Hence, priming an interdependence self-construal in a collectivist culture should not be as effective as priming independence and vice versa. Kemmelmeier and Cheng (2004) support this claim in a study using a language prime in a Hong Kong sample, although the findings are only consistent with the hypothesis if Chinese is the default cultural orientation in Hong Kong. In their meta-analysis, Oyserman and Lee (2008b) discuss the effectiveness of priming collectivism in Western and Asian countries and note a sizeable difference in effect sizes between the two regions. Specifically, they find a larger effect size for priming collectivism in the USA and Europe than in Asia ($d=0.44$ vs. $d=0.08$, respectively).[2] The authors do note that the Asian studies tend to use weaker primes and confound language and culture materials. In contrast, Oyserman and Lee (2008b) do not find the predicted difference in effect size between priming individualism in European-Americans and Asians ($d=0.39$ and $d=0.39$, respectively). More work is needed to understand the relationship between chronically active conceptual frames and temporarily induced orientations.

7.4.7 Theoretical Limitations

The studies we reviewed here, as well as those meta-analyzed by Oyserman and Lee (2008a), show that priming individual or collective social orientation has a robust effect on a variety of outcome variables. Despite important progress, the literature on culture priming has a number of critical gaps. These gaps relate to limited study populations, confounding distinct types of individualism and collectivism, and the slow progress of extending predictions of the situational model to decision-making and reasoning. Moreover, there currently is limited understanding of the underlying neural mechanisms that inform how the brain processes cultural primes (Chiao et al., 2009; Chiao, Mathur, Harada, & Lipke, 2009).

It is clear from Oyserman and Lee's (2008b) survey, and our own review, that the study populations are extremely limited. While this research is termed "cross-cultural," only a handful of countries are represented. The vast majority of the studies were conducted with U.S. undergraduates, although samples from Germany, the Netherlands, and Hong Kong were also included. There were no studies with populations in Scandinavia, Eastern or Southern Europe, Latin America, Middle East, or Africa. Moreover, there has been limited attention in cross-national theorizing or comparison of people in Islamic and African countries (Oyserman et al., 2002). Furthermore, continued reliance on convenient student samples is a serious obstacle

[2] As with any comparison of effect sizes, conclusions must be drawn with caution.

to theoretical and empirical progress in culture research, and in psychology in general (Henrich et al., 2005). The practice of drawing inferences about entire cultural groups from a sample of young, Western-educated students limits generalizability. Studies that look beyond the student population tend to focus on managers in high-profile international firms, a group that tends to be unrepresentative of an entire population (e.g., Ralston et al., 1995). Focus on such limited, homogenous, and atypical groups is likely to introduce serious confounds into most designs.

Another limitation of recent studies is the nearly exclusive reliance on the individualism and collectivism dichotomy as the theoretical foundation for the studies. First, individualism and collectivism are likely confounded with other variables, such as equality and hierarchy. Second, individualism and collectivism are concepts that describe a cultural level influence on individual behavior, but not social dynamics. We suggest that Relational Models Theory (Fiske, 2002) might provide a better foundation to account of cultural phenomena at the dyadic level than culture-level constructs. According to Fiske and Haslam (2005), "social relationships are distinct entities that must be analyzed at their own level, as forms of motivated coordination" (p. 267). Thinking about relationships is different from thinking about individuals or macro conceptions of culture; in interpersonal relationships people focus on "the structures and processes of interaction" (Fiske & Haslam, 2005, p. 282). Both the person and (macro) cultural prescriptions for behaviors inevitably influence these processes. Research on RMs is expected to reveal how people of certain cultures are likely to react different situations. Saner, Mathis, Blok, and Glazer (2013) began addressing the dearth of research on priming of relationship structures by studying whether preferences for RMs differ across adults of European, Iranian, and Chinese cultural backgrounds. Below we describe relational models and findings from Saner et al.'s study.

7.5 Relational Models Theory

The Relational Models Theory (Fiske, 1992, 2004) is a promising area for understanding social relationships because it is structured in terms of dyadic social interactions that occur across all cultures. Four distinct mental models (or schemas) are captured in the RMT, including Equality Matching, Market Pricing, Authority Ranking, and Communal Sharing (discussed further below). The four relational models are universal (Fiske, 1992) and help individuals coordinate and organize their social environment. People apply different RMs when they interact with others, plan social interactions, and make sense of others' social behavior (Fiske, 2004). Furthermore, the way in which RMs are expressed are likely to differ across cultures (Fiske, 2004). In this chapter, we present some evidence of cultural difference in the preference of RM's across cultures and suggest that distinct cognitive processing underlies each of the four types of mental models. We also suggest that priming RMs is a promising avenue for influencing how people interact with others from different cultures. Neuroimaging may be a useful tool to distinguish how the cognitive processing necessary to navigate various social interactions differs across these mental models. For example, fMRI research can identify which brain regions are central to each RM, and applying relevant knowledge about the functional tendencies of those central brain regions can illuminate how the mental models are similar and different from one another. Additionally, knowing the functional tendencies of key neural regions may be of service in designing and tailoring primes for each RM.

Thus, we propose that fMRI research on RMs will help distinguish prominent mechanisms through discovering parallels in neural substrates in other research domains. We discuss examples of such parallels in this section and suggest that this approach will bolster current cognitive hypotheses while simultaneously driving forward the understanding of how each mental model can be successfully primed. Only minimal research (e.g., Iacoboni et al., 2004) has examined RMs at the neural level despite such advantages. Hence, in the following section, we outline the four relational models and propose regions of interest (ROIs) that may reveal neural markers of the cognitive processing required by each model. Identifying such neural markers would enable other researchers to draw parallels in other research domains too. The neural pathways we propose as linked to RMs are based on related investigations in social and cultural neuroscience, primarily associated with perspective taking, empathy, and social perception, which are at the crux of relational models.

7.5.1 Communal Sharing

Communal Sharing (CS) refers to in-group equality and collective belonging, whereby members of in-groups provide and take resources as needed. It is based on the cognitive operation of grouping items, in this case people, into distinct categories based on similarities. "In this kind of relationship, the members of a group or dyad treat each other as all the same, focusing on commonalities and disregarding distinct individual identities" (Fiske, 1992, p. 690). Kinship bonds are a common example of Communal Sharing since close relatives are often perceived as being quite similar by both nature and nurture, having been raised in the same environment. As described by Fiske (1992), "people in a Communal Sharing relationship often think of themselves as sharing some common substance (e.g., "blood"), and hence think that it is natural to be relatively kind and altruistic to people of their own kind" (p. 691). In a Communal Sharing interaction, the needs of the other members are felt as one's own and can outweigh one's own personal goals and desires. For example, members of a work team might forgo their financial reward and opt to give it to a team member who needs it to cover medical expenses. Another example might be at a dinner party, the amount of food consumed by each guest relative to others is irrelevant, and keeping track of this quantity would violate the communal nature of the activity (Fiske & Haslam, 2005).

Because Communal Sharing is rooted in the perception of in-group belonging, we suggest that cortical midline structures, such as medial prefrontal cortex (MPFC), anterior cingulate cortex (ACC), and posterior cingulate cortex (PCC) are promising ROIs when the Communal Sharing model is active (Fig. 7.2). These cortical

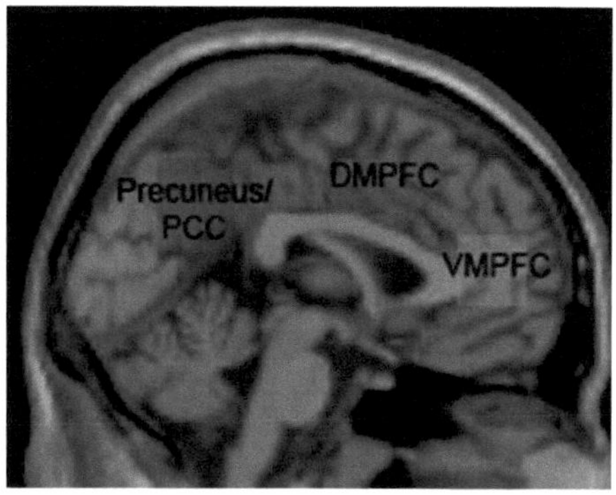

Fig. 7.2 Locations of self-representation areas. *PCC* posterior cingulate cortex, *DMPFC* dorsal medial prefrontal cortex, *VMPFC* ventral medial prefrontal cortex

midline structures are crucial for self-referential and social cognitive processing (Mitchell, Banaji, & MacRae, 2005; Ochsner et al., 2004). One specific neuroimaging study that illuminates how these regions are particularly important for identifying in-group members and responding to in-group members in a preferentially generous manner was in the domain of empathy (Mathur, Harada, & Chiao, 2012). African-Americans and Caucasian-Americans living in the United States generated increased MPFC response when viewing the pain of same-race strangers compared to other-race strangers. This increased empathy toward in-group members correlated with greater altruistic motivation for in-group members, as evidenced by increased giving of time and money. Mathur and colleague's (2012) study suggests neural activation that would similarly underlie the provision of resources to in-group members through Communal Sharing.

7.5.2 Authority Ranking

Authority Ranking (AR) refers to status differentials and hierarchical structures. It is fundamentally based on the cognitive operation of ranking. Fiske (1992) proposes that thinking about social interactions in this way results in a tendency to focus on ranking people by importance, although culture must determine the criterion by which the ranking is performed (as in age or wealth). For example, AR is observed when an expert in some domain directs a novice on how to accomplish a task or when subordinates wait for superiors to arrive or depart and then stand upon their arrival or departure (Fiske & Haslam, 2005). Another example might be observed in a school setting. In some societies, such as the USA and primary schools in the UK, students refer to their teachers in a formal manner that connotes the teacher has greater authority, such as Mr. [surname] or Mrs. [surname]. In primary schools in Hungary students refer to teachers as [First name] Aunt or [First name] Uncle, whereas in England and Ireland, secondary school aged students will refer to teachers as "Miss" or "Mister." In Italy, students will simply say "teacher" and in Japan "[Surname] Teacher." Finally, in countries where informality and egalitarian relationships are favored over hierarchy, such as Israel, students in primary school would say "Teacher [First Name]" and by secondary school "Teacher" is dropped and only the first name is used.

Iacoboni and colleagues (2004) found that viewing video clips depicting everyday social interactions in the form of Authority Ranking or Communal Sharing relational models activated the precuneus and dorsomedial prefrontal cortices (compared to rest or single person engaging in everyday activities). These areas are implicated in many cognitive functions, including the processing of social relationships (Iacoboni et al., 2004). Specifically, the precuneus, located on the outer portion of the back of the brain, is important for mental imagery, especially of episodic memories relevant to the self (Lou et al., 2004). Meanwhile, the dorsomedial prefrontal cortex, located behind the forehead, is highly involved in accurately judging the mental states of others (Saxe & Kanwisher, 2003). Moreover, evidence suggests that the precuneus and the medial prefrontal cortex are functionally coupled, meaning these regions often work together during social processing (Andrews-Hanna, Reidler, Sepulcre, Poulin, & Buckner, 2010).

Chiao, Mathur, and colleagues (2009) found, in a pain empathy experiment, that people higher on social dominance orientation were less likely to empathize with others they view in pain, as evident in less activation in the left anterior insula and anterior cingulate cortices. These areas are deeper within the brain behind the frontal lobe, and among many functions, they have been shown to activate when experiencing or observing suffering (Meyer et al., 2012). The above neuroscientific research results suggest that brain activity might relate with individual differences in relationship preferences, both in the form of person as unique to the group vs. similar to the group and as it pertains to hierarchy. Additionally, they suggest that the way people think about their relationship with others affects their social judgments.

More recent cross-cultural neuroimaging research suggests that the relational impact of social ranking differs based on particular cultural values. Specifically, cultural preferences for social hierarchy affect how strangers are judged, cared for, and aided within and across social groups. For example, in hierarchical cultures, such as Korea, individuals tended to demonstrate increased empathy toward others in their own group compared to members of other groups (Cheon et al., 2011). According to the researchers, this difference may be due in part, to heightened neural response within left temporoparietal junction (L-TPJ), a region important in mentalizing and perspective taking. However, in relatively more egalitarian cultures, such as the United States, individuals' L-TPJ response is the same across social groups. Importantly, at the individual level, greater endorsement of a social dominance orientation (i.e., preference for hierarchy) was higher among Korean participants and predicted a greater bias for selectively attuning to the needs of similar, relative to dissimilar, others (Cheon et al., 2011). These findings suggest that members of hierarchical cultures more greatly consider the thoughts and feelings of their group members compared to individuals in egalitarian cultures. More concretely, these findings suggest that culture level constructs influence how people process relationships. For example, people from egalitarian cultures, such as Americans, are more likely to expect others to express their thoughts and feelings directly, rather than bearing the social responsibility of needing to carefully infer these expectations or needs (Cheon et al., 2011). We suggest that these findings on the neural underpinnings of social hierarchy processing may reflect a similar cognitive mechanism as Authority Ranking (i.e., the role of mentalizing in deciphering appropriateness across situations, especially in one's primary social group where hierarchies are relevant on a daily basis). Hence, we speculate that L-TPJ is a likely candidate for being a neural structure relevant to Authority Ranking.

7.5.3 *Equality Matching*

Equality Matching (EM) refers to a relationship based on quid pro quo reciprocity, matched contributions, and equally divided distributions (Fiske, 2004). It is based on the cognitive operation of ensuring balances in the exchange of favors and payments. People that emphasize EM are often keeping score or tabs on who is ahead or whose turn it is (Fiske & Haslam, 2005). An example application of EM is a raffle. Each participant entering a raffle has the same chance of winning as anyone else. Another social example in the context of a dinner party is that if one is an invited guest, he or she will reciprocate the invitation in the same manner or with a gift worthy of the dinner. An example in the workplace might be observed in a team environment whereby each member is expected to contribute to the development of a product, but even if one person performed better than another, all members get the same reward.

In the majority of social interactions, people prefer when all individuals involved are treated equally (Civai, Crescentini, Rustichini, & Rumiati, 2012; Engelmann & Strobel, 2007; Skitka & Tetlock, 1992). Aligning with this notion, evidence from nonhuman primates, as well as large-scale cross-cultural research, suggests that humans have evolved a deep aversion to unequal treatment (Brosnan & de Waal, 2003; Henrich et al., 2006). As such, Equality Matching may be the default social norm in many scenarios. Recent neuroimaging research suggests that specific neural substrates are involved with identifying when equality is or is not present, as well as reacting to equal and unequal treatment.

An important ROI for the detection of inequality may be the anterior insula (AI), which activates in alarm when something in the environment is unpleasant or unexpected. Because equality is often the preferred distribution of resources, when individuals witness unequal treatment, activity in the AI heightens (Hsu, Anen, & Quartz, 2008; Tabibnia et al., 2008; Sanfey, Rilling, Aronson, Nystrom, & Cohen, 2003). In contrast, when the norms of equality are observed, individuals show increased activity in the nucleus accumbens, a central area in the reward network (Tabibnia et al., 2008). Because humans may be evolutionarily wired to prefer equality, people tend to exhibit seemingly antisocial thoughts and behavior when encountering a stranger who treats people in unequal ways without justification.

For example, brain activity differs in response to witnessing pain when the sufferer is viewed as a fair person versus an unfair person. Singer et al. (2006) designed characters that varied on the dimension of fairness based on their behavior in a standard economic game, a sequential Prisoner's Dilemma. Some characters reciprocated high transfers of money (equal treatment), while other characters decided to defect by transferring small amounts of money (unequal treatment). The unequal treatment was aversive and upsetting to participants. After this game, participants watched all the characters undergo physical pain. When male participants viewed strangers who treated others in either an equal (fair) or an unequal (unfair) manner, the participants showed distinctly different neural activity when the strangers underwent physical pain (Singer et al., 2006). If the stranger had been fair, the participants felt empathy; if the stranger had been unfair, the participants felt *schadenfreude*, a feeling of joy in response to another's suffering, as indicated by activity in the reward-indexing nucleus accumbens.

Thus, we speculate that EM likely activates regions associated with reward, like nucleus accumbens, when equality is reached and the pervasive social norm is upheld, and EM likely activates regions associated with negative appraisal, like AI, when such equality is not reached. However, this neural activity is likely modulated by other complicated factors, such as reputation and the perception of fairness.

7.5.4 Market Pricing

Finally, Market Pricing (MP) refers to proportional reciprocity and relationships based on equity distribution, as well as efficiency (Fiske & Haslam, 2005). It is founded on the cognitive operation of assigning values, including both positive and negative valences. Fiske (1992) theorizes that thinking about social interactions in this manner leads to making cost-benefit judgments. We suggest that RMT's MP focuses on the perceived value of a relationship with others. Attention is directed toward cues that denote the potential utility others possess that would enable a person or group to fulfill his/her/its goals. For example, if one is invited to a dinner party, one might determine if attending would have strategic benefits for one's personal or family goals. Another example is aligning oneself with a colleague who is highly respected, as the perception of relating to him or her might yield indirect benefits for oneself.

MP is distinct from EM, primarily on the basis of the role of equity versus equality, respectively. While equality is rooted in all parties being treated in the exact same way, equity refers to all parties being treated in systematically tailored ways depending on their differences to achieve a state of fairness. In addition to equity, the concept of expected value, the deciphered magnitude and probability of precise outcomes, is central to MP. Previous research suggests that the MPFC region is important for determining probabilities. Thus, this region may act as a neural underpinning of interpreting expected value (Knutson, Taylor, Kaufman, Peterson, & Glover, 2005), and perhaps MP more generally.

Because MP focuses on perceiving value in others, weighing costs and benefits of social investment, and treating others accordingly, extant neuroimaging research on social perception may illuminate ROIs that underlie MP at a more relational level. We speculate that caudate nucleus (CN), a region involved in social learning, may be a critical neural region associated with MP. Evidence suggests that people engage in a trial and error learning process upon interacting with strangers to decipher the value of future engagement, and the CN is highly activated during this process (Tricomi, Delgado, & Fiez, 2004).

For example, one particular neuroimaging study involved participants making risky choices about whether to trust hypothetical partners after reading vivid descriptions of life events depicting the partners as either highly moral, neutral, or immoral (Delgado, Frank, & Phelps, 2005). Although all partners behaved similarly in the experiment, the prior information on moral character was influential. As an MP perspective would predict, participants trusted the partners described as moral to a greater extent than those who were not considered moral and this might be attributed to the perceived benefits and high social value of associating with a moral partner. The CN was most activated during trials associated with the neutral partner, because deciphering the costs and benefits for this partner was relatively difficult and ambiguous. Thus, when individuals meet new people, especially with little information about their character and personality, deciding what the equitable treatment should be requires an attentive trial and error process likely rooted in CN neural activity.

7.5.5 Summary of Neural Correlates

As described in the sections above, we speculate that each of the four relational models may be associated with different cognitive mechanisms that are substantiated by various neural underpinnings. These speculations are based on neuroimaging evidence from social and cultural neuroscience that examines social perception and perspective-taking. Table 7.1 provides an overview of the regions of interest

Table 7.1 Proposed cognitive mechanisms and associated regions of interest (ROIs) for each relational model

Relational model	Cognitive mechanisms	ROIs	Rationale
Communal Sharing (CS)	Social grouping based on categorical similarities with the self	MPFC; ACC; PCC	These ROIs are crucial for self-processing and are associated with in-group benevolence
Authority Ranking (AR)	Social ranking; mentalizing; determining social appropriateness	Precuneus; DMPFC; L-TPJ	Precuneus and DMPFC are key ROIs in social perception; L-TPJ is important for mentalizing and deciphering social hierarchies
Equality Matching (EM)	Identifying and reacting to equal/unequal treatment	Nucleus Accumbens; AI	When equality is detected, reward areas are activated; when inequality is detected, aversive alarms are activated
Market Pricing (MP)	Proportional reciprocity; anticipation of expected value; experiential learning	MPFC; Caudate Nucleus	MPFC underlies determining expected value; CN is important for weighing costs and benefits to determine proportionality

Note: *MPFC* medial prefrontal cortex, *ACC* anterior cingulate cortex, *PCC* posterior cingulate cortex, *DMPFC* dorsomedial prefrontal cortex, *L-TPJ* left temporoparietal junction, *AI* anterior insula for each relational model. These suppositions would greatly benefit from future research, not only to substantiate our theorization, but also to develop our understanding of the critical differences between the relational models and how they can be uniquely primed.

7.5.6 Relational Models Across Cultures

RMT further asserts that using a RM cues a variety of detailed social prescriptions, propositions, precepts, and/or principles that might explain individuals' thinking (Fiske, 2004). Fiske (1992) explains that these prescriptions guide *how* people react in certain situations (Haslam, 2004); they are the foundation for social normative behaviors. Each social system reinforces norms that guide people's thoughts and behaviors (Markus & Kitayama, 1991). Through mimicry or reinforcement learning, members of a social system begin to internalize these different coordinating tools and modes of cognition and automatically respond to situations in ways that are consistent with the norms and ideals of the social system. These coordinating tools can also be cued through priming methods. However, a challenge in priming RMs is determining which cues suppress or draw out other coordinating tools. Moreover, to the extent that these representational systems highlight different ways of approaching various social situations, people across cultures are expected to differ in the extent to which they inculcate reliance on a given representational system and the extent to which they prioritize one system over another.

Studies on cultural priming, thus far, have isolated "individualism" or "collectivism" on assumptions of certain features and have demonstrated that they shift people's thinking about, for example, values and social obligations (Gardner et al., 1999). We suggest that by isolating specific social relationship preferences, we can also shift people's perspectives on social situations. In order to lay the groundwork for priming studies, we summarize preliminary results from Saner et al.'s (2013) scenario-based study that compares people of European, Chinese, and Iranian backgrounds on their ratings and rankings of RMs.

7.6 Preliminary Study Results

7.6.1 Scenario-Based Evidence That Relational Models Differ Across Cultures

As a precursor to testing the feasibility of priming RMs for the purposes of influencing responses to decision-making tasks, Saner et al. (2013) first explored whether people across cultures report different RM preferences on (five) different situations across seven domains. Saner's team conducted a scenario-based study to determine whether endorsement of the relational models differs by culture. This work was inspired by Woodhull and Louis's (2009) study design, in which they examined differences in relational model endorsement among U.S./Canadian and Mexican business managers; Woodhull and Louis examined relational model endorsement on seven contextual domains within a business environment, namely: (reciprocal) Exchange, Distribution and Use (of resources), (organization of) Work, Morals, (collective) Decisions, (social) Influence, and (constitution of) Identity. Applying a series of Mann-Whitney U tests with tests for each intersection of domain and RM, Woodhull and Louis found numerous differences between the two culture groups on all of the RMs and across all domains.

Saner et al. (2013) expanded on this work and examined endorsement of RMs in a wider range of contexts, comparing a wider selection of culture groups. They designed 35 hypothetical scenarios based on the seven domains Woodhull and Louis (2009) examined, and prepared questions based in five different contextual situations for each domain. The five situations for each domain went beyond a business context, and consisted of students in a psychology class working on a research project, participants attending an annual technology convention, residents of a suburban neighborhood, student reporters for a university's newspaper, and engineers building a bridge.

For example, study participants would first read the context of a scenario, such as the context of participants attending an annual technology convention, and then the study participants would read the scenario itself: "The participants are putting together a press conference. This is a complex undertaking with several different tasks that need to be performed. Rate the following statements in terms of your agreement." The agreement scale was five points labeled from left to right "strongly disagree," "somewhat disagree," "neutral," "somewhat agree," and "strongly agree." After rating each of the following items, study participants ranked each in order of preference. In this example, the four items and associated RMs were:

- "The convention president should devise a task completion plan for each participant and then expect each one to accomplish their assigned task." (AR)
- "They should each contribute in proportion to their company's revenues, since the event will benefit them proportionally." (MP)
- "Everyone can pitch in and work together to do whatever needs to be done." (CS)
- "The tasks should be divided equally among the participants so that each one does the same amount of work." (EM)

A self-report survey questionnaire was administered to nearly 200 US-resident adults of European, Chinese, and Iranian backgrounds, ranging in ages from 18 to 64. The non-Euro-American sample included first-generation immigrants, second-generation immigrants, and students temporarily residing in the United States.

Preliminary analyses of RM ratings using an ANCOVA (including sex as a factor and controlling for each participant's mean rating) indicate that while there are macro-level similarities in the RM preferences, there are significant cultural differences after controlling for sex and the effects of the 35 scenarios (Saner et al., 2013). Euro-Americans gave significantly higher ratings to AR, but lower ratings to MP than the other culture groups. Additionally Euro-Americans' EM rating was also higher than that for Chinese participants.

Although Saner et al. (2013) recognize that the rating differences between culture groups were fairly small on average, the fact that they were detected at all in such a diverse, but small sample across a range of scenario contexts, supports the contention that RM preferences differ across cultures, particularly in social decision-making scenarios. Saner and colleagues also found cultural differences within specific situations and domains. Therefore, studying RMs is a helpful exercise in understanding how people from different cultures prefer to organize and structure social interactions on different issues (i.e., domains), in different situations, and general tendencies across all situations and domains. It can reveal localized differences in behavior, in addition to broad trends of culture differences. With further targeting of particular contexts and larger samples, future studies will surely reveal even more differences in RM preferences between cultures. Additionally, results from the scenario-based study suggest that, in light of these cultural differences on RMs, cultural priming can be employed to increase people's abilities to detect, decode, and interpret others' communications through a relational orientation lens.

7.7 Applying Findings from Priming Research

Findings from priming and neuroimaging studies may have practical utility for improving cross-cultural awareness in the workplace context. Basic research findings may advance training and development of individuals engaged in work that has a cultural component, for example cultural analysis of a situation, negotiations with foreign or culturally different business associates, or even for travelers who want an easier adaptation into a host environment. However, on their own these findings might be insufficient for developing adequate training modules. Training developers would benefit from coupling the priming research findings with evidence-based practices rooted in theory of dynamic interpersonal interactions. Role theory is one that could serve as a basis for developing training.

7.7.1 Role Theory and Perspective-Taking

According to role theory (Katz & Kahn, 1978), people communicate by evaluating the consistency of message meaning with internalized models of social behavior. However, the message that one person, the role sender, intends to send, and the message as another person, the role receiver, interprets it, might differ due to cultural influences affecting how messages are encoded and decoded (Beehr & Glazer, 2005). More generally, communication processes rely on aligning communicators' goals and mutual understanding of the facts about the situation (e.g., Sperber & Wilson, 1995; Keysar, Barr, Balin, & Brauner, 2000). Through this alignment, communicators may be able to engage in perspective taking.

Correctly decoding another's message requires an ability to understand the social context in which it is conveyed and thus to be able take the perspective of the role sender (Heuer & Pherson, 2010). For instance, in a culture that discourages expression of disagreement in the presence of elders, a statement of agreement may have a different meaning depending on who else is also in the room. As another example, if a U.S. businessperson requests from a Japanese colleague to take the lead in preparing a contract and the Japanese colleague replies, "This might be possible," then the U.S. businessperson might anticipate compliance with the request. However, the Japanese person might actually mean, "No." The Japanese colleague gave his or her response as a possibility in order to politely decline the request without causing the U.S. person to "lose face." In short, by taking on a culturally-relevant perspective, the role receiver can consider how a role sender's cultural background and interpretation of social situations might influence his or her encoding method, and in doing so might be able to decode the sender's message with greater accuracy.

Although culture is only one of numerous factors that influence a role sender's message encoding (e.g., physical environment and situation are other factors), understanding its influence on *what* is communicated, *how* it is communicated, *when* it is communicated, *where* it is communicated, and *why* it is communicated can substantially improve communication decoding (i.e., understanding) and subsequent performance (e.g., Brew & Cairns, 2004). The more nuanced focus of RMs would help better situate communicators' interactions and enable more focused or tailored perspective-taking.

Perspective-taking is at the heart of being able to understand others' points of view and is a key approach to reducing conflict. Individuals who are unfamiliar with other cultural backgrounds are likely to use their own cultural filters to interpret behaviors of foreign others. Cultural filters and perceptual lens are synonymous terms to describe our tendencies to reduce information to fit our own worldviews. They refer to the way we perceive and interpret others' behaviors. Decoding messages through a role sender's perceptual lens can enable the receiver to better detect cultural nuances and, therefore, understand the intended meaning of the communications.

7.7.2 Application of RM Primes

It is possible that applying RM priming techniques to one's work activities can intensify the role receiver's awareness of another's social obligations and perspectives, where these differ cross-culturally. For example, an entrepreneurial leader assisting an industry competitor might seem irrational to a lay observer with an authority or hierarchical lens. However, with a Market Pricing cultural lens, one might observe a deeper cost vs. benefits approach that suggests that the entrepreneur's strategy ensures s/he maintains a competitive lead, while not becoming cornered into an industry monopoly (which would be organizational failure). In short, a cultural prime could cue role receivers to take on a nuanced cultural perspective and make more accurate inferences about intercultural interactions.

Table 7.2 Conceptual differences between training and priming paradigm

	Training	Priming
Effect duration	Long term	Temporary, short tem
Time course of intervention	Long	Brief
Participant level of awareness	High awareness	Low awareness (even subliminal)
Starting point for knowledge	No prior knowledge required	Some/lots of prior knowledge

Results from RM priming research are expected to facilitate improvements in intercultural training modules, but priming cannot replace training. Because many people would be expected to be able to activate some relational orientation, extensive training would likely make the RM connection easier to draw upon for making sense of a situation. Table 7.2 presents a brief description of the differences between priming and training. We believe that training and practice will increase aptitude for cultural perspective-taking and when given a cultural prime after training, we believe that the prime would activate linkages with other cultural factors to improve perspective-taking even more. In other words, cultural priming would take advantage of existing language and culture training, but would not replace it. Finally, the extent to which a cultural prime works will be determined, in part, by the specific communication skill or task that will utilize cultural knowledge.

Given Saner and colleagues' (2013) encouraging preliminary findings, researchers now can look toward testing specific RM primes. Neurological findings may assist in this process by highlighting which cognitive mechanisms should be targeted in the construction of the primes (e.g. mentalizing vs. reward processing vs. social learning as described earlier). Ultimately, the goals of RM priming research program would be: (1) to provide empirical evidence that people are capable of taking different RM perspectives when cued and (2) to develop a tool that would help shape culturally trained users' frame of mind when evaluating or engaging in communications with people from different cultures.

7.8 Conclusion

In this chapter, we presented findings from neuroimaging and cultural priming studies that suggest the need for studying RMs in order to support cultural perspective-taking training. We argued that the focus on individualism and collectivism is an inappropriate level of analysis to understand social relationships and interactions. Instead, we endorsed focusing on RMs when priming or training people to take on different cultural perspectives in a social dynamic and decision-making situation. We also amalgamated evidence from relevant areas of research to provide speculation about which neural regions may underlie the cognitive processing that distinguishes the four relational models from one another.

Our literature review suggests that priming of RMs would yield differences from unprimed responses on behavioral or attitudinal measures, but RM primes need to be developed. Additionally, the dearth of neuroimaging work that directly assesses RMs, suggests that future fMRI research on each mental model is needed to help identify prominent mechanisms through discovering parallels in neural substrates in other research domains. For example, as mentioned earlier, we suggest that the mechanism underlying Market Pricing may be a social learning process attuned to identifying costs, benefits, and values in others to determine interpersonal equity. Prior research in social perception has identified the caudate nucleus as a prominent region for social learning that guides subsequent interpersonal behavior. If future neuroimaging research on MP were to identify the caudate nucleus as a key region, this neural evidence would strongly support our mechanism hypothesis. Thus, we suggest that neuroimaging research be integrated into the area of RMs in order to validate current cognitive hypotheses as well as propel the understanding of underlying mechanisms of each mental model in new directions. Still, training on RMs may be valuable for individuals who need to assess relational orientations in a given situation. Until empirical studies can reveal empirical evidence of the utility of priming RMs, anecdotal studies and training program evaluations would be a reasonable and tangible, but temporary, substitute for demonstrating utility of RM primes.

References

Aarts, H., & Dijksterhuis, A. (2003). The silence of the library: Environment, situational norm, and social behavior. *Journal of Personality and Social Psychology, 84*(1), 18–28.

Al-Zahrani, S. S. A., & Kaplowitz, S. A. (1993). Attributional biases in individualistic and collectivistic cultures: A comparison of Americans with Saudis. *Social Psychology Quarterly, 56*(3), 223–233.

Anderson, J. R. (1983). A spreading activation theory of memory. *Journal of Verbal Learning and Verbal Behavior, 22*(3), 261–295.

Andrews-Hanna, J. R., Reidler, J. S., Sepulcre, J., Poulin, R., & Buckner, R. L. (2010). Functional-anatomic fractionation of the brain's default network. *Neuron, 65*, 550–562.

Bargh, J. A. (1989). Conditional automaticity: Varieties of automatic influence in social perception and cognition. In J. S. Uleman & J. A. Bargh (Eds.), *Unintended thought* (pp. 3–51). New York: The Guilford Press.

Bargh, J. A., & Chartrand, T. L. (2000). *The mind in the middle: A practical guide to priming and automaticity research*. New York: Cambridge University Press.

Baumeister, R. F., & Leary, M. R. (1995). The need to belong: Desire for interpersonal attachments as a fundamental human motivation. *Psychological Bulletin, 117*, 497.

Beehr, T. A., & Glazer, S. (2005). Organizational role stress. In J. Barling, K. Kelloway, & M. Frone (Eds.), *Handbook of work stress* (pp. 7–33). Thousand Oaks, CA: Sage.

Benet-Martínez, V., Leu, J., Lee, F., & Morris, M. W. (2002). Negotiating biculturalism: Cultural frame switching in biculturals with oppositional versus compatible cultural identities. *Journal of Cross-Cultural Psychology, 33*(5), 492–516.

Bond, M. H., & Leung, K. (2009). Cultural mapping of beliefs about the world and their application to a social psychology involving culture. In R. S. Wyer, C. Chiu, & Y. Hong (Eds.), *Understanding culture: Theory, research, and application* (pp. 109–126). New York: Psychology Press.

Bond, M. H., & Yang, K. S. (1982). Ethnic affirmation versus cross-cultural accommodation: The variable impact of questionnaire language on Chinese bilinguals in Hong Kong. *Journal of Cross-Cultural Psychology, 13*, 169–185.

Brew, F. P., & Cairns, D. R. (2004). Do culture or situational constraints determine choice of direct or indirect styles in intercultural workplace conflicts? *International Journal of Intercultural Relations, 28*, 331–352.
Brewer, M. B., & Gardner, W. (1996). Who is this we? Levels of collective identity and self-representations. *Journal of Personality and Social Psychology, 71*, 83–93.
Briley, D. A., & Wyer, R. S. (2001). Transitory determinants of values and decisions: The utility (or nonutility) of individualism and collectivism in understanding cultural differences. *Social Cognition, 19*, 197–227.
Briley, D. A., & Wyer, R. S. (2002). The effect of group membership salience on the avoidance of negative outcomes: Implications for social and consumer decisions. *Journal of Consumer Research, 29*, 400–415.
Brosnan, S. E., & de Waal, E. B. M. (2003). Monkeys reject unequal pay. *Nature, 425*, 297–299.
Bruner, J. S. (1957). On perceptual readiness. *Psychological Review, 64*, 123–152.
Bruner, J. S., & Goodman, C. C. (1947). Value and need as organizing factors in perception. *Journal of Abnormal and Social Psychology, 42*, 33–44.
Burris, C. T., & Rempel, J. K. (2004). "It's the end of the world as we know it": Threat and the spatial-symbolic self. *Journal of Personality and Social Psychology, 86*(1), 19–42.
Chartrand, T. L., & Bargh, J. A. (1996). Automatic activation of impression formation: Nonconscious goal priming reproduces effects of explicit task instructions. *Journal of Personality and Social Psychology, 71*, 464–478.
Chen, C. C. (1995). New trends in rewards allocation preferences: A Sino-U.S. comparison. *Academy of Management Journal, 38*, 408–428.
Cheon, B. K., Im, D., Harada, T., Kim, J., Mathur, V. A., Scimeca, J. M., et al. (2011). Cultural influences on neural basis of intergroup empathy. *NeuroImage, 57*, 642–650.
Chiao, J. Y., Harada, T., Oby, E. R., Li, Z., Parrish, T., & Bridge, D. J. (2009). Neural representations of social status hierarchy in human inferior parietal cortex. *Neuropsychologia, 47*(2), 354–363.
Chiao, J. Y., Mathur, V. A., Harada, T., & Lipke, T. (2009). Neural basis of preference for human social hierarchy versus egalitarianism. *Annals of the New York Academy of Sciences, 1167*, 174–181.
Choi, I., Nisbett, R. E., & Norenzayan, A. (1999). Causal attribution across cultures: Variation and universality. *Psychological Bulletin, 125*(1), 47–63.
Civai, C., Crescentini, C., Rustichini, A., & Rumiati, R. I. (2012). Equality versus self-interest in the brain: Differential roles of anterior insula and medial prefrontal cortex. *NeuroImage, 62*, 102–112.
Clark, H. H. (1985). *Language use and language users*. New York: Random House.
Delgado, M. R., Frank, R. H., & Phelps, E. A. (2005). Perceptions of moral character modulate the systems of reward during the trust game. *Nature Neuroscience, 8*, 1611–1618.
Dunning, D., Griffin, D. W., Milojkovic, J. D., & Ross, L. (1990). The overconfidence effect in social prediction. *Journal of Personality and Social Psychology, 58*, 568–581.
Engelmann, D., & Strobel, M. (2007). Preferences over income distributions. *Public Finance Review, 35*, 285–310.
Fischer, R., & Smith, P. B. (2003). Reward allocation and culture: A meta-analysis. *Journal of Cross-Cultural Psychology, 34*, 251–268.
Fiske, A. P. (1992). The four elementary forms of sociality: Framework for a unified theory of social relations. *Psychological Review, 99*(4), 689–723.
Fiske, A. P. (2002). Using individualism and collectivism to compare cultures—A critique of the validity and measurement of the constructs: Comment on Oyserman et al. (2002). [Editorial Material]. *Psychological Bulletin, 128*(1), 78–88.
Fiske, A. P. (2004). Relational models theory 2.0. In N. Haslam (Ed.), *Relational models theory: A contemporary overview* (pp. 3–25). Mahwah, NJ: Erlbaum.
Fiske, A. P., & Haslam, N. (2005). The four basic social bonds: Structures for coordinating interaction. In M. W. Baldwin (Ed.), *Interpersonal cognition*. New York: The Guildford Press.
Gardner, W. L., Gabriel, S., & Dean, K. K. (2004). The individual as melting pot: The flexibility of bicultural self-construals. *Current Psychology of Cognition, 22*, 181–201.

Gardner, W. L., Gabriel, S., & Lee, A. (1999). "I" value freedom but "we" value relationships: Self-construal priming mirrors cultural differences in judgment. *Psychological Science, 10*, 321–326.
Gardner, R. W., & Long, R. I. (1962). Cognitive controls of attention and inhibition: A study of individual consistencies. *British Journal of Psychology, 53*, 381–388.
Gilbert, D. T., & Malone, P. S. (1995). The correspondence bias: The what, when, how and why of unwarranted dispositional inference. *Psychological Bulletin, 117*, 21–38.
Haberstroh, S., Oyserman, D., Schwarz, N., Kühnen, U., & Ji, L. J. (2002). Is the interdependent self more sensitive to question context than the independent self? Self-construal and the observation of conversational norms. *Journal of Experimental Social Psychology, 38*(3), 323–329.
Han, S. (2010). *Cultural neuroscience approach to understanding of self. State of the art presentation*. Paper presented at the XXth International Congress of Cross-Cultural Psychology.
Haslam, N. (2004). *Relational models theory: A contemporary overview*. Mahwah, NJ: Erlbaum.
Henrich, J., Boyd, R., Bowles, S., Camerer, C., Fehr, E., Gintis, H., et al. (2005). "Economic man" in cross-cultural perspective: Behavioral experiments in 15 small-scale societies. *Behavioral and Brain Sciences, 28*(06), 795–815.
Henrich, J., McElreath, R., Barr, A., Ensminger, J., Barrett, C., Bolyanatz, A., et al. (2006). Costly punishment across human societies. *Science, 312*, 1767–1770.
Heuer, R. J., & Pherson, R. H. (2010). *Structured analytic techniques for intelligence analysts*. Washington, DC: CQ Press.
Higgins, E. (1996). Knowledge activation: Accessibility, applicability, and salience. In E. T. Higgins & A. W. Kruglanski (Eds.), *Social psychology: Handbook of basic principles* (pp. 133–168). New York: Guilford Press.
Higgins, E., & Brendl, C. (1995). Accessibility and applicability: Some "activation rules" influencing judgment. *Journal of Experimental Social Psychology, 31*(3), 218–243.
Higgins, E., & Chaires, W. (1980). Accessibility of interrelational constructs: Implications for stimulus encoding and creativity. *Journal of Experimental Social Psychology, 16*(4), 348–361.
Hofstede, G. (2001). *Culture's consequences: Comparing values, behaviors, institutions, and organizations across nations*. Thousand Oaks, CA: Sage Publications.
Hofstede, G. J., Pedersen, P. B., & Hofstede, G. H. (2002). *Exploring culture: Exercises, stories, and synthetic cultures*. Yarmouth, ME: Intercultural Press.
Hong, Y. (2009). A dynamic constructivist approach to culture: Moving from describing culture to explaining culture. In R. S. Wyer, C. Chiu, & Y. Hong (Eds.), *Understanding culture: Theory, research and application* (pp. 3–23). New York: Psychology Press.
Hong, Y., Benet-Martinez, V., Chiu, C., & Morris, M. (2003). Boundaries of cultural influence: Construct activation as a mechanism for cultural differences in social perception. *Journal of Cross-Cultural Psychology, 34*, 453–464.
Hong, Y.-Y., Chiu, C.-Y., & Kung, T. M. (1997). Bringing culture out in front: Effects of cultural meaning system activation on social cognition. In K. Leung, Y. Kashima, U. Kim, & S. Yamaguchi (Eds.), *Progress in Asian social psychology* (Vol. 1, pp. 135–146). Singapore: Wiley.
Hong, Y., Morris, M. W., Chiu, C.-Y., & Benet-Martinez, V. (2000). Multicultural minds: A dynamic constructivist approach to culture and cognition. *American Psychologist, 55*(7), 709–720.
Hsu, M., Anen, C., & Quartz, S. R. (2008). The right and the good: Distributive justice and neural encoding of equity and efficiency. *Science, 320*, 1092–1095.
Iacoboni, M., Lieberman, M. D., Knowlton, B. J., Molnar-Szakacs, I., Moritz, M., Throop, C. J., et al. (2004). Watching social interactions produces dorsomedial prefrontal and medial parietal BOLD fMRI signal increases compared to a resting baseline. *NeuroImage, 21*, 1167–1173.
Jacoby, L., & Dallas, M. (1981). On the relationship between autobiographical memory and perceptual learning. *Journal of Experimental Psychology: General, 110*(3), 306–340.
Ji, L., Peng, K., & Nisbett, R. (2000). Culture, control, and perception of relationships in the environment. *Journal of Personality and Social Psychology, 78*(5), 943–955.
Kashima, Y. (2009). Culture comparison and culture priming. In R. S. Wyer, C. Chiu, & Y. Hong (Eds.), *Understanding culture: Theory, research, and application* (pp. 53–77). New York: Psychology Press.
Katz, D., & Kahn, R. (1978). *The social psychology of organizations*. New York: Wiley.

Kemmelmeier, M. (2003). Individualism and attitudes toward affirmative action: Evidence from priming experiments. *Basic and Applied Social Psychology, 25*, 111–119.
Kemmelmeier, M., & Cheng, B. (2004). Language and self-construal priming: A replication and extension in a Hong Kong sample. *Journal of Cross-Cultural Psychology, 35*, 705–712.
Kemmelmeier, M., Wieczorkowska, G., Erb, H.-P., & Burnstein, E. (2002). Individualism, authoritarianism, and attitudes toward assisted death: Cross-cultural, cross-regional, and experimental evidence. *Journal of Applied Social Psychology, 32*, 60–85.
Keysar, B., Barr, D. J., Balin, J. A., & Brauner, J. S. (2000). Taking perspective in conversation: The role of mutual knowledge in comprehension. *Psychological Sciences, 11*, 32–38.
Kim, H., & Markus, H. R. (1999). Deviance or uniqueness, harmony or conformity? A cultural analysis. *Journal of Personality and Social Psychology, 77*(4), 785–800.
Knutson, B., Taylor, J., Kaufman, M., Peterson, R., & Glover, G. (2005). Distributed neural representation of expected value. *Journal of Neuroscience, 25*, 4806–4812.
Kühnen, U., & Oyserman, D. (2002). Thinking about the self influences thinking in general: Cognitive consequences of salient self-concept. *Journal of Experimental Social Psychology, 38*(5), 492–499.
Kunda, Z., & Nisbett, R. E. (1986). The psychometrics of everyday life. *Cognitive Psychology, 18*(2), 195–224.
Lee-Chai, A. Y., & Bargh, J. A. (Eds.). (2001). *The use and abuse of power: Multiple perspectives on the causes of corruption*. Philadelphia, PA: Psychology Press.
Lerner, J. S., Small, D. A., & Loewenstein, G. (2004). Heart strings and purse strings: Carryover effects of emotions on economic decisions. *Psychological Science, 15*(5), 337–341.
Leung, K., & Bond, M. (1984). The impact of cultural collectivism on reward allocation. *Journal of Personality and Social Psychology, 47*(4), 793–804.
Lin, Z., & Han, S. (2009). Self-construal priming modulates the scope of visual attention. *The Quarterly Journal of Experimental Psychology, 62*(4), 802–813.
Lou, H. C., Luber, B., Crupain, M., Keenan, J. P., Nowak, M., Kjaer, T. W., et al. (2004). Parietal cortex and representation of the mental self. *Proceedings of the National Academy of Sciences of the United States of America, 101*(17), 6827–6832.
Marian, V., & Kaushanskaya, M. (2004). Self-construal and emotion in bicultural bilinguals. *Journal of Memory and Language, 51*(2), 190–201.
Markus, H. R., & Kitayama, S. (1991). Culture and the self: Implications for cognition, emotion, and motivation. *Psychological Review, 98*(2), 224–253.
Masuda, T., & Nisbett, R. E. (2006). Culture and change blindness. *Cognitive Science, 30*, 381–399.
Mathur, V. A., Harada, T., & Chiao, J. Y. (2012). Racial identification modulates default network activity for same- and other- races. *Human Brain Mapping, 33*, 1883–1893.
Menon, T., Morris, M. W., Chiu, C.-Y., & Hong, Y.-Y. (2005). Culture and the construal of agency: Attribution to individual versus group dispositions. In D. L. Hamilton (Ed.), *Social cognition: Key readings in social psychology* (p. 333). New York: Psychology Press.
Meyer, M. L., Masten, C. L., Ma, Y., Wang, C., Shi, Z., Eisenberger, N. I., et al. (2012). Empathy for the social suffering of friends and strangers recruits distinct patterns of brain activation. *Social Cognitive and Affective Neuroscience*. doi:10.1093/scan/nss019.
Miller, J. (1984). Culture and the development of everyday social explanation. *Journal of Personality and Social Psychology, 46*(5), 961–978.
Miller, J. (1986). Early cross-cultural commonalities in social explanation. *Developmental Psychology, 22*(4), 514–520.
Miller, J. G., Bersoff, D. M., & Harwood, R. L. (1990). Perceptions of social responsibilities in India and in the United States: Moral imperatives or personal decisions? *Journal of Personality and Social Psychology, 58*(1), 33–47.
Mitchell, J. P., Banaji, M. R., & MacRae, C. N. (2005). The link between social cognition and self referential thought in the medial prefrontal cortex. *Journal of Cognitive Neuroscience, 17*, 1306–1315.

Morris, M. W., & Peng, K. (1994). Culture and cause: American and Chinese attributions for social and physical events. *Journal of Personality and Social Psychology, 67*(6), 949–971.
Navon, D. (1977). Forest before trees: The precedence of global features in visual perception. *Cognitive Psychology, 9*, 353–383.
Neely, J. H. (1977). Semantic priming and retrieval from lexical memory: Roles of inhibitionless spreading activation and limited-capacity attention. *Journal of Experimental Psychology: General, 106*, 226–254.
Nisbett, R. E., & Miyamoto, Y. (2005). The influence of culture: Holistic versus analytic perception. *Trends in Cognitive Sciences, 9*(10), 467–473.
Ochsner, K. N., Knierim, K., Ludlow, D. H., Ramachandran, T., Glover, G., & Mackey, S. C. (2004). Reflecting upon feelings: An fMRI study of neural systems supporting the attribution of emotion to self and other. *Journal of Cognitive Neuroscience, 16*, 1746–1772.
Oyserman, D. (2007). Social identity and self-regulation. In A. W. Kruglanski & E. T. Higgins (Eds.), *Social psychology: Handbook of basic principles* (2nd ed.). New York: Guilford Press.
Oyserman, D. (2011). Culture as situated cognition: Cultural mindsets, cultural fluency, and meaning making. *European Review of Social Psychology, 22*, 164–214.
Oyserman, D., Coon, H. M., & Kemmelmeier, M. (2002). Rethinking individualism and collectivism: Evaluation of theoretical assumptions and meta-analyses. [Review]. *Psychological Bulletin, 128*(1), 3–72.
Oyserman, D., & Lee, S. W. S. (2007). Priming 'culture': Culture as situated cognition. In S. Kitayama & D. Cohen (Eds.), *Handbook of cultural psychology* (pp. 255–279). New York: Guilford Press.
Oyserman, D., & Lee, S. W. S. (2008a). A situated cognition perspective on culture: Effects of priming cultural syndromes on cognition and motivation. In R. Sorrentino & S. Yamaguchi (Eds.), *Handbook of motivation and cognition across cultures*. New York: Elsevier.
Oyserman, D., & Lee, S. W. S. (2008b). Does culture influence what and how we think? Effects of priming individualism and collectivism. *Psychological Bulletin, 134*(2), 311–342.
Oyserman, D., & Sorensen, N. (2009). Understanding cultural syndrome effects on what and how we think: A situated cognition model. In R. Wyer, Y.-Y. Hong, & C.-Y. Chiu (Eds.), *Understanding culture: Theory, research and application* (pp. 25–52). New York: Psychology Press.
Posner, M. I., & Snyder, C. R. R. (1975). Attention and cognitive control. In R. Solso (Ed.), *Information processing and cognition: The Loyola symposium*. Hillsdale, NJ: Erlbaum.
Ralston, D. A., Cunniff, M. K., & Gustafson, D. J. (1995). Cultural accommodation: The effect of language on the responses of bilingual Hong Kong Chinese managers. *Journal of Cross-Cultural Psychology, 26*, 714–727.
Ross, L. (1977). The intuitive psychologist and his shortcomings: Distortions in the attribution process. *Advances in Experimental Social Psychology, 10*, 173–220.
Ross, L., & Nisbett, R. (1991). *The person and the situation: Perspectives of social psychology*. New York: McGraw-Hill.
Ross, M., Xun, W. Q. E., & Wilson, A. E. (2002). Language and the bicultural self. *Personality and Social Psychology Bulletin, 28*(8), 1040–1050.
Rossow, M. P. (2012). *Engineering ethics case study: The Challenger disaster*. Retrieved from http://www.cedengineering.com/upload/Ethics%20Challenger%20Disaster.pdf.
Saner, L., Mathis, A., Blok, S., & Glazer, S. (2013). Scenario norming study: Cross-cultural norming of scenarios. In S. Glazer & L. Saner (Eds.), *Validating the unique cultural space of relational models: A foundation for priming research*. College Park, MD: University of Maryland Center for Advanced Study of Language.
Sanfey, A. G., Rilling, J. K., Aronson, J. A., Nystrom, L. E., & Cohen, J. D. (2003). The neural basis of economic decision-making in the ultimatum game. *Science, 300*, 1755–1758.
Saxe, R., & Kanwisher, N. (2003). People thinking about thinking people: The role of the temporo-parietal junction in "theory of mind". *NeuroImage, 19*, 1835–1842.
Schwartz, S. H. (1992). Universals in the content and structure of values: Theoretical advances and empirical tests in 20 countries. In M. P. Zanna (Ed.), *Advances in experimental social psychology* (Vol. 25, pp. 1–65). New York: Psychology Press.

Schwartz, S. H. (2009). Culture matters: National value cultures, sources, and consequences. In C. Chiu, Y. Hong, S. Shavitt, & J. R. S. Wyer (Eds.), *Understanding culture: Theory, research, and application* (pp. 127–150). New York: Psychology Press.
Singer, T., Seymour, B., O'Doherty, J. P., Stephan, K. E., Dolan, R. J., & Frith, C. D. (2006). Empathic neural responses are modulated by the perceived fairness of others. *Nature, 439*(7075), 466–469.
Skitka, L. J., & Tetlock, P. E. (1992). Allocating scarce resources: A contingency model of distributive justice. *Journal of Experimental Social Psychology, 28*(6), 491–522.
Solley, C. M., & Murphy, G. (1960). *The development of the perceptual world.* New York: Basic Books.
Solomon, S., Greenberg, J., & Pyszczynski, T. (1991). A terror management theory of social behavior: The psychological functions of self-esteem and cultural worldviews. In M. P. Zanna (Ed.), *Advances in experimental social psychology* (Vol. 24, pp. 93–159). San Diego, CA: Academic Press.
Sperber, D., & Wilson, D. (1995). *Relevance: Communication and cognition.* Malden, MA: Wiley-Blackwell.
Srull, T. K. (2005). Person memory: Some tests of associative storage and retrieval models. In D. L. Hamilton (Ed.), *Social cognition: Key readings in social psychology* (Vol. 2, p. 372). New York: Psychology Press.
Srull, T. K., & Wyer, R. S. (1978). Category accessibility and social perception. *Journal of Personality and Social Psychology, 37*, 841–856.
Srull, T. K., & Wyer, R. S., Jr. (1979). The role of category accessibility in the interpretation of information about persons: Some determinants and implications. *Journal of Personality and Social Psychology, 37*, 1660–1672.
Strack, F., & Hannover, B. (1996). Awareness of influence as a precondition for implementing correctional goals. In P. M. Gollwitzer & J. A. Bargh (Eds.), *The psychology of action* (pp. 579–596). New York: Guilford Press.
Tabibnia, G., Satpute, A. B., & Lieberman, M. D. (2008). The sunny side of fairness preference for fairness activates reward circuitry (and disregarding unfairness activates self-control circuitry). *Psychological Science, 19*, 339–347.
Trafimow, D., Triandis, H. C., & Goto, S. (1991). Some tests of the distinction between private self and collective self. *Journal of Personality & Social Psychology, 60*, 649–655.
Tricomi, E. M., Delgado, M. R., & Fiez, J. A. (2004). Modulation of caudate activity by action contingency. *Neuron, 41*, 281–292.
Uskul, A. K., Kitayama, S., & Nisbett, R. E. (2008, June 24). *Ecocultural basis of cognition: Farmers and fishermen are more holistic than herders.* Paper published in the Proceedings of the National Academy of Sciences.
Utz, S. (2004). Self-construal and cooperation: Is the interdependent self more cooperative than the independent self? *Self and Identity, 3*(3), 177–190.
Wong, R. Y., & Hong, Y. (2005). Dynamic influences of culture on cooperation in the prisoner's dilemma. *Psychological Science, 16*, 429–434.
Woodhull, M. D., & Louis, D. J. (2009). Why did they do THAT??? A cross-cultural study of daily decision-making by mexican maquiladoras and US/Canadian managers. *Business Journal of Hispanic Research, 3*, 77–93.
Wyer, R. S., & Carlston, D. E. (1979). *Social cognition, inference, and attribution.* Hillsdale, NJ: Erlbaum.

Chapter 8
Intercultural Relations and the Perceptual Brain: A Cognitive Neuroscience Perspective

Robert Doole, Micaela Y. Chan, and Chih-Mao Huang

Abstract Doole, Chan, and Huang make the point that repeated experiences change the human brain, and enculturation is a prime example of this phenomenon. The authors point to the cognitive and behavioral differences in East Asians and Westerners with an emphasis on attention, reasoning, and self-concept. fMRI studies of visual processes have shown that Westerners show greater activation in object recognition-related brain regions. These studies appear to confirm that Westerners focus on objects while Easterners focus on the holistic. Further, structural MRI has demonstrated Westerners have greater cortical thickness in the frontal lobe and right parietal lobe. This provides further evidence that Westerners have more analytical cognitive patterns. The authors highlight how the understanding of the neuroscience of cultural differences can be used to make a positive impact in intercultural relations.

When people from different cultural backgrounds interact, misunderstandings arise not only from shortcomings in their ability with a shared language, but often from their failure to recognize that they do not possess all of the same implicit assumptions about what they are trying to communicate. Ideas about relationships, values, and logic can seem universal, embedded so deeply in consciousness that they go unvoiced in interpersonal interaction. Many of these concepts, which may seem representative of basic truths to us, are actually uniquely molded by our culture and

R. Doole
Department of Biological Science and Technology, National Chiao Tung University, 1001 University Road, Hsinchu 300, Taiwan
e-mail: robdoole333@gmail.com

M.Y. Chan
Center for Vital Longevity, School of Behavioral and Brain Sciences, The University of Texas at Dallas, 1600 Viceroy Drive, Suite 800, Dallas, TX 75235, USA
e-mail: mchan@utdallas.edu

C.-M. Huang (✉)
Department of Biological Science and Technology, National Chiao Tung University, 1001 University Road, Hsinchu 300, Taiwan

Laboratory for Cognitive Neuroscience, Institute of Linguistics, Academia Sinica, Taipei, Taiwan
e-mail: cmhuang@nctu.edu.tw

J.E. Warnick, D. Landis (eds.), *Neuroscience in Intercultural Contexts*,
International and Cultural Psychology, DOI 10.1007/978-1-4939-2260-4_8

may be alien to others. Overcoming false universalities of consciousness, as ethnographers have discovered, requires one to adopt—at least to a limited degree—a foreign perspective. Since no two individuals are precisely alike, is it even possible to think like the "other"? How accessible is a foreign perspective? People do make real and deep connections with one another on a daily basis, both within and between cultures. What is understood; what is lost in transmission, from one mind to another, from one cultural perspective to another? Investigation into how culture affects cognition and brain has yielded valuable information that both confirms our fundamental similarity, and points toward significant differences, a discovery of which could lead to improved intercultural relations. The burgeoning field of cognitive neuroscience of cultural differences has provided some evidence that cultural experiences sculpt behaviors, brain structure and neurocognitive functions (Chiao, 2009; Goh & Park, 2009; Han et al., 2013; Park & Gutchess, 2002, 2006; Park & Huang, 2010).

In this chapter, we focus primarily on two different cultural environments: East Asian (e.g., Chinese, Japanese, Korean) and Western (focus on American) societies, where a significant body of knowledge on this topic was developed over the past decades. We will begin by discussing the framework through which cognitive neuroscientists conceptualize the impacts of experience and culture on the brain. We will then introduce an area of research, perception and attention, which has provided the most mature data regarding cultural differences at the neurocognitive level. These cross-cultural studies cover experiments testing behavioral differences, primarily using eye-tracking methods, as well as functional and structural neuroimaging studies of the brain. Finally, we will discuss implications regarding the impact on intercultural relations which these cultural neuroscience studies elicit, as well as their shortcomings.

8.1 The Influence of Experiences on the Human Brain

Humans learn and adapt to pressures in their environment. Similar to improving muscle strength through repetitive use, many of our mental capacities are improved through training, practice, and sustained experiences. There is a wealth of evidence that sustained experience changes cognitive processes in the human brain. Cognitive states represent the pattern of activity throughout the brain as neurons communicate via synapses. For example, when one remembers an event in one's past, a specific pattern of neural activation occurs. This pattern will be repeated, with some variation, every time this particular memory is recalled. The probability that a neuron will communicate with another across a given synaptic bond, such as during a memory brain state, will be strengthened the more often that it is used. Like memory, cognitive capabilities, such as letter and number recognition, are mapped to specific, usually somewhat segregated, networks of brain activation. When Polk and Farah (1998) imaged the brains of Canadian postal workers, who sort mail based on postal codes which include both letters and numbers (e.g. L7G-4V8), they found that sustained exposure with a union of the two symbolic systems had remapped the normally separate regions of letters and numbers into a single unitary system. In addition, there is even evidence that sustained experience shapes the function and structure of the brain. In addition to strengthening with repeated use, synaptic bonds can also multiply. Cell bodies grow larger to accommodate the increased workload of heavier synaptic transmission. Structures in the brain that are known to correspond to specific mental capacities are found to increase in volume when those capacities are put to use more often. For example, Maguire et al. (2000) found that London taxi drivers, who engage in regular way finding, have larger gray matter of posterior hippocampi, brain regions believed to be critical for spatial navigation, than those of a control group. Moreover, Draganski et al. (2004) had experimental participants learn to juggle and reported that, after 3 months of practice, they had increased volume of brain regions associated with the processing of complex visual motion.

8.2 The Framework of Cultural Differences

One way to conceive of enculturation is as a sustained collection of related experiences that shape our cognition over time. Just as our basic cognitive processes and brain structures are altered by repetition of experiences like exercise, juggling, and way finding, so too are they sculpted by the systematic repetition of culturally mediated experiences. As social beings, all of our experiences are, in some sense, culturally situated. However, unlike learning to juggle, individual cultural experiences cannot be teased out and tested experimentally as independent variables. Instead, researchers have looked for different cognitive processes and brain structures between groups coming from dissimilar cultural backgrounds. The majority of research regarding cultural influences on cognition has compared people encultured in East Asia with people from Western backgrounds. Nisbett and colleagues proposed that, beginning in ancient times, Western thought (characterized by the Greeks) and Eastern thought (characterized by the Chinese) had fundamentally different philosophical views of the world. These differing views have persisted into the present day, and are still currently sculpting people's cognition and behavior. According to their framework (Nisbett & Masuda, 2003; Nisbett, Peng, Choi, & Norenzayan, 2001; Peng & Nisbett, 1999), Westerners, due to the individualistic, independent, self-based focus of their culture, have a tendency to process focal objects and organize information via rules and categories in an analytic way. In contrast, East Asians, based on their collectivist and interdependent representation, viewed themselves as part of a larger whole. This results in a holistic information-processing bias among East Asians, where object and contextual information are jointly encoded, and relational information is prioritized over categorical information (Goh & Park, 2009; Markus & Kitayama, 1991; Masuda & Nisbett, 2001; Nisbett & Masuda, 2003; Park & Huang, 2010). Thus far, there is broad agreement that cultural differences can be observed between East Asians and Westerners with respect to cognitive functions such as visual perception, attention, contextual processing, and reasoning (Nisbett & Masuda, 2003; Nisbett et al., 2001), as well as psychosocial processes such as motivation, relationality, and self-concept (Han & Northoff, 2008; Huang & Park, 2012; Markus & Kitayama, 1991). Such behavioral data in cultural psychology has provided innumerable demonstrations that there are subtle differences in the way individuals process information that appears to be a product of cultural experiences.

8.3 Cognitive Neuroscience of Cultural Differences

8.3.1 Eye-Tracking

The pattern of eye-movement permits measurement of both duration and location of fixations. Within the study of cognitive neuroscience of cultural differences, eye-tracking techniques have provided the means to observe where attention is directed and showed evidence as to how attentional focus acts as a mechanism to particular components of the visual information.

It has previously been thought that there may be culturally different processing and reasoning of information; yet, the fundamental perception of the world is universal (Nisbett & Masuda, 2003). In the past, facial expressions are thought to be universal in humans, and face perception studies have consistently shown that individuals perceive faces by fixating focus on the eyes and mouth, creating a triangular scanning pattern. However, these studies were conducted in Western parts of the world (e.g. USA, UK), and researchers questioned whether the results could be generalized to other populations.

By tracking eye movement in participants from different cultural backgrounds, several studies have found that Eastern and Western individuals adopted different strategies when processing visual information. For example, a study including both East Asians and Western Caucasians measured eye fixation patterns while participants learned, recognized, and categorized faces (Blais, Jack, Scheepers, Fiset, & Caldara, 2008). In contrast to the classic eye and mouth triangular fixation pattern observed in their Caucasian participants, their East Asians participants have a more centralized focus on the nose/center region of the face. Interestingly, although the two cultures differed in eyes fixation patterns, face recognition accuracy and reaction time did not differ (i.e., no performance differences). Blais et al. (2008) suggested that East Asians might prefer to avoid gazes to the eye regions because it is seen as a rude behavior in their culture. Following up this hypothesis, Kelly, Miellet, and Caldara (2010) conducted a study to explore whether East Asians will maintain avoidance of the eye regions even if the fixated target is a nonhuman face (sheep face), or a target that is not a face at all (greebles, face-like objects that are shown to be processed differently from faces). The results showed that for the sheep face, East Asians still preferred to fixate on the nose/center of the face, whereas Western Caucasians fixated on the eye and mouth regions, similar to how they fixate on human faces. Interestingly, fixations on greebles also showed cultural difference, where Western Caucasians focused more on the central head limb, and East Asians separated their focus on two prominent features on the greebles.

Based on the cross-cultural evidence above, it is shown that, although cultural differences were observed in the focus and perception of faces and nonhuman faces, the recognition and processing of faces seem to be equally well across cultures. Thus, perhaps culture serves the role of shaping how one perceives the world, in which strategy and processing method are developed to best match what is perceived, yielding similar outcomes in recognition accuracy and reaction time.

Researchers have further explored these cultural differences observed in face perception and expressions. Specifically, whether East Asians engage in a center-focused pattern in all situations, or only when peripheral vision could also capture the required information (i.e., still able to capture the eyes and mouth while focusing on the nose/center). Caldara, Zhou, and Miellet (2011) studied this question by restricting the vision of the participants, where, in certain conditions, the eyes and mouth were not visible when focusing on the nose. Under those conditions, East Asian participants utilized the eyes-mouth scanning pattern similar to the Western Caucasian participants. However, when vision restriction was not present, the East Asian participants revert to employing the nose/center-focused strategy. This shows that individuals from various cultures could utilize the same method to perceive faces, but there are culturally preferred ways. In addition to the basic encoding of faces, cultural difference was also found in another important aspect of face perception, the decoding of expressions. For example, when asked to categorize emotions of the faces, compared to Western Caucasian participants who focused both at the eye and mouth regions, East Asian participants focused more on the eye regions, and neglected the lower half of the face, such as the mouth area (Jack, Blais, Scheepers, Schyns, & Caldara, 2009). This increases the likelihood for East Asians to mislabel the expression of "fear" as "surprise"—or vice versa—because the major difference between these two expressions are in the mouth, upon which East Asians focused less.

In addition to face perception, different elements of information in presented visual stimuli shows that East Asian participants usually shift attention frequently between foreground objects and backgrounds, while Western participants tend to shift less often and spend most of their time gazing at the focal object (Chua, Boland, & Nisbett, 2005; Goh, Tan, & Park, 2009). For example, Chua et al. (2005) first examined the pattern of eye-movement in East Asians and Westerners while viewing scenes with embedded central objects. Westerners fixated longer and more often on focal objects whereas East Asians had shorter fixation durations and more saccades to background scenes, confirming basic predictions of culture and cognition models. In a later eye-tracking study, Goh et al. (2009) manipulated the objects and background combination within a scene to examine how the changing of combination (i.e., changing object, changing background, or both) would affect the participants. They found that eye-fixation of Caucasian Americans were more affected by the changes of objects when compared to East Asian participants. However, not all studies have found notable cultural differences. Rayner, Li,

Williams, Cave, and Well (2007) compared eye-fixation across a number of tasks in Chinese and American participants and found that Chinese participants did not spend more time fixating on backgrounds.

These findings from eye-movement paradigms provide evidence to suggest that the relative value of presented visual information in a scene does not hold the same weight across cultures. Whereas Westerners tend to fixate more on salient/focal information when confronted with a complex visual stimulus, East Asians seem more likely to integrate the relations between objects and contextual information.

8.3.2 Functional Neuroimaging of the Human Brain

Functional magnetic resonance imaging (fMRI) based on the blood oxygenation level-dependent (BOLD) contrast has become a widely used noninvasive method for localizing brain activity in humans (Kwong et al., 1992; Ogawa et al., 1992). Such studies are commonly performed on groups of human participants to identify the most consistent pattern of neural activation, either within or between groups, for the functional processes of interests. This cultural dichotomy for processing of visual information lead to a variety of investigations of how the cognitive biasing demonstrated in behavioral performance and eye-movement patterns are manifested in the brain (Han et al., 2013; Kitayama & Uskul, 2011; Park & Huang, 2010). Here, we consider functional neuroimaging studies (e.g., fMRI) of cognition that focus on visual perception of scenes, objects, and faces.

Functional neuroimaging studies have shown that there are functional differences between East Asian and Western people when viewing scenes, in terms of regions of the brain activating to varying degrees. Primarily, these differences are seen in the ventral visual cortex, a broad region encompassing a number of neural structures across the mediotemporal and occipital lobe that is associated with object representation and form recognition (Mishkin, Ungerleider, & Macko, 1983). For example, the fusiform region within the ventral visual cortex activates selectively to human faces (Kanwisher, McDermott, & Chun, 1997), and this structure is often referred to as the "fusiform face area." Selective responding to outdoor scenes, places, and houses occurs in the parahippocampal place area, and there is also evidence for specialization of object recognition in the lateral occipital complex or LOC (Epstein, Graham, & Downing, 2003; Grill-Spector et al., 1998).

Comparing East Asian participants to Westerners while viewing objects (e.g., elephant), backgrounds (e.g., jungle), or salient combinations (e.g., elephant in a jungle), Gutchess, Welsh, Boduroglu, and Park (2006) reported that Westerners showed more brain activity in object-related processing regions (i.e. LOC). Similarly, Goh et al. (2010) found greater activation of the fusiform face area in Westerners compared to Chinese Singaporeans when viewing faces, but less activation of lingual gyrus (a region selective to landmarks) when viewing scenes of houses. Additionally, activation of the fusiform face area was bilateral for Western participants, while Eastern participants had predominately right lateralized activation. These results support a model of holistic visual processing for East Asian people and analytic processing for Westerners. These findings clearly demonstrated cultural differences in perceptual processing, which are consistent with previous eye-movement data.

Another method used to isolate regions of neural specialization within the ventral visual cortex is through the use of a functional MRI adaptation paradigm (fMR-A; Malach et al., 1995). The fMR-A paradigm provides a means of measuring differences in selectivity and specialization in the ventral visual cortex based on the reduction in brain activations towards repeated stimuli. This phenomenon provides an index of the brain's ability to detect similarity between stimuli, and reflects the use of less neural resources to process information that is repeated (Goh & Huang, 2012). Using the fMR-A paradigm, Jenkins, Yang, Goh, Hong, and Park (2010) found that East Asians possess greater sensitivity to contextual information. They assessed the cultural differences in neural adaptation by asking American and Chinese participants to view repeated congruent (e.g., a deer in a forest) and incongruent scenes (e.g., a cow in a kitchen). They found that Chinese participants showed significantly greater adaptation in object processing regions of the brain than Americans when viewing incongruent pairs. The results suggest that the Chinese devoted more brain or cognitive resources to object processing when the scenes were incongruent due to their enhanced sensitivity to the entire scene. This finding provides clear brain evidence that cultural differences in ventral visual function are not only limited to variations in attentional focus on objects and backgrounds separately, but are also associated with variations in processing of semantic relationships between object and contextual information during visual perception and visual recognition (Goh & Huang, 2012; Goh & Park, 2009). Taken together, we have presented compelling data suggesting that cultural experience affects cognitive and neural function in systematic and fundamental ways, with the observation that East Asians are more sensitive to contextual information than Westerners, whereas Westerners have a tendency to process focal and discrete attributes of the visual environment.

8.3.3 Structural Neuroimaging of the Human Brain

In additional to changes in neural functions, cultural neuroscientists propose that prolonged exposure to cultural values and/or practices could also shape neural structures. This hypothesis does not seem far-fetched given the findings reviewed earlier on the influences of way finding (Maguire et al., 2000) and juggling (Draganski et al., 2004) on brain structure. Researchers have found differences in brain structures across several regions that correlate to culture. Using structural MRI, Chee, Zheng, Goh, and Park (2010) measured cortical thickness of 140 participants that were drawn from older and younger Singaporean Chinese and Americans. There were roughly equal participants in each group and the young adults from the two cultures were well matched for age, educational level, and neuropsychological assessment (e.g., speed of processing and working memory). After taking into consideration differences in head size, analysis of cortical thickness identified several cortical regions in both hemispheres as having significantly higher cortical thickness in young Americans relative to young Singaporean Chinese. Specifically, young Americans showed greater thickness in a number of frontal areas and right superior parietal lobule compared to the young Singaporean Chinese, whereas Singaporean Chinese had thicker left inferior temporal regions compared to the young Americans. However, these cultural differences disappeared when they compared the older groups (60–85 year old). The authors suggested that the increased thickness in the frontal areas of young Westerners could conceivably be due to the increased focus Western culture puts on reasoning, problem solving, and independent thinking, whereas the East Asian cultures rely more on following direction and rote memory. However, alternate explanations were also considered, such as dietary, genetic, and environmental differences unrelated to culture per se (Park & Huang, 2010).

8.4 Linking Cognitive Neuroscience of Cultural Differences to Intercultural Relations

Cognitive neuroscience of cultural differences is still in its infancy and requires much more basic research before its contributions to intercultural communication can reach their potential. With that in mind, based on observed cross-cultural differences in cognitive processing and its related functional/structural differences in the brain, several implications on intercultural relations can be drawn. First, since people from different cultures do not process visual information in an identical manner, it may be beneficial to adjust how such information is presented during intercultural communication. Let us take the example of a classroom in a country with an individualistic dominant culture (e.g., elementary school in the United States of America) with students from a collectivistic background (e.g., international students from China). We use this example because it is a common situation in American public schools; Quiroz, Greenfeld, and Altchech (1999) have identified American classrooms with primarily Latino students as having just such a cultural dynamic. Considering the eye-tracking studies mentioned above showing that individuals from collectivistic backgrounds tend to focus more on context and background, it may be helpful for teachers to adjust their visual presentation if they have a large group of students with collectivistic backgrounds. While it may seem that the simplest way to construct a poster trying to teach the concept of "animals" would be to display different animals on a blank background, thereby letting the students focus on the animals and not other background objects, collectivistic students may respond better if the poster depicts the animals in their habitats. So long as the animals are the focal point, it should remain as the focus for students from individualistic backgrounds, but one could expect the additional background information to benefit collectivistic students in building a more detailed mental representation of what "animals" do and how they live.

Not only is information processed differently in the visual domain, studies have shown that verbal information may also be processed differently. This provides added legitimacy to techniques that educators have already been using to "bridge the gap" between cultures. Trumbell, Rothstein-Fisch, and Greenfield (2000) has an example, "The Field Trip," which demonstrates how teachers can integrate the social experience which collectivistic encultured students use to understand the world with the content knowledge required by the American education system. They describe a teacher, whose goal is to teach about birds, allowing her students to tell their personal stories of experiences with birds and then extracting scientific knowledge from them. Therefore, instead of trying to focusing only on scientific knowledge, and possibly alienating them in the process, she used their social experiences to draw out the content that they needed to understand.

In addition, since unspoken assumptions could be some of the most impassable barriers to intercultural communication, discovering them could make a significant impact in improving relations. It is not just that individuals from across cultures behave differently, but their brains perceive and process the world differently on a very basic level (i.e., what and where they focus on). As a consequence, people from different cultures are often unaware of their attentional preferences and these expectations can result in coordination difficulties during intercultural collaboration. For example, divergent cultural preferences and expectations have been reported to strongly affect leadership and team functioning (Köhler & Berry, 2008) and may lead to intercultural conflict. We speculate that potential intercultural conflict concerning the coordination of intercultural collaboration may be stemming from the cultural dichotomy between context-inclusive vs. object-focused styles in the ventral visual processing stream of the brain. While looking at the same scenic picture, a Westerner may focus mostly on the focal object and use more brain resources for processing object-related information, whereas an East Asian will spend relatively longer time on the context and require more brain resources for processing contextual information. In a sense, their experience was different and their expectation would be different, even if they were looking at the same picture, in the same room. Eye-tracking and neuroimaging confirmed these perceptual differences, which individuals in an intercultural workplace should carefully take into account when interacting with others due to its effect on how mutual information is perceived, encoded, and retrieved differently across cultures. We believe that the evidence observed from the approach of cognitive neuroscience of cultural differences described above can help researchers to devise and develop appropriate interventions and trainings to improve intercultural communication.

We should note that cultures are not defined only by their place and/or nations along an individualism/collectivism spectrum. It is also influenced by an unknown number of dimensions such as 'Power Distance', 'Masculinity/Femininity', and 'Uncertainty Avoidance' affecting cultures (Hofstede, 1979, 1980; Hofstede & Bond, 1984). Since collectivism/individualism seems to be related to structural/functional differences in the brains of encultured individuals, it is possible that cross-cultural differences in the brain could be found relating to these dimensions. Setting aside the impact of natural environment and genetics, the relative impact of cultural factors could, one day, be identified through the study of cognitive neuroscience. Future studies could find, with greater accuracy than behavioral data, just where cultures tend to lie along these dimensions. In addition, neural differences could point towards cultural differences which have not yet been found through other kinds of research. Many aspects of culturally mediated cognition and behavior are so basic to our experience that they go unspoken; identifying fundamental differences in neural mechanisms across cultural groups in brain mechanism could be the best way to bring them to light.

Acknowledgements We gratefully acknowledge support from the Ministry of Science and Technology of Taiwan (MOST 103-2420-H-182 -002 -MY2 to C.M.H.).

References

Blais, C., Jack, R. E., Scheepers, C., Fiset, D., & Caldara, R. (2008). Culture shapes how we look at faces. *PLoS One, 3*, e3022.

Caldara, R., Zhou, X., & Miellet, S. (2011). Putting culture under the 'spotlight' reveals universal information use for face recognition. *PLoS One, 18*, e9708.

Chee, M. W., Zheng, H., Goh, J. O., & Park, D. C. (2010). Brain structure in young and old East Asians and Westerners: Comparisons of structural volume and cortical thickness. *Journal of Cognitive Neuroscience, 18*, 495–507.

Chiao, J. Y. (2009). Cultural neuroscience: A once and future discipline. *Progress in Brain Research, 178*, 287–304.

Chua, H. F., Boland, J. E., & Nisbett, R. E. (2005). Cultural variation in eye movements during scene perception. *Proceedings of the National Academy of Sciences of the United States of America, 102*, 12629–12633.

Draganski, B., Gaser, C., Busch, V., Schuierer, G., Bogdahn, U., & May, A. (2004). Neuroplasticity: Changes in grey matter induced by training. *Nature, 427*, 311–312.

Epstein, R., Graham, K. S., & Downing, P. E. (2003). Viewpoint-specific scene representations in human parahippocampal cortex. *Neuron, 37*, 865–876.

Goh, J. O., & Huang, C. M (2012). Images of the cognitive brain across age and culture. In P. Bright (Ed.), *Neuroimaging—Cognitive and clinical neuroscience*. InTech. doi:10.5772/23944.

Goh, J. O., Leshikar, E., Sutton, B. P., Tan, J. C., Sam, S., Hebrank, A., et al. (2010). Culture differences in neural processing of faces and houses in ventral visual cortex. *Social Cognitive and Affective Neuroscience, 5*, 227–235.

Goh, J. O., & Park, D. C. (2009). Culture sculpts the perceptual brain. *Progress in Brain Research, 178*, 95–111.

Goh, J. O., Tan, J. C., & Park, D. C. (2009). Culture modulates eye-movements to visual novelty. *PLoS One, 4*, e8238.

Grill-Spector, K., Kushnir, T., Hendler, T., Edelman, S., Itzchak, Y., & Malach, R. (1998). A sequence of object-processing stages revealed by fMRI in the human occipital lobe. *Human Brain Mapping, 6*, 316–328.

Gutchess, A. H., Welsh, R. C., Boduroglu, A., & Park, D. C. (2006). Cultural differences in neural function associated with object processing. *Cognitive, Affective, & Behavioral Neuroscience, 6*, 102–109.

Han, S., & Northoff, G. (2008). Culture-sensitive neural substrates of human cognition: A transcultural neuroimaging approach. *Nature Reviews Neuroscience, 9*, 646–654.

Han, S., Northoff, G., Vogeley, K., Wexler, B. E., Kitayama, S., & Varnum, M. E. (2013). A cultural neuroscience approach to the biosocial nature of the human brain. *Annual Review of Psychology, 3*, 335–359.

Hofstede, G. (1979). Value systems in forty countries. In L. Eckensberger, W. Lonner, & Y. Poortinga (Eds.), *Cross-cultural contributions to psychology* (pp. 389–407). Amsterdam: Swets and Zeitlinger.
Hofstede, G. (1980). *Culture's consequences: International differences in work-related values*. Beverly Hills, CA: Sage.
Hofstede, G., & Bond, M. (1984). Hofstede's culture dimensions. *Journal of Cross-Cultural Psychology, 15*, 417–433.
Huang, C. M., & Park, D. C. (2012). Cultural influences on Facebook photographs. *International Journal of Psychology, 48*, 334–343.
Jack, R. E., Blais, C., Scheepers, C., Schyns, P. G., & Caldara, R. (2009). Cultural confusions show that facial expressions are not universal. *Current Biology, 19*, 1543–1548.
Jenkins, L. J., Yang, Y. J., Goh, J. O., Hong, Y. Y., & Park, D. C. (2010). Cultural differences in the lateral occipital complex while viewing incongruent scenes. *Social Cognitive and Affective Neuroscience, 5*, 236–241.
Kanwisher, N., McDermott, J., & Chun, M. M. (1997). The fusiform face area: A module in human extrastriate cortex specialized for face perception. *Journal of Neuroscience, 17*, 4302–4311.
Kelly, D. J., Miellet, S., & Caldara, R. (2010). Culture shapes eye movements for visually homogeneous objects. *Frontiers in Psychology, 1*, 6. doi:10.3389/fpsyg.2010.00006.
Kitayama, S., & Uskul, A. K. (2011). Culture, mind, and the brain: Current evidence and future directions. *Annual Review of Psychology, 62*, 419–449.
Köhler, T., & Berry, M. (2008). Creating synergy for inter-cultural learning. In J. Salmons & L. Wilson (Eds.), *Handbook of research on electronic collaboration and organizational synergy* (pp. 142–158). Hershey, PA: Information Science Reference.
Kwong, K. K., Belliveau, J. W., Chesler, D. A., Goldberg, I. E., Weisskoff, R. M., Poncelet, B. P., et al. (1992). Dynamic magnetic resonance imaging of human brain activity during primary sensory stimulation. *Proceedings of the National Academy of Sciences of the United States of America, 89*, 5675–5679.
Maguire, E. A., Gadian, D. G., Johnsrude, I. S., Good, C. D., Ashburner, J., Frackowiak, R. S., et al. (2000). Navigation-related structural change in the hippocampi of taxi drivers. *Proceedings of the National Academy of Sciences of the United States of America, 97*, 4398–4403.
Malach, R., Reppas, J. B., Benson, R. R., Kwong, K. K., Jiang, H., Kennedy, W. A., et al. (1995). Object-related activity revealed by functional magnetic resonance imaging in human occipital cortex. *Proceedings of the National Academy of Sciences of the United States of America, 92*, 8135–8139.
Markus, H., & Kitayama, S. (1991). Culture and the self: Implication for cognition, emotion and motivation. *Psychological Review, 98*, 224–253.
Masuda, T., & Nisbett, R. E. (2001). Attending holistically versus analytically: Comparing the context sensitivity of Japanese and Americans. *Journal of Personality and Social Psychology, 81*, 922–934.
Mishkin, M., Ungerleider, L. G., & Macko, K. A. (1983). Object vision and spatial vision: Two cortical pathways. *Trends in Neurosciences, 6*, 414–417.
Nisbett, R. E., & Masuda, T. (2003). Culture and point of view. *Proceedings of the National Academy of Sciences of the United States of America, 100*, 11163–11170.
Nisbett, R. E., Peng, K., Choi, I., & Norenzayan, A. (2001). Culture and systems of thought: Holistic versus analytic cognition. *Psychological Review, 108*, 291–310.
Ogawa, S., Tank, D. W., Menon, R., Ellermann, J. M., Kim, S. G., Merkle, H., et al. (1992). Intrinsic signal changes accompanying sensory stimulation: Functional brain mapping with magnetic resonance imaging. *Proceedings of the National Academy of Sciences of the United States of America, 89*, 5951–5955.
Park, D. C., & Gutchess, A. H. (2002). Aging, cognition, and culture: A neuroscientific perspective. *Neuroscience and Biobehavioral Reviews, 26*, 859–867.
Park, D. C., & Gutchess, A. H. (2006). The cognitive neuroscience of aging and culture. *Current Directions in Psychological Science, 15*, 105–108.

Park, D. C., & Huang, C. M. (2010). Culture wires the brain: A cognitive neuroscience perspective. *Perspectives on Psychological Science, 5*, 391–400.

Peng, K., & Nisbett, R. E. (1999). Culture, dialectics, and reasoning about contradiction. *American Psychologist, 54*, 741–745.

Polk, T. A., & Farah, M. J. (1998). The neural development and organization of letter recognition: Evidence from functional neuroimaging, computational modeling, and behavioral studies. *Proceedings of the National Academy of Sciences of the United States of America, 95*, 847–852.

Quiroz, B., Greenfeld, P. M., & Altchech, M. (1999). Bridging cultures with a parent-teacher conference. *Educational Leadership, 56*, 68–70.

Rayner, K., Li, X., Williams, C. C., Cave, K. R., & Well, A. D. (2007). Eye movements during information processing tasks: Individual differences and cultural effects. *Vision Research, 47*, 2714–2726.

Trumbell, E., Rothstein-Fisch, C., & Greenfield, P. M. (2000). *Bridging cultures in our schools: New approaches that work.* West Ed. Retrieved October 15, 2002, from the West Ed Web site: http://web.wested.org/online_pubs/bridging/welcome.shtml

Chapter 9
Cultural Influences on Social and Self-Relevant Memory

Sarah Huff, Laura Ligouri, and Angela Gutchess

9.1 Introduction

Culture can impact the way in which an individual perceives the world around him or her. While definitions of culture vary widely between and within the varying social sciences, a cross-disciplinary definition might encompass culture as the values, ideas, and systems of belief that are shared amongst a network of interacting individuals who form a societal structure from which persons derive a sense of identification and interpret the world around them (Bruner, 1990; D'Andrade, 1984; DiMaggio, 1997; Geertz, 1973; Kashima, Woolcock, & Kashima, 2000; Sperber, 1996). Thus, culture can influence cognition related to social, emotional, and self-relevant processes. These processes correspondingly can shape the construction of culture, through the effects of personally held values and ideas as well as through interactions with others. A host of research indicates that Westerners tend to focus on objects, categories, and the self as an independent entity, whereas Easterners attend more to contexts, functional relationships, and group-relevant information. The lens imparted by one's culture can direct attention, filtering which aspects of one's environment are noted and encoded into memory (Gutchess & Indeck, 2009). In terms of memory retrieval, the cultural lens can shape which details are stored in memory and which cues serve as effective elicitors of information from memory.

One domain in which culture has been shown to impact memory is in attention to central objects vs. background contexts. Literature comparing Easterners (e.g., Chinese, Japanese) to Westerners (e.g., Americans, Canadians) indicates that while

S. Huff • L. Ligouri • A. Gutchess (✉)
Department of Psychology, Brandeis University, 415 South Street,
MS 062, Waltham, MA 02454-9110, USA
e-mail: gutchess@brandeis.edu

J.E. Warnick, D. Landis (eds.), *Neuroscience in Intercultural Contexts*,
International and Cultural Psychology, DOI 10.1007/978-1-4939-2260-4_9

Westerners largely focus on central objects in a complex scene, Easterners are more attuned to background and contextual information. Differences in social contexts across cultures could lead to these cognitive differences, in that heightened attention to others and the broader social context could provoke East Asians to generally attend to the context, even in nonsocial settings (Nisbett & Masuda, 2003; Nisbett, Peng, Choi, & Norenzayan, 2001). Eye-tracking data reveal that Westerners fixate more to a focal object during the first 300 msec of presentation than Easterners (Chua, Boland, & Nisbett, 2005 but see Evans, Rotello, Li, & Rayner, 2009), suggesting that these attentional differences emerge early in the viewing of complex scenes. These attentional biases can also affect memory for objects and backgrounds in that East Asians recall more information about background elements of animated vignettes compared to Americans (Masuda & Nisbett, 2001). Furthermore, East Asians exhibit poorer memory for objects when the context has been removed from a picture, whereas Americans are relatively insensitive to the presence or absence of background context (Masuda & Nisbett, 2001). Functional neuroimaging data also show that Americans engage neural regions implicated in object processing more than East Asians during the viewing of complex scenes (Gutchess, Welsh, Boduroglu, & Park, 2006). A study comparing younger and older adults across cultures found that while older adults exhibit reduced adaptation, that is, a reduced response to viewing pictures as a function of repetition, in object processing regions, the effect of aging was magnified for East Asians relative to Westerners (Goh et al., 2007). Thus, these functional neuroimaging studies are consistent with results from behavioral investigations in demonstrating greater sensitivity to objects for Westerners than Easterners.

Memory is a particularly interesting domain in which to study cross-cultural differences in that memory has the potential to shape, as well as reflect, one's experiences in the world. Whether or not a certain event in fact occurred or not can often be objectively verified. However, the *details* of these memories play an integral part in how an individual constructs the world and his or her own sense of self. Because memory is a constructive process (Schacter, 1999), information is encoded and retrieved based on those aspects that are most salient to the individual, such as conceptual, perceptual, or relational features. Cultural differences in the ways in which individuals remember the qualities of information could contribute to substantial differences in the subjective experiences of individuals, which could have implications for cross-cultural communication and the potential for misunderstanding in global relations.

The study of social and self-relevant memory could be particularly important for how individuals construct a model of the world around them, given the importance of social interactions. We review evidence for cross-cultural differences in these domains of memory in this chapter. In order to create a framework for how culture can shape memory for social and self-relevant information, we first present three different mechanisms through which culture has been suggested to impact cognitive processes (Gutchess, Schwartz, & Boduroglu, 2011). Cultural differences could reflect: (1) the engagement of different cognitive *processes*, such that individuals from different cultures adopt different strategies (e.g., relating information about others to oneself) or process distinct elements of information (e.g., focal person versus social context), (2) differences in the *content* of which facets of information are stored and accessed by individuals from different cultures (e.g., information from one's own perspective vs. the perspective of others), or (3) differences in the degree of *difficulty* across cultures, with one task being more challenging and resource-intensive for members of one culture than another. While the first mechanism, differences in cognitive processes, has received the most focus thus far in the broader culture and cognition literature, the social domain may often blur together the first and second processes such that which distinct elements of information receive focus may be closely linked to what *content* is encoded into memory. For example, viewing a situation from the perspective of the independent self, that is, self as an entity separable from others, may evoke more self-relevant processing which can also impact the extent to which information is encoded into memory from the perspective of oneself (e.g., in a complex social context, remembering events that impacted oneself rather than others or remembering from one's own viewpoint). The third mechanism, involving difficulty, to date has rarely impacted the study of social cognition across cultures. This is because comparisons of more or less effortful social processes (e.g., use of automatic stereotypes based on group membership vs. controlled processes to override stereotypes or individuate a target) have not been the focus of cross-cultural research thus far.

9.2 Self-Reference Effect

To date, the self constitutes the social process with relevance to memory that has been most studied in terms of cultural differences. The self-reference effect (Rogers, Kuiper, & Kirker, 1977) has been widely studied and replicated in behavioral and neuroimaging studies, primarily in the USA. When given the task to rate adjectives on four dimensions (structural, phonemic, semantic, self-reference), participants demonstrate significantly better memory for the adjectives in the self-relevant condition (Rogers et al., 1977). Further work established that referencing the self improved memory even relative to referencing other people (see Symons & Johnson, 1997 for a review). This effect in memory indicates that the self engages unique cognitive processes that evoke deep, elaborative processing and organization of information in memory (Klein & Kihlstrom, 1986; Klein & Loftus, 1988). The self reference effect is also consistent with the notion of an independent self-construal, such that one distinguishes the self from others as a distinct and separable entity. Recently research on the self-reference effect has been extended to neuroimaging research as well (Kelley et al., 2002; Macrae, Moran, Heatherton, Banfield, & Kelley, 2004). In these studies, participants show differential activation in regions implicated in self-referential processing (medial prefrontal cortex (mPFC) and posterior cingulate cortex (PCC)) while judging self-relevant adjectives that are subsequently remembered. Such studies were critical in substantiating the claims that self-referencing engaged unique processes and were not simply an extension of levels of processing (Craik & Lockhart, 1972) that merely reflected a particularly meaningful semantic condition (Heatherton, Macrae, & Kelley, 2004).

Most striking about these findings is that it is not the hippocampus that supports the memory-related process. Rather, cortical midline structures (mPFC and PCC) are activated (Macrae et al., 2004), therefore suggesting that self-referential memory is a unique memory process, distinct from other types of explicit memory, which typically rely on the hippocampus (Gabrieli, 1998; Paller & Wagner, 2002).

9.3 Cultural Differences in Self-Construal

A discussion of cultural influences on self-relevant memory necessitates an understanding of cultural differences in self-concept and identity. As previously mentioned, the self-reference effect has been studied in Western contexts where the majority of people are independent or self-focused, meaning that preferential memory for self-relevant information is not too surprising. An increasing body of both behavioral and neuroimaging research suggests broad cultural differences in self-concept; one of the most widely researched distinctions is between the independent and interdependent self-construal (Markus & Kitayama, 1991). A person with an independent self-construal derives his or her identity from uniqueness and separation from others. Alternatively, a person with an interdependent self-construal emphasizes connectedness and relations with others in understanding his or her own identity. The independent self-construal is most common in Western societies, such as the USA, whereas the interdependent self-construal is more indicative of individuals in East Asian countries, such as China (Markus & Kitayama, 1991; Sui, Zhu, & Chiu, 2007; Triandis, 1995).

9.4 Cultural Differences in the Neural Underpinnings of Self-Referential Processing

While the self-reference effect has been demonstrated many times in Western cultures, it has only recently been investigated in other cultures. Although there is minimal cross-cultural research on the self-reference effect in terms of memory, cross-cultural investigations of broader aspects of self-referential processing seem to be much more common. Self-referential processing is typically measured by asking participants to think about themselves, or make judgments about whether adjectives describe the self, a close other (e.g., mother), or a distant other (e.g., the president). Neuroimaging evidence suggests that there is increased activation in the mPFC and PCC while judging relevance to the self, as contrasted against judging the relevance to others, suggesting the existence of a network that is uniquely involved in processing information about oneself (Craik et al., 1998; Gutchess, Kensinger, & Schacter, 2007, 2010; Han & Northoff, 2009; Heatherton et al., 2004; 2006; Kelley et al., 2002; Macrae et al., 2004; Northoff, Heinzel, de Greck, Bermpohl, & Dobrowolny, 2006).

Cross-cultural investigations suggest that cultural identity may modify the engagement of this self-referential network. Recent evidence has shown that both American and Chinese participants engage the self-referential processing network (mPFC and PCC) while making judgments about self-relevant information. However, these same regions are activated in Chinese participants for mother-relevant information as well (Zhang et al., 2006; Zhu, Zhang, Fan, & Han, 2007). For Westerners, there appears to be a distinct neural representation of self processing, while in Chinese participants, self representation is inclusive of close others, such as the mother (Zhu et al., 2007) (see Fig. 9.1). These findings are consistent with the previously mentioned behavioral research, which demonstrates an independent self-construal in Westerners and an interdependent self-construal in East Asians, and a potential explanation for cultural differences in self-relevant memory.

In addition to distinct neural representations of the self versus others across cultures, there is also evidence of differences in the self-reference memory effect, based on both behavioral and neural results. Recent research demonstrates that both Westerners and Chinese participants exhibit the self-reference effect such that information related to oneself is better remembered than information related to another person (Zhang et al., 2006; Zhu & Zhang, 2002; Zhu et al., 2007). However, similar to the findings for self-referential processing, Chinese participants also show memory enhancement for mother-relevant items, thus providing further evidence for the inclusion of the other in East Asians' representation of the self (Zhang et al., 2006; Zhu & Zhang, 2002; Zhu et al., 2007).

To date, the self-reference effect in memory and other self-referential processes have been studied with the methods previously described, wherein participants typically make judgments about whether a personality trait is descriptive of the self or another person and in some tasks they complete a subsequent memory task. However, behavioral research suggests that one of the most important distinctions between Easterners and Westerners is in the independent (i.e., unique and person) self-construal and the interdependent (i.e., interconnected and social) self-construal (Cousins, 1989; Heine, 2001; Markus & Kitayama, 1991, 2010). Therefore, it is important to investigate self-referential processing beyond simple personality traits by including information about the social self. Recently, researchers have addressed the limitation of focusing only on the personal self (without also considering the social self) by including social roles and physical attributes, in addition to personality traits, in self-referential tasks. In an investigation of Chinese and Dutch college students, Ma et al. (2014) found greater mPFC activation in Danes than in Chinese during self-reflection (vs. reflection about a public figure) across all three dimensions (personality traits, physical traits, social roles/identities). Alternatively, during self-reflection on social attributes, Chinese showed significantly greater activation in the temporoparietal junction (TPJ), which has been implicated in mentalizing and thinking about others' mental states and beliefs (Jenkins & Mitchell, 2010; Lombardo et al., 2010; Saxe & Kanwisher, 2003). Additionally, individual differences in self-construal (independent vs. interdependent) mediated the patterns of neural activity. MPFC activation during self-reflection on social attributes was negatively correlated with interdependence and TPJ activation was positively correlated with interdependence

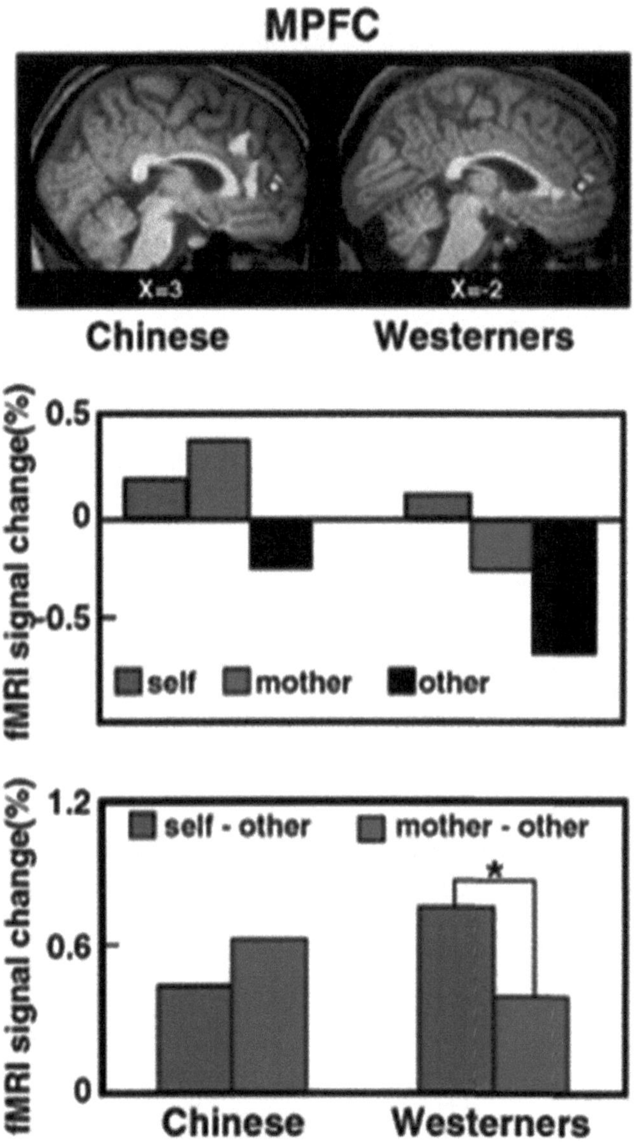

Fig. 9.1 Cross-cultural differences have been shown during mother-referencing, such that Chinese participants engage mPFC for self- and mother-referencing whereas Westerners engage the region for referencing the self more than mother. Reprinted from *Neuroimage*, 34, Zhu, Y., Zhang, L., Fan, J., and Han, S., Neural basis of cultural influence on self-representation, 1310–1316, Copyright (2007), with permission from Elsevier

during self-reflection on social attributes. In other words, the more interdependent an individual is, the less likely she is to engage the mPFC, and the more likely she is to engage the TPJ during self-reflection of social attributes.

Sul, Choi, and Kang (2012) performed a similar study but they added a memory component, which allows for examination of cultural differences in the self-reference effect for personal and social information. These researchers were interested in the self-reference effect across personal and social domains in Korean college students, including a measure of individual differences in individualism and collectivism (INDCOL index; Singelis, Triandis, Bhawuk, & Gelfand, 1995), which has often been used to assess independent and interdependent self-construals. In terms of neural activity the results were consistent with the Chinese sample in the Ma et al. (2014) study, such that the Koreans showed higher activity in the mPFC while judging self-relevant personality traits and higher TPJ activation while processing self-relevant social identity information. In addition, the self-reference effect was moderated by cultural orientation, as measured by individualism and collectivism. Specifically, participants who reported being more individualistic experienced greater memory enhancement for personality traits, whereas individuals who tended towards collectivism showed greater memory enhancement for social traits (Sul et al., 2012).

9.5 Bicultural Identity and Priming as Moderators for Neural Activity

Another recent avenue of research has incorporated bicultural individuals into the study of self-referential processing. Individuals with a bicultural identity are an interesting population because they have had exposure to at least two cultural meaning systems. As a result, they are able to switch between different cultural frames to varying degrees based on their cultural identity. In our lab, we investigated how having a bicultural identity influences self-referential (vs. mother-referential or other-referential) processing and memory using an adjective trait judgment task and a surprise recognition memory task. Based on the Bicultural Identity Integration Scale (BII), individuals were identified as having either a *blended* (i.e., compatible identities and competence in both cultures) or an *alternating* (i.e., viewing cultural identities as oppositional or conflicting and/or identifying with only one culture) bicultural identity (Benet-Martinez & Haritatos, 2005; Phinney & Devich-Navarro, 1997). Contrary to expectations, results indicate that the dmPFC was more engaged for mother-referencing than self-referencing in our sample of bicultural Asian Americans. In terms of subsequent memory, the region that supported successful encoding into memory was the posterior cingulate cortex (PCC), showing greater activation for mother-relevant and other-relevant (compared to self-relevant) information. Finally, we observed reversals in the pattern of activity in the dmPFC implicated in subsequent memory for those with a blended bicultural identity versus those with alternating bicultural identity. Specifically, those with alternating bicultural identities engaged canonical self-referential regions for encoding of self-relevant information, while those with blended bicultural identities recruited these same areas for the encoding of mother-relevant information. These findings suggest that cultural effects, specifically individual differences in bicultural identity, modulate neural activity during judgment and encoding of information relevant to the self and others (Huff, Yoon, Lee, Mandadi, & Gutchess, 2013).

Since culture represents context for the individual, manipulating the context using techniques such as priming can further elucidate cultural differences in the distinction between self versus other and subsequent memory for information relevant to these individuals. Techniques such as priming are also particularly advantageous in that they allow for causal inferences about the role of culture. Whereas people cannot be randomly assigned to different cultural groups (e.g., Eastern vs. Western), the role of culture is indirectly inferred, although a number of other confounding factors could actually explain the pattern of results, such as cohort effects (e.g., an effect of living in a particular context at a given point in time, rather than something pervasive about the culture itself; see Gutchess and Goh [2013] for further discussion). This concern about culture is in contrast to the use of primes to directly manipulate culture in that participants can be randomly assigned to one cultural group or another. This is achieved through priming by orienting individuals to different aspects of the self-concept (e.g., independence or interdependence), which can potentially influence memory and neural activity differently during self-reference vs. other-reference judgments (Gutchess & Indeck, 2009). In order to demonstrate the contribution of priming to cross-cultural differences in self-referential processing and how self-identity is considered in relation to others Chiao et al. (2010) exposed bicultural individuals to either an individualistic or a collectivistic prime prior to a self-judgment task. In this task there were two categories of personality trait judgments (general or contextual) and participants were asked to determine whether the trait described them. In the general condition, the participant was asked to respond to whether a trait described her "in general." In the contextual condition, the participant was given a specific context, such as "when talking to my mother," and asked to decide whether the trait described her in that context. Those exposed to the individualism prime were characterized by greater activation in mPFC and PCC for general relative to contextual judgments, whereas those in the collectivism prime condition demonstrated greater mPFC and PCC activation for contextual relative to general self-descriptions (Chiao et al., 2010). Another study employed priming prior to a general trait judgment task (including self-relevant, father-relevant, unfamiliar other person-relevant judgments) and demonstrated the effect that cultural priming modulates activity in the dorsal, but not the ventral, portion of the mPFC in Asian Americans (Harada, Li, & Chiao, 2010). More specifically, neural activity differed in the dorsal mPFC as a function of prime when making judgments related to the self and others. Researchers in Hong Kong found additional evidence for cultural influence on neural activation in the ventral medial prefrontal cortex (vmPFC), such that those exposed to a Western prime demonstrate increased activation during self-relevant (vs. other relevant) judgments. In contrast, after seeing the Chinese prime, participants showed no difference in vmPFC activation for self-relevant (vs. other-relevant) judgments. Research to date provides convincing evidence that cultural context influences self-referential processing, identity, and memory using behavioral and neuroimaging methods. These cultural differences emerge across studies relying on naturally occurring distinct cultural groups as well as studies employing priming to activate distinct cultural identities in what was initially a single group of participants. In contrast to the relatively large body of research employing neuroimaging to investigate cross-cultural differences in self-relevant processes, other areas of social memory are marked by a dearth of neural data and are characterized primarily by behavioral methods.

9.6 Cultural Influences on Memory for Social Information

Research investigating cross-cultural differences in memory for social information and information about others has yet to be fully developed, making this area of research another promising domain for future work. One existing study shows that recall of social interactions differs across Americans and East Asians (Chua, Leu, & Nisbett, 2005). Americans recall more information about a central character than Taiwanese participants, relative to recollection about information regarding other characters. In their recall of narratives and videos, Americans also attribute more intentionality to the characters. Judgments of emotion are also impacted by cultural differences in attention to social contexts. Taiwanese recall more emotional content than Americans. Another study integrates the study of East Asians' superior memory for emotional information with memory for non-central characters. Japanese were better at recognizing changes in facial expression in characters in the background than Americans, whereas Americans and Japanese remembered similar amounts of information about the central character (Masuda et al., 2008). Similarly, young Koreans are better than young Americans at recognizing background contexts (that were not explicitly social per se) presented behind emotional faces. Surprisingly, however, the effect does not simply go away with age but the bias actually reverses such that older Americans remember more backgrounds than older Koreans (Ko, Lee, Yoon, Kwon, & Mather, 2011).

Neuroimaging methods have yet to be applied to the study of cross-cultural differences in memory for social information. However, research conducted predominantly with Westerners suggests that explicit memory for social information may rely more on distinct neural systems than explicit memory for nonsocial information (for a review, see Kensinger & Gutchess, 2015). For example, medial prefrontal cortex supports the encoding of self-relevant information and impressions of others (Gilron & Gutchess, 2012; Macrae et al., 2004; Mitchell, Macrae, & Banaji, 2004), and amnesic patients with widespread disruption of explicit memory resulting from hippocampal damage are able to successfully form and retrieve impressions of others (Johnson, Kim, & Risse, 1985; Todorov & Olson, 2008). Given that many of the cross-cultural differences in cognition are thought to result from social differences across groups, we might expect that the neural systems devoted to remembering social information would be particularly prone to effects of culture (see Freeman, Rule, & Ambady, 2009; Rule, Freeman, & Ambady, 2013 for reviews). This could also lead to robust interactions with inhibitory and other processes implicated in the perception and remembering of stereotyped and outgroup members (e.g., Amodio, Devine, & Harmon-Jones, 2008; Bruneau & Saxe, 2010).

9.7 Cultural Influences on Autobiographical Memory

Autobiographical memory has consistently been defined as memory for a personal event that is remembered from an individual's own life (Conway & Pleydell-Pearce, 2000; Nelson & Fivush, 2004), making the self an inherent focus of this literature as well. Given interpersonal content of many autobiographical memories, these memories largely involve social information and social contexts as well. Nelson and Fivush (2004) argue that autobiographical memory is explicit, declarative, and unique because the perspective is that of the self in relation to others. Autobiographical memory does not emerge automatically and evidence suggests that it develops throughout the preschool years, as facilitated by processes including understanding of the self and others, language, narrative, and socialization, among others that are less relevant to our current focus on social memory (see Nelson & Fivush, 2004). In addition, these processes also are likely to underlie the development of cultural differences in autobiographical memory, as well as other types of memory. Socialization, language acquisition, and parental interaction can reinforce attention to focal objects or relationships (as discussed by Nisbett, 2003), with American mothers, for example, emphasizing the properties of objects and Asian mothers teaching relationships during play (Fernald & Morikawa, 1993).

Of primary importance to the discussion of autobiographical memory differences across cultures is an understanding of the self and its relation to others. Since the focal individual in autobiographical memory is the self, creation and recall should be dependent on an individual's understanding of the self. Nelson and Fivush (2004) argue that autobiographical memories cannot be created until a child demonstrates self-recognition and that this must be measured in terms of the sociocultural context. As previously noted, there is substantial evidence of considerable cultural differences in self-construal, such that Easterners tend to hold a more interdependent self-concept, whereas Westerners are more likely to possess an independent self-concept (Markus & Kitayama, 1991, 2010). Based on differences in self-construal, Western culture may encourage a more detailed image of the self in the past, whereas Eastern culture may place more emphasis on the interdependent self as a member of a communal past (Nelson & Fivush, 2004). In fact, cultural variation in self-construal is evident in research on autobiographical memory. In a study of Chinese and American adults, Wang and Conway (2004) found that Chinese individuals were more likely to recall social and historical events and to focus on the involvement of others. Alternatively, Americans recalled more specific, discrete personal events with an emphasis on their role and feelings. These authors argue that these cultural differences help to solidify the self as either interdependent or independent. For Chinese participants, this style of remembering encourages thinking about the self as interconnected with others, whereas the style in Americans facilitates thinking about the self as self-contained and different from others. Many other studies have shown similar findings (Jobson & O'Kearney, 2008; Wang, 2001, 2004), such that individuals from Western cultures (i.e., the USA and Australia) have more autobiographical memories that are more specific and detailed than East Asians. Additionally, European American children and adults report having their first memory earlier than East Asians (Peterson, Wang, & Hou, 2009; Wang, 2006). This difference likely occurs because European Americans are encouraged from a very early age to develop an independent self-construal, which facilitates self-remembering at an earlier point of development (Wang, 2006). For individuals with an independent self-construal, autobiographical memory serves the function of helping them distinguish themselves. In contrast, for those with an interdependent self construal, autobiographical memory reaffirms their relationship with others and social interactions (Wang, 2011). Research comparing groups raised with different childrearing practices also supports the idea that the concept of an independent self critically determines the development of autobiographical memory. Children raised collectively in reformed kibbutzim report first memories that occur at a later age than children raised in more individualistic settings (Harpaz-Rotem & Hirst, 2005).

Another primary process in the emergence of autobiographical memory from a developmental perspective is language. Language development is important for understanding the organization of autobiographical memory. Additionally, having language capacity allows children to discuss past experiences and form organized memories. Discussion facilitates an understanding of the different perspectives from which memories can be seen (Nelson & Fivush, 2004). Consistent with this argument, evidence from memory studies using bicultural individuals has supported this claim (Marian & Kaushanskaya, 2004; Wang, Shao, & Li, 2010). In one study, children in Hong Kong were asked to recall memories in both English and Chinese. When the interview was conducted in English, children recalled lengthier, more self-focused autobiographical memories than when they were interviewed in Chinese. These findings support both the role of language in the development of autobiographical memory and in the dynamic construction of the self, which can be influenced by the cultural context, as defined by language in the current study (Wang et al., 2010). In addition, Gutchess and Siegel (2011) review that language can shape some aspects of thought, which could have broader implications for the ways in which cultural differences in language impact autobiographical memory.

Narrative style and socializing with others in the culture are other important aspects of the development of autobiographical memory. Western parents are more likely to ask their children about their feelings and thoughts about an event. In contrast, East Asian parents are more likely to tell children how they should feel and think about an event (Ross & Wang, 2010). The Western style of parenting encourages the child to think about their uniqueness and their individual role in a given event, while the East Asian parenting style facilitates thinking about the child's role within the situation and context. These early interactions with parents regarding the cultural norms have profound influence on how children actually think about and recall events from their past. For example, when discussing personal experiences, European American children provide more details and focus on their feelings, preferences, and opinions. Alternatively, East Asian children focus less on detail and more on social interactions (Ross & Wang, 2010). Early interactions between children and their parents facilitate the development of autobiographical memories that are either more self-focused, encouraging individuality, or more focused on interconnectedness and relationships. Additionally, these interactions encourage the development of cultural narratives and the continuation of cultural norms.

A more recent line of investigation has been on within-culture variation in autobiographical memory. Humphries and Jobson (2012) looked at individual trauma history as a factor that influences how cultural background influences autobiographical memory. They used tests of autobiographical memory and history of trauma with Chinese and British students. The prediction that British students would recall more specific memories than Chinese students was supported; however, there was also a significant difference in the number of specific memories recalled as a function of trauma. Individuals with high trauma exposure recalled fewer specific memories than those with low trauma exposure, regardless of cultural background. This study suggests that there are factors other than culture that influence autobiographical memory and should be the focus of future research.

Future investigations into the cultural influences on autobiographical memory would benefit from focusing on the neural processes that underlie both encoding and recall of this specific type of memory. In recent years several researchers have presented reviews and meta-analyses investigating the neural network underlying autobiographical memory. Most studies suggest that that autobiographical memory relies on a left lateralized network including prefrontal, medial and lateral temporal, and cingulate cortex, as well as the temporoparietal junction and the cerebellum (Martinelli, Sperduti, & Piolino, 2013; Svoboda, McKinnon, & Levine, 2006). In a review, Cabeza and St. Jacques (2007) associate specific subcomponents of autobiographical memory with neural regions within this network, suggesting that memory search is facilitated by activation of the lateral prefrontal cortex (PFC), self-referential processes draw on the medial PFC, recollection is facilitated by the hippocampus and retrosplenial cortex, emotional processing occurs in the amygdala, and visual imagery is supported by the occipital lobe, cuneus, and precuneus. While there has been an extensive focus on the neural networks implicated in memory in Western cultures, there has been little to no focus on the function of neural processes of autobiographical memory across cultures. As previously mentioned, there are robust cultural differences in neural activity while thinking about the self and thinking about others (Zhang et al., 2006; Zhu et al., 2007), so presumably there are cultural differences in autobiographical memory, as it is a fundamentally self-relevant memory process. The rich behavioral literature indicating robust differences in the content and features of autobiographical memories across cultures (Jobson & O'Kearney, 2008; Marian & Kaushanskaya, 2004; Peterson et al., 2009; Ross & Wang, 2010; Wang, 2001, 2004, 2006, 2011; Wang & Conway, 2004; Wang et al., 2010) also bolsters the call for further research using neuroimaging measures to investigate autobiographical memory across cultures.

9.8 Cultural Influences on Memory for Emotional Information

Memories for each of the above types of information—self-relevant, social, and autobiographical—may often involve emotional content. Researchers have often found that emotional events are more likely to be remembered than events lacking an emotional component (Reisberg & Hertel, 2005). In 1977, Brown and Kulik published their seminal paper on "flashbulb memory," a term referring to the enhanced encoding and retrieval of memory that results after experiencing an emotionally arousing event. The hallmark of flashbulb memories is their distinctive and highly detailed nature, described by proponents of the theory to be well remembered and accurate. Studies suggest that universal memory enhancement may not be the best description for how affective responses influence memory. Later research challenged the notion that the totality of an emotional scene would be wholly detailed and accurately remembered, referring to emotional memory as "memory narrowing" or "tunnel memory" where memory is narrowed onto select aspects of an emotional event, lessening the focus of memory for surrounding information (reviewed by Holland & Kensinger, 2010; Levine & Edelstein, 2009; Mather & Sutherland, 2011; Reisberg & Heuer, 2004). Research by Kensinger and colleagues (e.g., Kensinger, Gutchess, & Schacter, 2007; Kensinger, Piguet, Krendl, & Corkin, 2005; Waring & Kensinger, 2009) suggests that that affective responses may lead to memory tradeoffs, enhancing memory for select features of an event while acting to impair memory for temporally or spatially proximate details. The amygdala and orbitofrontal cortex seem to contribute to these effects, with activity associated with memory for objects but not the associated backgrounds (Waring & Kensinger, 2011).

Thus far, little is known about how cultural differences might impact the formation and retrieval of emotional memory. Work devoted to investigating the influence of culture on emotion and memory has largely fallen into two main categories: the influence of culture on emotion and the influence of culture on memory.

We first focus on the ways in which culture influences emotional processing. Moving away from early work focusing on emotions as universal, recent work has instead focused upon the ways in which sociocultural dynamics impact emotion (Kitayama & Markus, 1994; Mesquita & Frijda, 1992). Specifically, these studies have focused upon the ways in which culture directs attention to affective stimuli and to differing conceptions of what constitutes positive and negative emotions cross-culturally.

Recent studies have contributed to the understanding of how culture influences attention-related orientation to affective stimuli. This idea was explored by Grossman, Ellsworth, and Hong (2012), whose studies provided evidence for the influence of culture on attention to positive and negative stimuli. Drawing on previous ethnographic evidence that assigned a positive value to brooding, or the immersion of oneself within negative feelings, in Russian culture, the authors hypothesized that Russians would be more likely than Americans to focus on negative stimuli. The results showed that when presented with a series of images, Americans did not preferentially discriminate between pleasant and unpleasant ones whereas Russian participants spent a significantly greater amount of time studying the unpleasant images. This finding led the researchers to suggest that cultural differences in emotion may be driven by attention-related tendencies linked to information processing.

However, what this and other similar studies draw into question is whether or not there is a universal understanding across cultures of what it means to have a "positive" and "negative" emotion. Cross-national comparisons show cultural variation in the definition of what constitutes a positive emotion. For example, in many Western cultural contexts positive feelings are associated with individual success, high self-esteem, and good health (Heine, Lehman, Markus & Kitayama, 1999; Kitayama, Markus, & Kurokawa, 2000; Taylor & Brown, 1988). However, positive emotions like happiness are not viewed as unequivocally "good" in many Asian cultural contexts. As one example of this, Japanese are more likely than European Americans to associate happiness with negative social consequences, such as jealousy in others and disharmony in social relationships (Uchida & Kitayama, 2009). Furthermore, cultures differ in their "ideal affect," as in preferences for high arousal (e.g., excited) vs. low arousal (e.g., calm) emotions (Tsai, Knutson, & Fung, 2006), which could impact what information is preferentially attended to in one's environment or how incidents are interpreted and encoded into memory.

Literature bridging the three topics together (culture, emotion, and memory) is extremely limited. One notable exception can be found in the work of Scollon et al. (2004) examining whether cross-cultural variation in emotional experience is a factor of the way in which emotions are themselves experienced or whether the difference lies instead in the memory reconstruction of the event that is later recalled to the researcher. That is to say, do differences in the ways that cultures value emotion, or hold certain emotions in higher esteem than others, influence the selection of information and rate at which memories are encoded? Further, if memory for emotion is a reconstructive process, how might culture influence variations in memory? In this study, researchers drew samples from five ethnic groups: three from within the USA (European American, Asian American, and Hispanic) and two from non-Western cultures (India and Japan). The authors hypothesized that due to a cultural emphasis on positive feelings found within Hispanic and American cultures, participants originating from these groups would have a higher frequency of recall for pleasant feelings than the remaining three cultural groups. In contrast, because literature on Asian culture describes an equality of value for both positive and negative emotions, the authors predicted that Indian and Japanese respondents would show equal levels of recall for both pleasant and unpleasant feelings. The study found that global ratings of affect figured prominently in people's memories of emotion across all five groups. However, interesting cultural differences did emerge when observing the degree to which intensity of the emotional experience predicted the *frequency* of recall. In this case pleasant emotional memories were recalled more frequently for Hispanic Americans who weighed the intensity of pleasant emotions more strongly and gave very little consideration to the intensity of unpleasant emotions. Indian participants emphasized the impact of intensity for pleasant emotions but gave equal weight to the frequency and intensity of unpleasant emotions.

However, questions addressing how culture impacts information selection for emotional memory were not directly answered by the Scollon et al. (2004) study. Instead, the study focused on the appropriateness of measures used across cultures. While measurement issues are very important to consider when conducting cross-cultural research, questions remain as to the ways culture differentially impacts selection of information as well as the level of accuracy maintained during retrieval. These questions illuminate a number of interesting potential directions for future research on cross-cultural differences in emotion and memory.

9.9 Autobiographical Memory, Emotion, and International Relations

In previous sections of this chapter, we discussed the self-reference effect and the ways in which memory improves for items and events that evoke a self-relevant condition. Further, we discussed the importance of cross-cultural differences in autobiographical memory as contributing to one's understanding of the self and its relation to others. This line of reasoning brings us to the next line of inquiry, which is to ask how autobiographical memory affects human interaction on an increasingly global scale.

As rendered above, recent discussions on autobiographical memory have focused upon cultural differences in self-concept, drawing distinctions between independent and interdependent self-construal. However, investigations are limited that address how these distinctions in the self-concept impact interactions between groups of individuals, such as at the level of interactions between cultures, or nations, that are predominantly independent vs. interdependent. As Renshon and Duckitt (2000) posed the challenge to future investigators in his discussion of culture's contribution to political psychology, "An essential question, therefore, for any modern cultural and cross-cultural political psychology… is how diverse cultural traditions can be integrated into political units that transcend them" (p. 10). Relevant to this discussion, how do the psychological processes studied largely at the level of the individual translate into policies or other organizational aspects of nations? This is important to understand, as it impacts how subgroups are ultimately subsumed within the structure of a single nation or how nations with very different perspectives, as a result of their cultures, work together.

Let us expand upon this question and apply our current discussion to concepts relevant to memory. How do cross-cultural variations in autobiographical memory ultimately transcend the individual and come to impact the culture writ large, whose social, economic, and political institutions ultimately reflect that differentiation? Said simply, how do autobiographical and emotional memory impact international relations? To address such a complex question will require the interaction of multiple disciplines to understand how specific psychological processes, such as autobiographical memory, lead to the creation of collective memories at the level of the group, which ultimately influence processes relevant to politics and international relations. See work by Assmann and Shortt (2011), which takes such an approach of bringing experts from multiple disciplines together to investigate the topic of collective memory.

In considering the interconnection between autobiographical and emotional memory with international relations, psychologists may be well served to look to other disciplines that have considered related issues. As anthropologists have grappled with for the last century, psychologists now must ask a similar question: is culture merely a collective sum of its parts? In this instance, can understanding questions regarding social and self-relevant memory at the level of the individual, and even appreciating the differences across individuals as a function of culture, eventually build to an understanding of these processes at the level of groups or nations? An important question when considering what makes psychology distinct from other forms of analysis is the search for explanation, description, and prediction at the individual level of analysis (Jervis, (1976). That individual level of analysis is then extrapolated to that of the regional and descriptively utilized to conceptualize a culture as independent or interdependent.

Anthropological debate on the topic may help to inform psychological conception of the way in which individual memory is transferred to a collectively motivated, political body. Noted anthropological founding father, sociologist Maurice Halbwachs (1992), claimed that collective memory is not a socially constructed notion nor a mysterious kind of group mind, but is formed through a collection of singular persons, the individuals as group members both forming and retaining societal memories and conventions thusly disseminating said notions to future members. While this seems consistent with the psychological approach, Emile Durkheim (1953)'s perspective is not. He argued that collective representations originate not from the individuals themselves, but via the *association* of the individuals, and that it is from these associations that a separate entity, culture, is comprised.

Applying this later concept to the study of psychology, memory, and international relations, the analysis of individual behavioral and neural processes has largely overlooked the transcendent processes that result when individuals interact with one another. Individualistic expressions of autobiographical memory and emotion converge, and are transmuted through the associative process that occurs across individuals. Ultimately, the effects of autobiographical memory and emotion are expressed via cultural institutions and political actions.

In much of the literature, political psychology oriented towards international relations has, at its foundation, a focus on rational choice theory (e.g., de Mesquita & Root 2000; McDermott, 2004; Mercer, (2005). That is, the models assume that rational, unitary actors seek to maximize their circumstances given predetermined and established patterns of preference. Psychological models present individuals as complex actors whose behavior at times diverges from that which has been established as rational. For example, the behavior can reflect nonrational acts, attributed to a confluence of cognitive and affective processes. However, two problems immediately arise with the rational models. First is the focus on the individual to the exclusion of the communal. Second is that nonrational behavior may not actually be irrational at all, but rather reflect the intersection of a cultural group's extended historical experience with that of an individual's psychological experience.

It is in the expression of the so-called nonrational that we can see the communal. When expressed en masse, the way in which autobiographical memory is experienced through the individual's cultural prism produces a collective representation that is distinct from the original autobiographical memory. It is this collective representation that is at work in international relations. It will be useful at this point to consider an extended case study of the ways in which autobiographical and emotional memory collectively coalesce to influence international relations between two cultures. One such example is recent work on the Israeli–Palestinian conflict. Study of this conflict has focused on the ways in which following violence, multiple experiences of individualized trauma can coalesce into a field of collective traumatization, which impacts Israeli–Palestinian relations.

The ongoing political violence sustained throughout the continuation of the Israeli–Palestinian conflict has left many susceptible to high rates of trauma and increasing instances of posttraumatic stress disorder. For cultures subject to such events, it will be important to understand the ways in which cumulative effects of exposure of trauma, and perhaps even the development of PTSD, affect emotion and cognition. It will be particularly important to understand the ways in which the effects differ across cultures, as well as the cultural universals. We will focus on the Israeli–Palestinian conflict as an example case through which to consider the effects of trauma on the processing of emotional information and the potential repercussions for memory. After sixty years of conflict, both Israeli as well as Palestinian populations have an extended and continuous relationship with trauma. Authors such as psychologists Landau (1997), Weinberg and Nuttman-Shwartz (2006), and Wistrich and Ohana (1995) describe Israel's association with trauma stemming from experiences including, but not limited to, the Zionist reading of the precariousness of Jewish existence as a defenseless minority in the Diaspora nightmarishly substantiated by the *Shoa* or Holocaust, the war to establish the state of Israeli in 1948, the 1967 War, the 1987 Intifada, the 2006 war with Lebanon, Intifada Al-Aqsa, terrorist campaigns within Israel, and so on.

Given the situation of protracted military conflict in which Israel has been involved since its inception, many Israelis routinely process military and "security" considerations as integral aspects of social reality (Friedman-Peleg & Bilu, 2011). But the cultural meaning system of "cognitive militarism" that evolved in the state's formative years (Kimmerling, 1993) and the myth of heroism it canonized fostered a collective mood that made light of the psychological toll of war and military violence and only until recently stigmatized its emotional manifestations.

Similarly Palestinians too have experienced years of violence and trauma as a direct result of the Israeli–Palestinian conflict. The 1948 war, known by Israelis as the War of Independence, and by Palestinians as *Al-Nakba*, or the Catastrophe, caused over 750,000 Palestinians to flee their homes and to take refuge in camps hastily erected by the Red Cross and other humanitarian agencies in the West Bank, Gaza, Lebanon, Syria, Jordan, and Egypt (Chatty, 2007). While living in exile, violence has been a continual and uninterrupted experience. During the first and second *Intifadas* (uprisings), more than 6,200 Palestinians were killed (B'Tselem, 2008a, 2008b), more than 60,000 wounded (JMCC, 2008; PCHR, 2008), and more than 65,000 have been detained (B'Tselem, 2008a, 2008c; JMCC, 2008). Palestinians have also experienced a virtual civil war, as Palestinian factions fight for the leadership of Palestine with very different visions of what they hope to accomplish and how they hope the Palestinian State will be governed. These internal political tensions have led to gun battles, assassinations, and a division of the Palestinian Authority, with separate governing authorities in Gaza and the West Bank. Palestinian Authority security forces in the West Bank reportedly arrest over 150 Palestinians each month and in Gaza over 250 people are being detained each month by the security forces of the de facto Hamas government. Over 590 individuals have been killed in internecine warfare between Palestinian political factions (B'Tselem, 2008a, 2008b).

A plethora of studies have shown the high rate of trauma and onset of PTSD symptoms among representative samples of Palestinians (e.g., Afana, Dalgard, Bjertness, Grunfeld, & Hauff, 2002; Hobfoll et al., 2011; Srour & Srour, 2006). Representative samples show as high as 73 % of Palestinians (sampled from Gaza, the West Bank, and East Jerusalem) exhibit avoidance criteria (avoidance of thoughts or feelings associated with the traumatic experience) and 35.5 % fall within the Post-Traumatic Stress Disorder (PTSD) criteria of the DSM-III diagnostic categories (Hobfoll et al., 2011). As a result, psychological studies seeking to compare trauma and the development of PTSD among Palestinians have called into question the effects of trauma on Israeli–Palestinian relations (Awwad, 2004; Lustick, 1994).

Therefore, what effect does collective, prolonged, and repeated experiences of trauma have on culture itself and how might this impact individual memory? One finding reviewed in our previous section on autobiographical memory is that individuals experiencing high levels of trauma tend to have less specificity to memories than those with lower levels of trauma exposure (Humphries & Jobson, 2012). In addition, Gutchess et al. (2011) suggest a number of ways memory may differ based on cultural experiences. If ways of processing information are encouraged and reified through one's development within a particular cultural context, what effect might a "traumatized culture" have on information processing and memory?

In order to begin to understand these questions at the very basic level of what can be studied in a controlled laboratory study, we have begun a cross-cultural experiment in the USA and Israel addressing the formation of emotional memory studying the emotion-induced memory trade-off effect, in which where select information is encoded at the expense of temporally or spatially proximate details (Ligouri & Gutchess, 2013). For example, after studying an image of a snake in the forest, participants retain memory for the snake but poorer memory for the forest relative to other backgrounds. In this example, memory for the forest is worse if it had been paired with an emotional item, such as a snake, compared to if it had been paired with a neutral item, such as a squirrel, in the forest. Researchers have referred to this memory pattern as a trade-off because the memory for the background context (the forest) is traded in favor of memory for the emotional item (the snake).

Recent examinations have sought to understand the impact of trauma exposure on the memory trade-off effect by comparing three groups: emotion-induced memory trade off for participants with current PTSD, participants who had experienced trauma but did not have current PTSD, and a control group who had neither experienced significant trauma nor met criteria for current PTSD (Mickley-Steinmetz, Scott, Smith, & Kensinger, 2012). Contrary to expectations, the results indicated a reduced trade-off effect for trauma-exposed subjects without a current diagnosis of PTSD only. Interestingly, both the control group and participants who had developed symptoms of PTSD produced comparable results, both showing an enhanced trade-off effect from the affective stimuli that were consistent with prior studies (e.g., Kensinger et al., 2005; Waring & Kensinger, 2009). This was puzzling that the control group (with no prior traumatic experience) and the PTSD group not only display enhanced memory trade-off effects, but exhibited comparable results, while the trauma-only group showed variation in the trade-off effect. Researchers theorized that this pattern occurred due to the trauma-only exposed group having developed positive coping mechanisms or an enhanced capacity to process emotional events in a similar fashion to that involved in experiencing neutral information.

Using the research by Mickley-Steinmetz et al. (2012) as a model, the study by Ligouri and Gutchess (2013) examines potential cross-cultural variation in emotional memory by comparing American and Israeli participants. Similarly to the trauma-exposed group without PTSD, a reduced emotion-induced memory trade-off is predicted to occur for Israeli participants due to coping strategies developed as a result of repeated exposure to violence as a result of the sixty-year protracted Israeli–Palestinian conflict. The results of this study will allow us to determine whether emotional memory trade-off effects are culturally mediated and if so, the types of processes and experiences that may lead to differences across cultures. Further, work begun here will begin to illuminate the ways in which individual expressions of autobiographical memory and emotion could directly impact larger cultural institutions, ultimately influencing and affecting political actions linked to the Israeli–Palestinian conflict.

9.10 Role of Memory Across Groups

Why is memory important to investigate across cultures? For one, Psychological research has been conducted predominantly in Western cultures. Thus, our understanding of human memory may not characterize subgroups within a culture or non-Western populations. Broadening our understanding of cultural differences will lead to more accurate models of memory and a better understanding of which principles of memory are universals.

Potential applications of the proposed work could be to explore cultural differences in processes for which memory is particularly important or prone to impairment. Studying the role of memory in educational settings is one example of a potential application. Given the importance of understanding the factors that lead to effective learning and retention of information would have profound implications for "best practices" to adopt universally or would indicate culture-specific modifications or programs that are tailored to the strengths and weaknesses of the predominant learning style of members of a particular culture. An understanding of cross-cultural differences in learning could suggest modifications to be made to teaching styles and assessment measures. Another potential application is to the study of the memory strategies associated with successful aging. Providing effective strategies can reduce age-related declines on some tasks (Kausler, 1994). Because we argue that cultures may possess their own sets of preferred strategies and biases for information processing, different cultures could vary in the strategies that they emphasize, which are more or less effective in overcoming age-related cognitive changes. This would allow us to identify strategies or cognitive lenses that are most adaptive for successful aging. Alternatively, findings of similar memory declines with aging across cultures, despite differences in strategies and perspectives, may suggest universal effects of aging. Such findings would indicate limits to the malleability of cognition and memory with aging.

By studying a basic cognitive process, we can better appreciate ways that individuals perceive and remember distinct aspects of their environment, which can be exacerbated in complex environments and social situations. Better understanding of cognition across cultures can facilitate cross-cultural interactions in a variety of settings. For example, erroneous memories that are biased in culturally specific ways may lead to misunderstandings due to the reconstructive nature of memory. Memories that emphasize different aspects of a complex event based on cultural group could lead to disagreements based on differences in the qualities and types of details remembered, although both parties agree that the same overall event occurred. For example, Americans and Chinese could offer starkly different accounts of the same event, as an effect of emphasizing different details in memory and relying on distinct inferences to connect gaps in memory. These problems could be especially pronounced for the interpersonal aspects of events. For example, Westerners' greater focus on self-relevant information at the expense of information relevant to others or social interactions is the converse of Easterners' greater focus on others and social interactions rather than self-relevant information. Thus, individuals from these two groups may remember distinctly different accounts of social events and experience difficulty remembering or appreciating an account that emphasizes an opposing perspective. Individuals spending time in a foreign country may benefit from appreciating the ways in which the collective memory of their host culture differs from their own, and understanding the values and conventions that can be conveyed and emphasized through memory. Research that advances our understanding of cultural differences may help to ensure smoother international relations through culturally sensitive interactions and ways of conveying information, particularly by understanding systematic differences in memory across cultures.

9.11 Conclusions

The study of social memory is only just beginning to suggest ways in which social information may draw upon a unique memory system that is, at least to some degree, distinguishable from other types of long-term explicit memory (for a review, see Kensinger & Gutchess, 2015). Studying this system across cultures only compounds the potential to unveil a number of different properties and features of this memory system, some of which should operate as cultural universals and some making culturally specific contributions. The complexity of the social world, and the breadth of literature revealing cross-cultural differences in social processes suggest that this type of memory could be particularly prone to being shaped by the effects of culture, compared to other types of memory.

Another interesting perspective to bear in mind about social information is the number of embedded levels that contribute to one's social identity. While an individual might have her own unique perspective on memory, including what is prioritized and emphasized in attention and information processing, this self-identity is shaped by a number of factors, likely including aspects of one's identity (e.g., being female, American) and unique experiences (e.g., relationships with family members; exposure to trauma). In the complex case of bicultural identity (see Nguyen & Benet-Martinez, 2007 for a nuanced discussion of this topic), identity can be influenced, at the very least, by the degree of assimilation into those cultures and whether those cultural identities are blended or alternating, such that one individual can shift between multiple different cultural perspectives. For example, an Asian American might have two unique frames available for processing information, such that she can shift between different perspectives. Depending whether the individual is momentarily in an "Asian" or "American" frame, different aspects of information may be encoded or retrieved. If one has a blended "Asian American" identity, he may have created a unique integrated identity that overlaps little with others who are remembering from an "Asian" or an "American" perspective. While our discussion of biculturalism thus far emphasizes direct contact with cultural forces, indirect contact is rapidly increasing in today's society through globalization and greater access to shared media. Moving outside of the individual, that person is embedded in a number of social groups, and exposed to specific aspects of collective memory based on those groups and identities. When one individual interacts with and contemplates other individuals, the observer likewise perceives others and makes assessments based on a rich set of observable individual and group-based identities. Reminiscence from autobiographical memory represents one process through which sharing and socializing about past experiences aids in developing and expressing one's self-concept. More importantly, these processes facilitate the development and understanding of cultural norms, fostering the continuation of cultural narratives.

In conclusion, we hope that we have achieved our aim with this chapter of not only highlighting the ways in which one's subjective identity, shaped by culture, can mold memory, but that we have also illustrated ways in which one's subjective memory, shaped by culture, can influence one's perspective on self and others. We have also motivated the importance of studying the iterative process through which collective memory and cultural identity shape each other, despite the challenges of studying such a complex and multifaceted process.

Acknowledgements The authors gratefully acknowledge support from the National Science Foundation (BCS-1147707), supporting their research in this area. We thank Pete Millar for comments and assistance with the preparation of this chapter.

References

Afana, A., Dalgard, O. S., Bjertness, E., Grunfeld, B., & Hauff, E. (2002). The prevalence and associated socio-demographic variables of post-traumatic stress disorder among patients attending primary health care centres in the Gaza Strip. *Journal of Refugee Studies, 15*(3), 283–295.

Amodio, D. M., Devine, P. G., & Harmon-Jones, E. (2008). Individual differences in the regulation of intergroup bias: The role of conflict monitoring and neural signals for control. *Journal of Personality and Social Psychology, 94*, 60–74.

Assmann, A., & Shortt, L. (Eds.). (2011). *Memory and political change*. New York: Palgrave Macmillan.

Awwad, E. (2004). Broken lives: Loss and trauma in Palestinian-Israeli relations. *International Journal of Politics, Culture, and Society, 17*(3), 405–414.

B'Tselem. (2008a). *Fatalities in the first Intifada*. Retrieved from http://www.btselem.org/english/statistics/First_Intifada_Tables.asp

B'Tselem. (2008b). *Intifada fatalities*. Retrieved from http://www.btselem.org/english/statistics/Casualties.asp

B'Tselem. (2008c). *Statistics on Palestinians in the custody of the Israeli security forces*. Retrieved from http://www.btselem.org/english/statistics/Detainees_and_Prisoners.asp

Benet-Martinez, V., & Haritatos, J. (2005). Bicultural identity integration (BII): Components and psychosocial antecedents. *Journal of Personality, 73*, 1015–1050.

Brown, R., & Kulik, J. (1977). Flashbulb memories. *Cognition, 5*, 73–99.

Bruneau, E. G., & Saxe, R. (2010). Attitudes toward the outgroup are predicted by activity in the precuneus in Arabs and Israelis. *Neuroimage, 52*, 1704–1711.

Bruner, J. (1990). *Acts of meaning*. Cambridge, MA: Harvard University Press.

Cabeza, R., & St. Jacques, P. (2007). Functional neuroimaging of autobiographical memory. *TRENDS in Cognitive Sciences, 11*, 219–227.

Chatty, D. (2007). Researching refugee youth in the Middle East: Reflections on the importance of comparative research. *Journal of Refugee Studies, 20*(2), 265–280.

Chiao, J. Y., Harada, T., Komeda, H., Li, Z., Mano, Y., Saito, D., … Iidaka, T. (2010). Dynamic cultural influences on neural representations on the self. *Journal of Cognitive Neuroscience, 22*, 1–11.

Chua, H. F., Boland, J. E., & Nisbett, R. E. (2005). Cultural variation in eye movements during scene perception. *Proceedings of the National Academy of Sciences of the United States of America, 102*, 12629–12633.

Chua, H. F., Leu, J., & Nisbett, R. E. (2005). Culture and diverging views of social events. *Personality and Social Psychology Bulletin, 31*, 925–934.

Conway, M. A., & Pleydell-Pearce, C. W. (2000). The construction of autobiographical memories in the self-memory system. *Psychological Review, 107*, 261–288.

Cousins, S. D. (1989). Culture and self-perception in Japan and the United States. *Journal of Personality and Social Psychology, 56*, 124–131.

Craik, F. I. M., & Lockhart, R. S. (1972). Levels of processing: A framework for memory research. *Journal of Verbal Learning and Verbal Behavior, 11*, 671–684.

Craik, F. I. M., Moroz, T. M., Moscovitch, M., Stuss, D. T., Winocur, G., Tulving, E., et al. (1998). In search of self: A positron emission tomography study. *Psychological Science, 10*, 26–34.

D'Andrade, R. G. (1984). Cultural meaning systems. In R. A. Shweder & R. A. LeVine (Eds.), *Culture theory: Essays on mind, self, and emotion* (pp. 88–119). New York: Cambridge University Press.

de Mesquita, B. B., & Root, H. L. (Eds.). (2000). *Governing for prosperity*. Yale University Press.

DiMaggio, P. J. (1997). Culture and cognition. *Annual Review of Sociology, 23*, 263–287.

Durkheim, É. (1953). The determination of moral facts. *Sociology and philosophy*, pp. 35–63.

Evans, K., Rotello, C. M., Li, X., & Rayner, K. (2009). Scene perception and memory revealed by eye movements and receiver-operating characteristic analyses: Does a cultural difference truly exist? *The Quarterly Journal of Experimental Psychology, 62*, 276–285.

Fernald, A., & Morikawa, H. (1993). Common themes and cultural variations in Japanese and American mothers' speech to infants. *Child Development, 64*, 637–656.

Freeman, J. B., Rule, N. O., & Ambady, N. (2009). The cultural neuroscience of person perception. *Progress in Brain Research, 178*, 191–201.

Friedman-Peleg, K., & Bilu, Y. (2011). From PTSD to "national trauma": The case of the Israel Trauma Center for victims of terror and war. *Transcultural Psychiatry, 48*, 416–436.

Gabrieli, J. D. (1998). Cognitive neuroscience of human memory. *Annual Review of Psychology, 49*, 87–115.

Geertz, C. (1973). *The interpretation of cultures: Selected essays* (p. 89). New York: Basic Books.

Gilron, R., & Gutchess, A. (2012). Remembering first impressions: Effects of intentionality and diagnosticity of subsequent memory. *Cognitive, Affective, and Behavioral Neuroscience, 12*, 85–98.

Goh, J. O., Chee, M. W., Tan , J.C., Venkatraman, V., Hebrank, A., Leshikar, E. D., … Park, D. C. (2007). Age and culture modulate object processing and object-scene binding in the ventral visual area. *Cognitive, Affective, and Behavioral Neuroscience, 7*, 44–52.

Grossman, I., Ellsworth, P. C., & Hong, Y. (2012). Culture, attention, and emotion. *Journal of Experimental Psychology: General, 141*, 31–36.

Gutchess, A. H., & Goh, J. O. S. (2013). Refining concepts and uncovering biological mechanisms for cultural neuroscience. *Psychological Inquiry: An International Journal for the Advancement of Psychological Theory, 24*, 31–36.

Gutchess, A. H., & Indeck, A. (2009). Cultural influences on memory. *Progress in Brain Research, 178*, 137–150.

Gutchess, A. H., Kensinger, E. A., & Schacter, D. L. (2007). Aging, self-referencing and medial prefrontal cortex. *Social Neuroscience, 2*, 117–133.

Gutchess, A. H., Kensinger, E. A., & Schacter, D. L. (2010). Functional neuroimaging of self-referential encoding with age. *Neuropsychologia, 48*, 211–219.

Gutchess, A. H., Schwartz, A. J., & Boduroglu, A. (2011). *The influence of culture on memory.* Proceedings of international workshop on human-computer interaction, lectures notes in computer science. Springer

Gutchess, A. H., & Siegel, M. (2011). Memory across cultures. In A. Assmann & L. Shortt (Eds.), *Memory and political change* (pp. 201–215). New York: Palgrave Macmillan.

Gutchess, A. H., Welsh, R. C., Boduroglu, A., & Park, D. C. (2006). Cross-cultural differences in the neural correlates of picture encoding. *Cognitive, Affective, and Behavioral Neuroscience, 6*, 102–109.

Halbwachs, M. (1992). *On collective memory*. Chicago, IL: University of Chicago Press.

Han, S., & Northoff, G. (2009). Understanding the self: A cultural neuroscience approach. *Progress in Brain Research, 178*, 203–212.

Harada, T., Li, Z., & Chiao, J. Y. (2010). Differential dorsal and ventral medial prefrontal representations of the implicit self modulated by individualism and collectivism: An fMRI study. *Social Neuroscience, 5*, 257–271.

Harpaz-Rotem, I., & Hirst, W. (2005). The earliest memory in individuals raised in either traditional and reformed kibbutz or outside the kibbutz. *Memory, 13*, 51–62.

Heatherton, T. F., Macrae, C. N., & Kelley, W. M. (2004). What the social brain sciences can tell us about the self. *Current Directions in Psychological Science, 13*, 190–193.

Heatherton, T. F., Wyland, C. L., Macrae, C. N., Demos, K. E., Denny, B. T., & Kelley, W. M. (2006). Medial prefrontal activity differentiates self from close others. *Social Cognitive and Affective Neuroscience, 1*, 18–25.

Heine, S. J., Lehman, D. R., Markus, H. R., & Kitayama, S. (1999). Is there a universal need for positive self-regard? *Psychological Review, 106*, 766–794.

Heine, S. J. (2001). Self as a product of culture: An examination of East Asian and North American selves. *Journal of Personality, 69*, 881–906.

Hobfoll, S. E., Canetti, D., Hall B. J., Brom, D., Palmieri, P. A., Johnson, R. J., … Galea, S. (2011). Are community studies of psychological trauma's impact accurate? A study among Jews and Palestinians. *Psychological Assessment, 3*, 599–605.

Holland, A. C., & Kensinger, E. A. (2010). Emotion and autobiographical memory. *Physics of Life Review, 7*, 88–131.

Huff, S. E., Yoon, C., Lee, F., Mandadi, A., & Gutchess, A. H. (2013). Self-referential processing and encoding in bicultural individuals. *Culture and Brain, 1*(1), 16–33.

Humphries, C., & Jobson, L. (2012). Short report: Influence of culture and trauma history on autobiographical memory specificity. *Memory, 20*(8), 915–922.

Jenkins, A. C., & Mitchell, J. P. (2010). Mentalizing under uncertainty: Dissociated neural responses to ambiguous and unambiguous mental state inferences. *Cerebral Cortex, 20*, 404–410.

Jervis, R. (1976). *Perception and misperception in international politics*. Princeton, NJ: Princeton University Press.

Jerusalem Media and Communication Center (JMCC). (2008). *The intifada: An overview, the first two years*. Retrieved from http://www.jmcc.org/research/reports/intifada.htm#statist

Jobson, L., & O'Kearney, R. (2008). Cultural differences in retrieval of self-defining memories. *Journal of Cross-Cultural Psychology, 39*, 75–80.

Johnson, M., Kim, J., & Risse, G. (1985). Do alcoholic Korsakoff's syndrome patients acquire affective reactions? *Journal of Experimental Psychology: Learning, Memory, and Cognition, 11*, 22–36.

Kashima, Y., Woolcock, J., & Kashima, E. S. (2000). Group impressions as dynamic configurations: The tensor product model of group impression formation and change. *Psychological Review, 107*, 914–942.

Kausler, D. H. (1994). *Learning and memory in normal aging*. San Diego, CA: Academic Press.

Kelley, W. M., Macrae, C. N., Wyland, C. L., Caglar, S., Inati, S., & Heatherton, T. F. (2002). Finding the self? An event-related fMRI study. *Journal of Cognitive Neuroscience, 14*, 783–794.

Kensinger, E. A., Gutchess, A. H., & Schacter, D. L. (2007). Effects of aging and encoding instructions on emotion-induced memory trade-offs. *Psychology and Aging, 22*, 781–795.

Kensinger, E. A., Piguet, O., Krendl, A. C., & Corkin, S. (2005). Memory for contextual details: Effects of emotion and aging. *Psychology and Aging, 20*, 241–250.

Kensinger, E. A., & Gutchess, A. H. (2015). Memory for emotional and social information in adulthood and old age. In A. Duarte, M. Barense, & D. R. Addis (Eds.), *The Wiley Handbook on The Cognitive Neuroscience of Human Memory* (pp. 393–414). Chichester, UK: Wiley Blackwell.

Kimmerling, B. (1993). Patterns of militarism in Israel. *European Journal of Sociology, 2*, 1–28.

Kitayama, S., & Markus, H. R. (1994). *Emotion and culture: Empirical studies of mutual influence*. Washington, DC: American Psychological Association.

Kitayama, S., Markus, H. R., & Kurokawa, M. (2000). Culture, emotion, and well-being: Good feelings in Japan and the United States. *Cognition and Emotion, 14*, 93–124.

Klein, S. B., & Kihlstrom, J. F. (1986). Elaboration, organization, and the self-reference effect in memory. *Journal of Experimental Psychology: General, 115*, 26–38.

Klein, S. B., & Loftus, J. (1988). The nature of self-referent encoding: The contribution of elaborative and organizational processes. *Journal of Personality and Social Psychology, 55*, 5–11.

Ko, S. G., Lee, T. H., Yoon, H. Y., Kwon, J. H., & Mather, M. (2011). How does context affect assessments of facial emotion? The role of culture and age. *Psychology and Aging, 26*, 48–59.

Landau, S. (1997). Conflict resolution in a highly stressful society: The case of Israel. In D. P. Fry & K. Björkqvist (Eds.), *Cultural variation in conflict resolution: Alternatives to violence* (pp. 123–136). Mahwah, NJ: Lawrence Erlbaum Publishers.

Levine, L. J., & Edelstein, R. S. (2009). Emotion and memory narrowing: A review and goal-relevance approach. *Cognition and Emotion, 23*, 833–875.

Ligouri, L., & Gutchess, A. (2013). *The effects of cross-cultural variation on the emotion-induced memory trade-off*. [Unpublished raw data]

Lombardo, M. V., Chakrabarti, B., Bullmore, E. T., Wheelwright, S. J., Sadek, S. A., Suckling, J., et al. (2010). Shared neural circuits for mentalizing about the self and others. *Journal of Cognitive Neuroscience, 22*, 1623–1635.

Lustick, I. (1994). *Religion, culture, and psychology in Arab-Israeli relations*. New York: Garland Pub.

Ma, Y., Bang, D., Wang, C., Allen, M., Frith, C., Roepstorff, A., et al. (2014). Sociocultural patterning of neural activity during self-reflection. *Social Cognitive and Affective Neuroscience, 9*(1), 73–80.

Macrae, C. N., Moran, J. M., Heatherton, T. F., Banfield, J. F., & Kelley, W. M. (2004). Medial prefrontal activity predicts memory for self. *Cerebral Cortex, 14*, 647–654.

Marian, V., & Kaushanskaya, M. (2004). Self-construal and emotion in bicultural bilinguals. *Journal of Memory and Language, 51*, 190–201.

Markus, H. R., & Kitayama, S. (1991). Culture and the self: Implications for cognition, emotion, and motivation. *Psychological Review, 98*, 224–253.

Markus, H. R., & Kitayama, S. (2010). Cultures and selves: A cycle of mutual constitution. *Perspectives on Psychological Science, 5*, 420–430.

Martinelli, P., Sperduti, M., & Piolino, P. (2013). Neural substrates of the self-memory system: New insights from a meta-analysis. *Human Brain Mapping, 34*(7), 1515–1529.

Masuda, T., Ellsworth, P. C., Mesquita, B., Leu, J., Tanida, S., & De Veerdonk, E. V. (2008). Placing the face in context: Cultural differences in the perception of facial emotion. *Journal of Personality and Social Psychology, 94*, 365–381.

Masuda, T., & Nisbett, R. E. (2001). Attending holistically versus analytically: Comparing the context sensitivity of Japanese and Americans. *Journal of Personality and Social Psychology, 81*, 922–934.

Mather, M., & Sutherland, M. R. (2011). Arousal-biased competition in perception and memory. *Perspectives in Psychological Science, 6*, 114–133.

McDermott, R. (2004). The feeling of rationality: The meaning of neuroscience for political science. *Perspectives on Politics, 2*(4), 691–706.

Mercer, J. (2005). Rationality and psychology in international politics. *International Organization, 59*(01), 77–106.

Mesquita, B., & Frijda, N. H. (1992). Cultural variations in emotions: A review. *Psychological Bulletin, 112*, 179–204.

Mickley-Steinmetz, K. R., Scott, L. A., Smith, D., & Kensinger, E. A. (2012). The effects of trauma exposure and posttraumatic stress disorder (PTSD) on the emotion-induced memory trade-off. *Frontiers in Integrative Neuroscience, 6*, 34.

Mitchell, J., Macrae, C., & Banaji, M. (2004). Encoding-specific effects of social cognition on the neural correlates of subsequent memory. *Journal of Neuroscience, 24*, 4912–4917.

Nelson, K., & Fivush, R. (2004). The emergence of autobiographical memory: A social cultural developmental theory. *Psychological Review, 111*, 486–511.

Nguyen, A.-M. D., & Benet-Martinez, V. (2007). Biculturalism unpacked: Components, measurement, individual differences, and outcomes. *Social and Personality Psychology Compass, 1*, 101–114.

Nisbett, R. E. (2003). *The geography of thought: How Asians and Westerners think differently ... and why*. New York: The Free Press.

Nisbett, R. E., & Masuda, T. (2003). Culture and point of view. *Proceedings of the National Academy of Sciences of the United States of America, 100*, 11163–11170.

Nisbett, R. E., Peng, K., Choi, I., & Norenzayan, A. (2001). Culture and systems of thought: Holistic versus analytic cognition. *Psychological Review, 108*, 291–310.

Northoff, G., Heinzel, A., de Greck, M., Bermpohl, F., & Dobrowolny, H. (2006). Self-referential processing in our brain. *NeuroImage, 31*, 440–457.

Palestinian Center for Human Rights (PCHR). (2008). Statistics related to Al Aqsa intifada: 29 September 2000, updated 05 January, 2010. Retrieved from http://www.pchrgaza.org/alaqsaintifada.html

Paller, K. A., & Wagner, A. D. (2002). Observing the transformation of experience into memory. *Trends in Cognitive Sciences, 6*, 93–102.

Peterson, C., Wang, Q., & Hou, Y. (2009). "When I was little": Childhood recollections in Chinese and European Canadian grade-school children. *Child Development, 80*, 506–518.

Phinney, J. S., & Devich-Navarro, M. (1997). Variations in bicultural identification among African American and Mexican American adolescents. *Journal of Research on Adolescence, 7*, 3–32.

Reisberg, D., & Hertel, P. (2005). *Memory and emotion*. New York: Oxford University Press.

Reisberg, D., & Heuer, F. (2004). Memory for emotional events. In D. Reisberg & P. Hertel (Eds.), *Memory and emotion* (pp. 3–41). New York: Oxford University Press.

Renshon, S. A., & Duckitt, J. (2000). Culture and crosscultural political psychology: Revitalizing a founding tradition for a new subfield. In S. A. Renshon & J. Duckitt (Eds.), *Political psychology: Cultural and crosscultural foundations* (pp. 3–17). New York, NY: New York University Press.

Rogers, T. B., Kuiper, N. A., & Kirker, W. S. (1977). Self-reference and the encoding of personal information. *Journal of Personality and Social Psychology, 35*, 677–688.

Ross, M., & Wang, Q. (2010). Why we remember and what we remember: Culture and autobiographical memory. *Perspectives on Psychological Science, 5*, 401–409.
Rule, N. O., Freeman, J. B., & Ambady, N. (2013). Culture in social neuroscience: A review. *Social Neuroscience, 8*, 3–10.
Saxe, R., & Kanwisher, N. (2003). People thinking about thinking people: fMRI investigations of theory of mind. *Neuroimage, 19*, 1835–1842.
Schacter, D. L. (1999). The seven sins of memory: Insights from psychology and cognitive neuroscience. *American Psychologist, 54*, 182–203.
Scollon, C. N., Diener, E., Oishi, S., & Biswas-Diener, R. (2004). Emotions across cultures and methods. *Journal of Cross-Cultural Psychology, 35*(3), 304–326.
Singelis, T. M., Triandis, H. C., Bhawuk, D., & Gelfand, M. (1995). Horizontal and vertical dimensions of individualism and collectivism: A theoretical and measurement refinement. *Cross-Cultural Research, 29*, 240–275.
Sperber, D. (1996). *Explaining culture: A naturalistic approach*. Malden, MA: Blackwell.
Srour, R. W., & Srour, A. (2006). Communal and familial war-related stress factors: The case of the Palestinian child. *Journal of Loss and Trauma, 11*, 289–309.
Sui, J., Zhu, Y., & Chiu, C.-Y. (2007). Bicultural mind, self-construal, and self- and mother-reference effects: Consequences of cultural priming on recognition memory. *Journal of Experimental Social Psychology, 43*, 818–824.
Sul, S., Choi, I., & Kang, P. (2012). Cultural modulation of self-referential brain activity for personality traits and social identities. *Social Neuroscience, 7*, 280–291.
Svoboda, E., McKinnon, M. C., & Levine, B. (2006). The functional neuroanatomy of autobiographical memory: A meta-analysis. *Neuropsychologia, 44*, 2189–2208.
Symons, C. S., & Johnson, B. T. (1997). The self-reference effect in memory: A meta-analysis. *Psychological Bulletin, 121*, 371–394.
Taylor, S. E., & Brown, J. D. (1988). Illusion and well-being: a social psychological perspective on mental health. *Psychological bulletin, 103*(2), 193–210.
Todorov, A., & Olson, I. (2008). Robust learning of affective trait associations with faces when the hippocampus is damaged, but not when the amygdala and temporal pole are damaged. *Social Cognitive and Affective Neuroscience, 3*, 195–203.
Triandis, H. C. (1995). *Individualism and collectivism*. Boulder, CO: Westview.
Tsai, J. L., Knutson, B., & Fung, H. H. (2006). Cultural variation in affect valuation. *Journal of Personality and Social Psychology, 90*, 288–307.
Uchida, Y., & Kitayama, S. (2009). Happiness and unhappiness in east and west: themes and variations. *Emotion, 9*(4), 441–456.
Wang, Q. (2001). Cultural effects on adults' earliest childhood recollection and self- description: Implications for the relation between memory and the self. *Journal of Personality and Social Psychology, 81*, 220–233.
Wang, Q. (2004). The emergence of cultural self-constructs: Autobiographical memory and self-description in European American and Chinese children. *Developmental Psychology, 40*, 3–15.
Wang, Q. (2006). Earliest recollections of self and others in European American and Taiwanese young adults. *Psychological Science, 17*, 708–714.
Wang, Q. (2011). *Autobiographical memory and culture*. Online readings in psychology and culture, Unit 5. Retrieved from http://scholarworks.gvsu.edu/orpc/vol5/iss2/2
Wang, Q., & Conway, M. A. (2004). The stories we keep: Autobiographical memory in American and Chinese middle-aged adults. *Journal of Personality, 72*, 911–938.
Wang, Q., Shao, Y., & Li, Y. J. (2010). "My way or Mom's way?" The bilingual and bicultural self in Hong Kong Chinese children and adolescents. *Child Development, 81*, 555–567.
Waring, J. D., & Kensinger, E. A. (2009). Effects of emotional valence and arousal upon memory trade-offs with aging. *Psychology and Aging, 24*, 412–422.
Waring, J. D., & Kensinger, E. A. (2011). How emotion leads to selective memory: Neuroimaging evidence. *Neuropsychologia, 49*, 1831–1842.

Weinberg, H., & Nuttman-Shwartz, O. (2006). Group work and therapy in Israel: Mirroring a regressed-traumatized society. *Organisational and Social Dynamics, 6*, 95–110.
Wistrich, R. S., & Ohana, D. (1995). *The shaping of Israeli identity: Myth, memory, and trauma.* London: F. Cass.
Zhang, Y., Zhou, T., Zhang, J., Liu, Z., Fan, J., & Zhu, Y. (2006). In search of Chinese self: An fMRI study. *Science in China, 49*, 89–96.
Zhu, Y., & Zhang, L. (2002). An experimental study on the self-reference effect. *Science in China Series C: Life Sciences, 45*, 120–128.
Zhu, Y., Zhang, L., Fan, J., & Han, S. (2007). Neural basis of cultural influence on self-representation. *NeuroImage, 34*, 1310–1316.

Chapter 10
How Social Dynamics Shape Our Understanding of Reality

Anna Abraham

In the relative sense, then, the sense in which we contrast reality with simple unreality, and in which one thing is said to have more reality than another, and to be more believed, reality means simply relation to our emotional and active life.

(William James, 1890, pp. 295)

Abstract Abraham examines the intrinsic effortless capacity for human beings to create and immerse themselves in multiple fictional worlds yet still not lose sight of reality. The reality–fiction distinction is discussed both as a developmental phenomenon and as an emergent product of our social experience. It is argued that the reality–fiction distinction is facilitated by spontaneous attributions of personal relevance, which is mainly defined by a variety of cultural factors.

Abraham discusses the experimental neuroimaging evidence for the brain response when making reality–fiction distinctions and demonstrates that the findings are consistent with other culture-related phenomena. The medial prefrontal cortex is highlighted as a key brain region that modulates the determinations of personal relevance. Multiple studies are highlighted to corroborate this postulation.

Abraham concludes with a discussion of the implications of the reality–fiction distinction for intercultural relations. It is argued that an understanding of the reality–fiction distinction could be applied to studies of acculturation, community identity, and prejudice.

Human beings expend a great deal of time and energy in their daily lives engaging in multiple fictional worlds through films, television, books, computer games, theater, pretend play, and even while fantasizing. Despite the abounding possibilities that exist through such mediums to confuse our senses about the borders of our reality, we are rarely perplexed about our real world relative to these fictional

A. Abraham, Ph.D. (✉)
Department of Psychology, School of Social, Psychological and Communication Sciences, Leeds Beckett University, Leeds, UK
e-mail: annaabr@gmail.com

J.E. Warnick, D. Landis (eds.), *Neuroscience in Intercultural Contexts*, International and Cultural Psychology, DOI 10.1007/978-1-4939-2260-4_10 worlds. What makes the reality–fiction distinction so resistant to assault from our daily engagement in fantasy worlds especially given that this distinction has to be continually updated and flexibly maintained? What cognitive and neural factors contribute in facilitating our implicit understanding of the reality–fantasy distinction? This chapter will expound the idea that socially modulated personal relevance arising through familiarity and experience is a key factor that underlies this understanding. Avenues through which such insights can inform research on the neuroscience of intercultural relations will also be explored.

10.1 Reality Versus Fantasy

Storytelling or narratives form an integral part of our lives. It is a fundamental means by which we acquire knowledge, about not just fictional worlds being described, but also our own world (Mar, 2004). Narratives form the basis on which culture is transmitted, particularly in preliterate cultures, and they can be transmitted by means of verbal or nonverbal mediums (e.g., in the use of the hula dance to tell the creation myth of Hawaiian culture). We derive enjoyment from engaging in narratives as they enable us to explore alternative worlds from a safe vantage point. No extra cognitive effort is required to understand the dynamics of these fictional worlds. In fact, we may find ourselves so immersed within the happenings of these worlds so as to be emotionally affected by them (Nell, 1988; Oatley, 2002). Narratives can make us laugh with joy, sob with sadness, cringe with embarrassment, fear the sound of a creaking door, tremble with anger at an injustice, and experience pure excitement at the mere anticipation of events. What is therefore remarkable is how easily healthy individuals can, nonetheless, tell fiction apart from reality, and how rarely we confuse the two.[1]

This is not to say that we always keep reality and fiction distinct. Indeed, we are routinely guilty of errors of omission and commission when we think about reality. Experiments of selective attention in the form of "change blindness" (Simons & Chabris, 1999), where we fail to notice sizeable changes in scenes, are clear indications of lack of awareness of large-scale fluctuations in reality. Our propensity to be fooled by magic tricks, thereby believing the impossible, is also an indicator of our inability under certain contexts to tell apart reality and fantasy.

As fantasy is evoked in a range of differing contexts, what is necessary is a categorization of the same. One framework was proposed by Pascal Boyer (1997) which distinguished between two orthogonal factors to characterize different manifestations of the reality–fantasy distinction: (a) real-unreal in objective terms, and (b) intuitive-counterintuitive to expectations. Under this classification, magical

[1] Healthy individuals exhibit a generally stable reality–fantasy distinction, but in several psychiatric and neurological disorders, such as schizophrenia and delusional misidentification syndromes, this division between reality and fantasy can be aberrant or ambiguous under specific conditions.

illusions, such as those orchestrated by Derren Brown, would fall into the "counter-intuitive but real" category, whereas the world of Hogwarts in the Harry Potter series would fall into the "intuitive but unreal" category. The term "fiction" is employed in the current chapter as opposed to the wider concept of "fantasy" because the explored literature is limited to topics that are relevant to the understanding of our real world relative to "intuitive but unreal" worlds, which stem from fictional works such as comics, children's literature, and cartoons.

Boyer's classification was proposed as a response to the seminal article by Jacqueline Woolley (1997) in which children's sense of the reality–fantasy distinction was compared to that of adults (Woolley, 1997). The developmental trajectory of the understanding of the reality–fiction distinction will be discussed in the next section.

10.2 Understanding the Reality–Fiction Distinction: Development

The ability to tell apart fantasy from reality emerges early during development (Wellman & Wellman, 1990). Children customarily show the first signs of engaging in fantasy by the age of 2, when they begin pretend play. Around the age of 4, children understand that there are physical rules that determine reality and they detect violations of these rules (Rosengren & Hickling, 1994). They can even discriminate between impossible events, which cannot take place in the real world, and improbable events, which are unfamiliar but could occur in reality (Weisberg & Sobel, 2012). And, by the age of 5, a rather sophisticated understanding of the reality–fiction distinction is already in place (Skolnick & Bloom, 2006).

Deena Skolnick and Paul Bloom (2006) employed an elegant paradigm to investigate children's explicit and implicit knowledge of the reality–fiction distinction by assessing not only fantasy/reality distinctions (Can you touch Batman?), but also within-world distinctions (Can Robin touch Batman?) and between-world or fantasy/fantasy distinctions (Can SpongeBob touch Batman?). They found that 5-year-olds successfully differentiated reality from fiction, and distinguished between different fictional worlds.

So our ability to determine what is real versus unreal begins very early during child development and this understanding is quite profound even prior to commencing primary school. What is still unclear though is *how* this ability develops.

10.3 Making the Reality–Fiction Distinction: What Are the Modulating Factors?

Functional neuroimaging provides a unique avenue through which potential answers to this question can be revealed. Using neuroimaging techniques, it is possible to assess which parts of the brain are engaged when performing any perceptual, cognitive, or behavioral task. As each brain region (or network of brain areas) is accompanied by a particular functional profile in that a circumscribed range of mental operations elicit activity in that brain area, such investigations can inform us about the information processing demands of the psychological task in question.[2]

Investigating the manner in which the brain is engaged when we are telling apart reality from fiction could inform us about what underlies our implicit knowledge of this distinction. This was the objective of a series of fMRI investigations on the topic (Abraham & von Cramon, 2009; Abraham, von Cramon, & Schubotz, 2008). In the first of these studies, participants were presented with sentences in which a real person called Peter engaged with either a known real entity (e.g., George Bush, Angela Merkel) or a fictional character (e.g., Cinderella, Batman) in informative contexts (e.g., heard about) or interactive contexts (e.g., spoke to). Following this, subjects had to determine whether this scenario was possible or not given the constraints of our real world. Using this experimental paradigm, it was possible to assess whether our understanding of the reality–fiction distinction is context-dependent (interacting with a fictional character is impossible whereas interacting with a real entity is possible) or character-dependent (qualitative and/or quantitative differences in the conceptual representations of real versus fictional entities within our brain networks).

The former "context-dependent" hypothesis would resonate with the principle of minimal departure (Ryan, 1980) in the narrative comprehension literature. This refers to the idea that when we faced with information regarding a fictional world, we consider the rules of this world as being entirely analogous to our reality except in the context of exceptional circumstances, which are specifically outlined in the narrative. So in Harry Potter's world, the non-Muggle world of wizardry with its own exceptional rules exists in parallel with the Muggle world, which is similar on all counts to our real world. Other evidence, however, where different types of fictional worlds were compared in terms of their distance from reality and the type of facts being incorporated, indicated that participants generate intuitions even about unspecified facts and follow a decidedly more nuanced approach when considering fictional worlds relative to reality (Weisberg & Goodstein, 2009). Such findings indicate that these distinctions follow from the rather complex and differential integration of various kinds of contextual information.

The question therefore still remains about whether this implicit reality–fiction distinction between is context-dependent (e.g., I can dream about SpongeBob Squarepants, but I cannot arm-wrestle him) or character-dependent (e.g., My knowledge about SpongeBob's world is organized in my brain in a distinct manner compared to information about my own world).

So investigating the brain response when making reality–fiction distinctions was targeted at developing a better understanding of the dynamics of this implicit ability

[2] It is to be noted that deduction through reverse inference needs to be exercised with considerable caution (Poldrack, 2008). However, if applied in the right manner and in the right context, it can be extremely useful (Hutzler, 2013), particularly if the goal is for those insights to direct future empirical work (Poldrack, 2008).

(Abraham et al., 2008). The brain response when processing any kind of context involving a real or fictional protagonist (compared to an unrelated control task) commonly resulted in activations of regions known for their involvement in declarative memory retrieval (e.g., hippocampus) and mental state reasoning (e.g., dorsomedial prefrontal cortex). Interestingly, contrasting different types of contexts (interactive versus informative) did not lead to any significant findings. But processing any kind of context concerning real entities relative to those containing fictional characters led to a significant difference in the ensuing pattern of brain activity. Reading information about real people lead to activations in the anterior medial prefrontal cortex (PFC) (see Fig. 10.1) and precuneus/posterior cingulate (PCC), whereas processing information about fictional characters lead to activations along the lateral inferior frontal gyrus (IFG).

What are the essential differences in the functional profiles of these brain regions? The role of the lateral IFG is well-established in the field of semantic cognition (Badre & Wagner, 2007) particularly in semantic selection and semantic retrieval. The medial PFC and the PCC, on the other hand, are commonly engaged during evaluative judgment, autobiographical or episodic memory retrieval and self-referential processing (Zysset, Huber, Ferstl, & von Cramon, 2002). Moreover, the anterior-most regions of the PFC are held to be specifically recruited when a higher-order behavioral goal requires the integration of information from two or more separate cognitive operations (Ramnani & Owen, 2004). So while brain areas relevant for semantic cognition are engaged during the processing of fiction, brain regions involved in episodic cognition are implicated in the processing of reality.

There are two factors to keep in mind regarding the interpretation of these brain activation distinctions. First, this semantic/episodic dissociation suggests that fiction (relative to reality) is coded as fact in the brain, whereas reality (relative to fiction) is processed in terms of subjectively coded representations. Second, there is more informational integration during the processing of information concerning real entities compared to that of fictional characters. How can these conclusions best be explained?

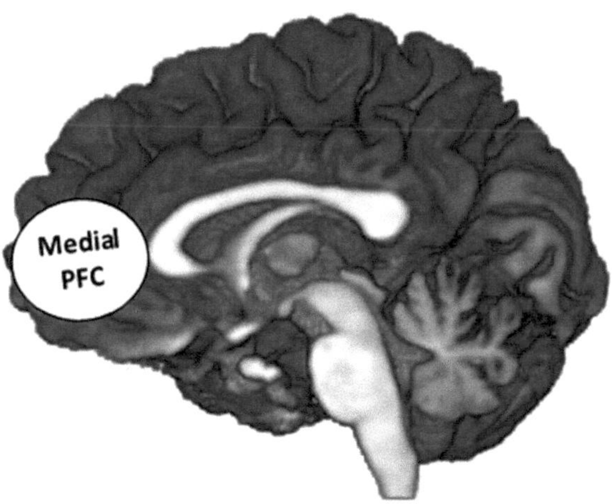

Fig. 10.1 The location of the medial prefrontal cortex (PFC) in the human brain

Among the major differences between known real entities and fictional characters is the amount of information that we can readily draw upon in reference to each protagonist and the frequency with which we encounter information (new/old) concerning them. We are regularly bombarded with information concerning real entities, such as famous people, through the media. Even if we are never likely to encounter these people in reality, they, unlike fictional characters, nevertheless occupy a significant space in our social world.

Moreover, although we can arrive at quite a detailed understanding of a fictional world (such as that of Cinderella), we still have, relatively speaking, very limited information about her world in comparison to what we know about our own world. With a real entity, such as Barack Obama, one has access to different types of information about him: the degree of perceived attractiveness, his position in the social hierarchy, the degree of influence his politics has on one's own life and that of others, what morals/values he stands for, one's personal feelings toward him (e.g., like/dislike, respect/irreverence), the last time one saw him on television or read about him in the newspaper, etc. So reading about a familiar entity leads to the spontaneous access, integration, and coordination of many different kinds of information (e.g., semantic, episodic, emotional, self-referential, evaluative, interoceptive). And this occurs even in the absence of any explicit behavioral goal that imposes such cognitive or behavioral demands.[3]

These findings have important implications for our understanding of the reality–fiction distinction as they essentially indicate that this distinction between the processing of fiction as factual knowledge in contrast to reality as subjectively coded representations is a relative one. If the degree of associated self-relevance to a representation is one of the factors that affects what we classify as real, the question of how we process fiction versus reality needs to be approached in terms of the degree of personal relevance associated with the protagonist in question as opposed to a simple dichotomy (objective reality versus unreality). A continuum-based approach may be crucial to understanding various fascinating aspects of human behavior, particularly those that are culture-specific manifestations, such as religiosity and compulsive gaming, in which the reality–fiction distinction can be blurred as such fictional contexts tend to be coded in highly self-relevant terms.

10.4 Culture and the Reality–Fiction Distinction

Behavioral and neuroimaging evidence indicate that cultural factors, such as ethnicity and socialization, exert considerable influence on several facets of human psychological functioning including self-perception, self-concept, fundamental

[3] While the reality/fiction distinction may have certain parallels with the familiarity/unfamiliarity distinction, they cannot be considered equivalent per se. Relevance goes beyond familiarity or awareness to also include associated importance or significance (Abraham, 2013).

attribution error, experience of emotion, self-esteem and life satisfaction, cognitive dissonance, motivation, and memory (Fiske & Taylor, 2007). For instance, Zhu and colleagues (2007) discovered that, compared to people of Western origin, the medial PFC in Chinese participants was strongly engaged not only during self-referential processing but also during information processing related to a close "other," i.e., one's mother (Zhu, Zhang, Fan, & Han, 2007). These studies are reviewed in another chapter in this book (Han, Chap. 2). The rationale offered for this pattern of findings was that China represents an interdependent culture where the conceptual representations of one's self and close others would be expected to be more tightly coupled than in the case of independent cultures, such that of Western Europe.

This idea relates well to other work within the field of social neuroscience which have shown that ventral and anterior regions of the medial PFC are engaged when making judgments about other people who are similar to us in terms of sociopolitical views (Mitchell, Macrae, & Banaji, 2006), and who are close or socially relevant to us (Krienen, Tu, & Buckner, 2010). Indeed, even considering the perspectives of one's own preferred candidate relative to that of the opponent prior to the 2008 US presidential elections was found to be reflected in heightened medial PFC activity (Falk, Spunt, & Lieberman, 2012). Engagement of the medial PFC has generally been documented in related research domains on salience processing and valuation, particularly in the presence of some degree of personal involvement (Abraham, 2013; Roy, Shohamy, & Wager, 2012; Somerville, Kelley, & Heatherton, 2010). While the medial prefrontal cortex as a whole is considered to be involved in self-relevance appraisal, the ventral aspects in particular are held to mediate "identification and appraisal of stimulus-induced self relevance" (Schmitz & Johnson, 2006, 2007).

What is critical to note here is that the engagement of this brain structure is "stimulus-induced." This means that this region is not only involved within explicit contexts, where subjects have to make conscious judgments of oneself or close others, but also in implicit contexts, where the self-relevant stimuli are provided but no self-referential judgment has to be made (Abraham & von Cramon, 2009; Moran, Heatherton, & Kelley, 2009). This affirms not only that information processing in the brain is proactive and predictive (Bubic, von Cramon, & Schubotz, 2010), but also that stimulus-induced spontaneous modulations of the brain can be used to understand different aspects of brain function.

The findings of the Abraham et al. (2008) study suggested that the reality–fiction distinction is potentially mediated by the degree of associated personal relevance with the entity/character in question. A clearer demonstration of personal relevance-based mediation would come from showing a graded effect such that the medial PFC becomes increasingly more activated as a function of personal relevance. Familiar individuals existing within our sociocultural world, such as famous celebrities or cultural icons, would be expected to be more relevant to us compared to fictional characters because they occupy a space in our shared social world. But individuals who are part of our intimate circle of family and friends, and to whom we feel particularly connected, would be even more personal significant for us because their actions have a direct bearing on our lives. If the medial prefrontal cortex codes for personal relevance, the activation profile seen in this brain region when processing information concerning friends or family (high relevance), famous people (medium relevance), and fictional characters (low relevance) should vary accordingly. A follow up fMRI study confirmed these predictions as the medial PFC was most strongly engaged during high relevance contexts (e.g., involving one's mother), moderately engaged in medium relevance contexts (e.g., involving Barack Obama) and least engaged in low relevance contexts (e.g., involving Cinderella) (Abraham & von Cramon, 2009).[4]

The findings thereby confirmed that the anterior medial PFC is modulated by the degree of stimulus associated personal relevance. This bolsters the idea that one of the factors that guide our implicit knowledge of the reality–fiction distinction is the degree of coded personal relevance associated with a particular entity/character representation. Recent views have highlighted that the constructive processes orchestrated by anterior regions of the ventral medial PFC "is one of combining elemental units of information—from sensory systems, interoceptive cues, long-term memory—into a gestalt representation of how an organism is situated in its environment, which then drives predictions about future events" (Roy et al., 2012).

So how does this happen? How does information about real people get coded with a higher degree of personal relevance, whereas reading a work of fiction about vampires living amongst us does not? One possibility is that when encountering information about a new protagonist, the conceptual knowledge that we possess that is directly or indirectly associated with this person is spontaneously activated. In contrast, background knowledge would be nonexistent or very limited for a new fictional character from a previously unknown world. So the concepts formed about this new world and its characters are freshly generated and integrated to existing schemas whenever there is an overlap. For instance, I recently came across a comic book series, "The 99 - Superheroes from the Muslim world," of which I had no prior knowledge. New concepts had to therefore be automatically generated within my brain to represent this new fictional world. After reading a few issues, I came across a story where "The 99" crosses paths with the "Justice League of America," a comic series with which I am very familiar. An association between the two fictional worlds (through direct interactions between the characters and their pursuit of common goals) was thereby spontaneously formed.

In contrast, when learning about a new real entity, such as a newly elected president of a country like the USA, there is far more existing knowledge in the semantic network of my mind (e.g., about the country, previous presidents, etc.) within which this new information can be readily embedded. Our conceptual knowledge in relation to real people is not only far more extensive but also exceptionally multifaceted compared to our knowledge of fictional characters. For instance, the kind of associations most people have for a fictional character such as Cinderella (e.g., evil stepfamily, glass slipper, fairy godmother, the significance of midnight) are limited to the context of the story in which we learnt about her. In comparison, as mentioned earlier, our associations about a real person, such as Barack Obama are far more

[4] Differing levels of personal relevance (friends-real > famous-real > fiction) were assumed a priori. No behavioral measures have been devised thus far to estimate degrees of relevance associated with known entities/characters.

wide-ranging and heterogeneous (episodic, semantic, evaluative, interoceptive, and so on). And this is exponentially more so for people to whom we feel very close or who are highly personally relevant for us, such as close family members and friends.

The reason why a real person feels more "real" to us than a fictional character at a phenomenological level may be because we are automatically primed to access far more comprehensive and diverse types of conceptual knowledge in relation to the real people than fictional characters. This may also explain why, relatively speaking, a real person we know personally (e.g., a friend) feels more real or actual to us than a real person who we do not know personally (e.g., a news anchor). Studies have shown, for instance, that levels of physiological arousal and emotional reactivity, as measured by skin conductance response (SCR) for instance, are enhanced upon viewing familiar faces than unfamiliar faces. Findings from clinical populations are particularly telling in this regard (Young, 2009). Although people with prosopagnosia (or face-blindness) fail to correctly identify pictures of familiar faces as belonging to someone they know, they nonetheless exhibit heightened emotional reactivity, as evidenced by elevated SCR, when presented with these pictures. In contrast, this emotional familiarity effect is absent in people with Capgras syndrome who suffer from the delusion that a close family member or friend has been replaced by a doppelganger-imposter. When presented with a familiar face of someone they know but do not believe to be the person in question, people with Capgras disorder correctly identify the faces but do not exhibit elevated SCR. This evidence points to importance of interoceptive cues in determining our reality.

Other factors also play a key role. Children, for instance, have been shown to evaluate the factual nature of fictional events based on how they fit with their own world knowledge. 4-year olds consider fictional characters that are associated with specific regular events in one's life, such as Santa Claus and the Tooth Fairy, to be more real than fictional characters that are not related to real-life events, such as dragons and fairies (Rosengren & Hickling, 1994). 5-year-olds judge novel entities to be more real when they encounter them in every day or scientific contexts compared to fantastical contexts (Woolley & Van Reet, 2006).

Adults can also be expected to use contextual information in the same manner when making a decision about the reality status of a novel entity. Barack Obama may appear more "real" to Americans just as Angela Merkel may seem more "real" to Germans or Xi Jinping more "real" to the Chinese because each of these leaders carry more personal relevance within their own socio-cultural contexts than outside. So cultural factors most certainly play a key modulatory role in how we attribute personal significance to people, events, and objects in our lives.

10.5 Implications for Intercultural Relations

How can these insights from the neuroscientific study of the reality–fiction distinction be applied within the domain of intercultural relations? One of the key ideas propounded within this chapter has been that the degree to which personal relevance or significance is experienced in a particular context has a viable impact on how real we experience the world to be. Factors like socialization and enculturation have a tremendous capacity to exert a substantial influence on the manner in which we classify aspects of our worlds to be personally significant or not given that a cultural group can preserve its behavioral attributes among subsequent generations through different avenues of cultural transmission (vertical, oblique, horizontal). Indeed, this can also happen through acculturation, which occurs when this process is influenced from other cultures that are not one's primary culture (Berry, Poortinga, Segall, & Dasen, 2002).

Having to adapt to changing cultural contexts is inevitable in plural societies, which is the reality of the vast majority of large cities the world over. This adaptation can take on different forms as it can be internal/psychological (e.g., sense of well-being) or sociocultural (e.g., how connected one is to others in the new society). The usual case is that one culture (the dominant one) customarily exerts more influence on the other culture (the nondominant one) and the dynamics of that relationship has an impact on which strategy of acculturation is adopted: assimilation, integration, separation, or marginalization (Berry et al., 2002). Intercultural relations in terms of cultural identity and attitudinal reactions (such as prejudice or discrimination) are particularly affected as a result.

The picture is, however, far more complicated in truly multicultural contexts, such as in a country like Canada or a metropolitan city like London, where there is more than one dominant culture and/or hybrid ethnic identities are commonplace. Not only do other strategies, such as individualism, come into play, several other contextual factors, such as the influence of colonial histories, the sociopolitical orientation of the dominant group, and the immediate conditions of everyday life, need to be taken into consideration to appreciate the full picture (Bourhis, Moise, Perreault, & Senecal, 1997; Doucerain, Dere, & Ryder, 2013). To evaluate how research on the understanding of the reality–fiction distinction can inform us about the mechanisms underlying acculturation relevant processes and strategies, it would be important to first address this question in contexts where the division between the dominant and nondominant groups are relatively clear-cut. This would enable the accrual of foundational knowledge from which more nuanced analyses of the influence of other contextual factors can be explored.

One of the more obvious avenues for exploration would be to evaluate how separated (or marginalized) a nondominant group is within a specific sociocultural context by applying insights from the neuroscientific study of the reality–fiction distinction to the implicit knowledge of how community identity is shaped by in-group versus out-group relations. Assessing the degree to which the anterior medial PFC in members of a dominant group is responsive when reading information about a nondominant group would be one such strategy. If the nondominant out-group were seen as a socially significant (and consequently personally relevant) group, this brain region would be expected to be strongly engaged. However, if the nondominant group were seen as a socially insignificant (and consequently personally irrelevant) group, the same pattern of brain activity would not be expected. While related hypotheses have been explored in cross-cultural work in fields such as emotional recognition (Chiao et al., 2008), they have rarely been investigated in terms of intercultural relations (Elfenbein & Ambady, 2002).

Another path for exploration would be to assess the neurobiological effects of strategies aimed at reducing prejudice between groups by increasing intercultural contact, as per the contact hypothesis (Allport, 1979). Brain activation patterns as a function of processing information concerning dominant and nondominant groups before and after strategy interventions would help determine whether the behavioral effects of reduced prejudice among intergroup relations (Pettigrew & Tropp, 2008) are accompanied by concomitant brain related changes—such as increased medial PFC activation. As such a pattern would indicate a higher degree of coded personal relevance in relation to the corresponding out-group, the findings could be used to suggest that some degree of attitudinal change, in terms of assimilation or integration, had actually taken place as a result of the interventions.

The effect of the internet on intercultural relations affords yet another opportunity for understanding the dynamics behind acculturation-related adaptive processes. This is a rarely explored field that offers fertile ground for exploration as novel types of cultural contact and change are coming to the forefront with increasingly more individuals turning to the internet to express and identify with themselves with new communities through blogs, social networks, social media, and online forums (Bentley & O'Brien, 2012). Despite the lack of direct physical contact, these cannot be considered fictional realms and they can also be associated with a high degree of personal relevance. For instance, it is possible that an avid member of the World of Warcraft gaming community may exhibit greater medial PFC activity when processing information concerning a fictional entity of high personal relevance (e.g., a character in the game) or a real but unknown entity of high personal relevance (e.g., an anonymous member of that online gaming community) than to a real known person of low personal relevance (e.g., his math teacher). Exploring the dynamics behind how novel cultural factors impact our psychological function is among the many promising avenues for future investigation.

A first step in such a venture would be to develop a scale to assess personal relevance that can be applied across situations (persons, characters, objects, events, topics). So far, personal relevance has been assessed in a very limited fashion by asking subjects to provide a single rating for the degree of concern or self-relatedness they experience towards different types of stimuli (e.g., Northoff et al., 2009; Tomaszczyk, Fernandes, & MacLeod, 2008; Ülkümen & Thomas, 2013). A comprehensive scale for assessing the degree of associated personal relevance to any given stimulus therefore needs to be developed which incorporates different dimensions of relevance, such as intensity of feeling (none-strong), depth of knowledge (superficial-deep), judgment (like-neutral-dislike), affect (positive-neutral-negative), significance to one's self concept (none-high), significance to one's life (none-high), sense of identification (none-high), sense of concern (none-high), and so on. Just as with pan-cultural investigations of affective meaning of concepts (see Heise, 2014), an instrument to assess the degree of personal relevance associated with a person (or object or idea) would readily lend itself to cross-cultural investigations of the reality–fiction distinction.

Conclusions

The rapidly burgeoning field of cross-cultural psychology has made evident that the effects of different cultures on human cognitive and behavioral function are more profound than previously thought. But the picture is actually far more complicated than we even imagine. Our social world is increasingly becoming more multicultural in terms of geography and more accessible via growing access to the internet throughout the world. The challenges that face the field of intercultural relations in understanding the intercultural dynamics that guide human cognition behavior are vast. This chapter explored some of the means through which knowledge could be gained in this respect—by exploring our implicit understanding of what makes the world real, significant, and relevant to us.

References

Abraham, A. (2013). The world according to me: Personal relevance and the medial prefrontal cortex. *Frontiers in Human Neuroscience, 7*, 341. doi:10.3389/fnhum.2013.00341.

Abraham, A., & von Cramon, D. Y. (2009). Reality = relevance? Insights from spontaneous modulations of the brain's default network when telling apart reality from fiction. *PloS One, 4*(3), e4741. doi:10.1371/journal.pone.0004741.

Abraham, A., von Cramon, D. Y., & Schubotz, R. I. (2008). Meeting George Bush versus meeting Cinderella: The neural response when telling apart what is real from what is fictional in the context of our reality. *Journal of Cognitive Neuroscience, 20*(6), 965–976. doi:10.1162/jocn.2008.20059.

Allport, G. W. (1979). *The nature of prejudice: 25th anniversary edition* (Unabridged ed.). Basic Books.

Badre, D., & Wagner, A. D. (2007). Left ventrolateral prefrontal cortex and the cognitive control of memory. *Neuropsychologia, 45*(13), 2883–2901. doi:10.1016/j.neuropsychologia.2007.06.015.

Bentley, R. A., & O'Brien, M. J. (2012). Cultural evolutionary tipping points in the storage and transmission of information. *Frontiers in Psychology, 3*, 569. doi:10.3389/fpsyg.2012.00569.

Berry, J. W., Poortinga, Y. H., Segall, M. H., & Dasen, P. R. (2002). *Cross-cultural psychology: Research and applications* (2nd ed.). Cambridge, UK: Cambridge University Press.

Bourhis, R. Y., Moise, L. C., Perreault, S., & Senecal, S. (1997). Towards an interactive acculturation model: A social psychological approach. *International Journal of Psychology, 32*(6), 369–386. doi:10.1080/002075997400629.

Boyer, P. (1997). Further distinctions between magic, reality, religion, and fiction. *Child Development, 68*(6), 1012–1014. discussion 1027–1030.

Bubic, A., von Cramon, D. Y., & Schubotz, R. I. (2010). Prediction, cognition and the brain. *Frontiers in Human Neuroscience, 4*, 25. doi:10.3389/fnhum.2010.00025.

Chiao, J. Y., Iidaka, T., Gordon, H. L., Nogawa, J., Bar, M., Aminoff, E., … Ambady, N. (2008). Cultural specificity in amygdala response to fear faces. *Journal of Cognitive Neuroscience, 20*(12), 2167–2174. doi:10.1162/jocn.2008.20151.

Doucerain, M., Dere, J., & Ryder, A. G. (2013). Travels in hyper-diversity: Multiculturalism and the contextual assessment of acculturation. *International Journal of Intercultural Relations, 37*(6), 686–699.

Elfenbein, H. A., & Ambady, N. (2002). On the universality and cultural specificity of emotion recognition: A meta-analysis. *Psychological Bulletin, 128*(2), 203–235.

Falk, E. B., Spunt, R. P., & Lieberman, M. D. (2012). Ascribing beliefs to ingroup and outgroup political candidates: neural correlates of perspective-taking, issue importance and days until the election. *Philosophical Transactions of the Royal Society of London. Series B, Biological Sciences, 367*(1589), 731–743. doi:10.1098/rstb.2011.0302.

Fiske, S., & Taylor, S. (2007). *Social cognition, from brains to culture* (1st ed.). New York: McGraw-Hill Humanities/Social Sciences/Languages.

Heise, D. R. (2014). Cultural variations in sentiments. *SpringerPlus, 3*, 170. doi:10.1186/2193-1801-3-170.

Hutzler, F. (2013). Reverse inference is not a fallacy per se: Cognitive processes can be inferred from functional imaging data. *NeuroImage*. doi:10.1016/j.neuroimage.2012.12.075.

Krienen, F. M., Tu, P.-C., & Buckner, R. L. (2010). Clan mentality: Evidence that the medial prefrontal cortex responds to close others. *The Journal of Neuroscience, 30*(41), 13906–13915. doi:10.1523/JNEUROSCI. 2180-10.2010.

Mar, R. A. (2004). The neuropsychology of narrative: Story comprehension, story production and their interrelation. *Neuropsychologia, 42*(10), 1414–1434. doi:10.1016/j.neuropsychologia.2003.12.016.

Mitchell, J. P., Macrae, C. N., & Banaji, M. R. (2006). Dissociable medial prefrontal contributions to judgments of similar and dissimilar others. *Neuron, 50*(4), 655–663. doi:10.1016/j.neuron.2006.03.040.

Moran, J. M., Heatherton, T. F., & Kelley, W. M. (2009). Modulation of cortical midline structures by implicit and explicit self-relevance evaluation. *Social Neuroscience, 4*(3), 197–211. doi:10.1080/17470910802250519.

Nell, V. (1988). The psychology of reading for pleasure: Needs and gratifications. *Reading Research Quarterly, 23*(1), 6–50.

Northoff, G., Schneider, F., Rotte, M., Matthiae, C., Tempelmann, C., Wiebking, C., … Panksepp, J. (2009). Differential parametric modulation of self-relatedness and emotions in different brain regions. *Human Brain Mapping, 30*(2), 369–382. doi:10.1002/hbm.20510.

Oatley, K. (2002). Emotions and the story worlds of fiction. In M. Green, J. Strange, & T. Brock (Eds.), *Narrative impact: Social and cognitive foundations* (pp. 39–70). Mahwah, NJ: Lawrence Erlbaum.

Pettigrew, T. F., & Tropp, L. R. (2008). How does intergroup contact reduce prejudice? Meta-analytic tests of three mediators. *European Journal of Social Psychology, 38*(6), 922–934. doi:10.1002/ejsp.504.

Poldrack, R. A. (2008). The role of fMRI in cognitive neuroscience: Where do we stand? *Current Opinion in Neurobiology, 18*(2), 223–227. doi:10.1016/j.conb.2008.07.006.

Ramnani, N., & Owen, A. M. (2004). Anterior prefrontal cortex: Insights into function from anatomy and neuroimaging. *Nature Reviews: Neuroscience, 5*(3), 184–194. doi:10.1038/nrn1343.

Rosengren, K. S., & Hickling, A. K. (1994). Seeing is believing: Children's explanations of commonplace, magical, and extraordinary transformations. *Child Development, 65*(6), 1605–1626.

Roy, M., Shohamy, D., & Wager, T. D. (2012). Ventromedial prefrontal-subcortical systems and the generation of affective meaning. *Trends in Cognitive Sciences, 16*(3), 147–156. doi:10.1016/j.tics.2012.01.005.

Ryan, M.-L. (1980). Fiction, non-factuals, and the principle of minimal departure. *Poetics, 9*(4), 403–422. doi:10.1016/0304-422X(80)90030-3.

Schmitz, T. W., & Johnson, S. C. (2006). Self-appraisal decisions evoke dissociated dorsal-ventral aMPFC networks. *NeuroImage, 30*(3), 1050–1058. doi:10.1016/j.neuroimage.2005.10.030.

Schmitz, T. W., & Johnson, S. C. (2007). Relevance to self: A brief review and framework of neural systems underlying appraisal. *Neuroscience and Biobehavioral Reviews, 31*(4), 585–596. doi:10.1016/j.neubiorev.2006.12.003.

Simons, D. J., & Chabris, C. F. (1999). Gorillas in our midst: Sustained inattentional blindness for dynamic events. *Perception, 28*(9), 1059–1074. doi:10.1068/p2952.

Skolnick, D., & Bloom, P. (2006). What does Batman think about SpongeBob? Children's understanding of the fantasy/fantasy distinction. *Cognition, 101*(1), B9–18. doi:10.1016/j.cognition.2005.10.001.

Somerville, L. H., Kelley, W. M., & Heatherton, T. F. (2010). Self-esteem modulates medial prefrontal cortical responses to evaluative social feedback. *Cerebral Cortex, 20*(12), 3005–3013. doi:10.1093/cercor/bhq049.

Tomaszczyk, J. C., Fernandes, M. A., & MacLeod, C. M. (2008). Personal relevance modulates the positivity bias in recall of emotional pictures in older adults. *Psychonomic Bulletin and Review, 15*(1), 191–196.

Ülkümen, G., & Thomas, M. (2013). Personal relevance and mental simulation amplify the duration framing effect. *Journal of Marketing Research, 50*(2), 194–206. doi:10.1509/jmr.10.0172.

Weisberg, D. S., & Goodstein, J. (2009). What belongs in a fictional world? *Journal of Cognition and Culture, 9*(1–2), 69–78. doi:10.1163/156853709X414647.

Weisberg, D. S., & Sobel, D. M. (2012). Young children discriminate improbable from impossible events in fiction. *Cognitive Development, 27*(1), 90–98. doi:10.1016/j.cogdev.2011.08.001.

Wellman, H. M., & Woolley, J. D. (1990). From simple desires to ordinary beliefs: The early development of everyday psychology. *Cognition, 35*, 245–275.

Woolley, J. D. (1997). Thinking about fantasy: Are children fundamentally different thinkers and believers from adults? *Child Development, 68*(6), 991–1011.

Woolley, J. D., & Van Reet, J. (2006). Effects of context on judgments concerning the reality status of novel entities. *Child Development, 77*(6), 1778–1793. doi:10.1111/j.1467-8624.2006.00973.x.

Young, G. (2009). In what sense "familiar"? Examining experiential differences within pathologies of facial recognition. *Consciousness and Cognition, 18*(3), 628–638. doi:10.1016/j.concog.2009.06.006.

Zhu, Y., Zhang, L., Fan, J., & Han, S. (2007). Neural basis of cultural influence on self-representation. *NeuroImage, 34*(3), 1310–1316. doi:10.1016/j.neuroimage.2006.08.047.

Zysset, S., Huber, O., Ferstl, E., & von Cramon, D. Y. (2002). The anterior frontomedian cortex and evaluative judgment: An fMRI study. *NeuroImage, 15*(4), 983–991. doi:10.1006/nimg.2001.1008.

Index

J.E. Warnick, D. Landis (eds.), *Neuroscience in Intercultural Contexts*,
International and Cultural Psychology, DOI 10.1007/978-1-4939-2260-4

O

P

R

S

MIX
Papier aus verantwortungsvollen Quellen
Paper from responsible sources
FSC® C105338

If you have any concerns about our products,
you can contact us on
ProductSafety@springernature.com

In case Publisher is established outside the EU,
the EU authorized representative is:
Springer Nature Customer Service Center GmbH
Europaplatz 3, 69115 Heidelberg, Germany

Printed by Libri Plureos GmbH
in Hamburg, Germany